Poincaré's Legacies, Part I

pages from year two
of a mathematical blog

Terence Tao

AMERICAN MATHEMATICAL SOCIETY

2000 *Mathematics Subject Classification.* Primary 00A99.

For additional information and updates on this book, visit
www.ams.org/bookpages/mbk-66

Library of Congress Cataloging-in-Publication Data

Tao, Terence, 1975–
    Poincaré's legacies : pages from year two of a mathematical blog / Terence Tao.
      p. cm.
    Includes bibliographical references and index.
    ISBN 978-0-8218-4883-8 (pt. 1 : alk. paper)—ISBN 978-0-8218-4885-2 (pt. 2 : alk. paper)
    1. Poincaré conjecture. 2. Ricci flow. 3. Differential equations, Partial. I. Title.

QA613.2.T36   2009
514′.72—dc22

                                      2009009832

To Garth Gaudry, who set me on the road;
To my family, for their constant support;
And to the readers of my blog, for their feedback and contributions.

Contents

Preface

In February of 2007, I converted my "What's new" web page of research updates into a blog at `terrytao.wordpress.com`. This blog has since grown and evolved to cover a wide variety of mathematical topics, ranging from my own research updates, to lectures and guest posts by other mathematicians, to open problems, to class lecture notes, to expository articles at both basic and advanced levels.

With the encouragement of my blog readers, and also of the AMS, I published many of the mathematical articles from the first year (2007) of the blog as [**Ta2008b**], which will henceforth be referred to as *Structure and Randomness* throughout this book. This gave me the opportunity to improve and update these articles to a publishable (and citeable) standard, and also to record some of the substantive feedback I had received on these articles from the readers of the blog. Given the success of the blog experiment so far, I am now doing the same for the second year (2008) of articles from the blog. This year, the amount of material is large enough that the blog will be published in two volumes.

As with *Structure and Randomness*, each part begins with a collection of expository articles, ranging in level from completely elementary logic puzzles to remarks on recent research, which are only loosely related to each other and to the rest of the book. However, in contrast to the previous book, the bulk of these volumes is dominated by the lecture notes for two graduate courses I gave during the year. The two courses stemmed from two very different but fundamental contributions to mathematics by Henri Poincaré, which explains the title of the book.

This is the first of the two volumes, and it focuses on ergodic theory, combinatorics, and number theory. In particular, Chapter 2 contains the lecture notes for my course on *topological dynamics and ergodic theory*, which origi- nated in part from Poincaré's pioneering work in chaotic dynamical systems. Many situations in mathematics, physics, or other sciences can be modeled by a discrete or continuous *dynamical system*, which at its most abstract level is simply a space X, together with a shift $T : X \to X$ (or family of shifts) acting on that space, and possibly preserving either the topological or measure-theoretic structure of that space. At this level of generality, there are a countless variety of dynamical systems available for study, and it may seem hopeless to say much of interest without specialising to much more concrete systems. Nevertheless, there is a remarkable phenomenon that dy- namical systems can largely be classified into "structured" (or "periodic") components, and "random" (or "mixing") components,[1] which then can be used to prove various *recurrence theorems* that apply to very large classes of dynamical systems, not the least of which is the *Furstenberg multiple re- currence theorem* (Theorem 2.10.3). By means of various *correspondence principles*, these recurrence theorems can then be used to prove some deep theorems in combinatorics and other areas of mathematics, in particular yielding one of the shortest known proofs of *Szemerédi's theorem* (Theorem 2.10.1) that all sets of integers of positive upper density contain arbitrarily long arithmetic progressions. The road to these recurrence theorems, and several related topics (e.g. ergodicity, and Ratner's theorem on the equidis- tribution of unipotent orbits in homogeneous spaces) will occupy the bulk of this course. I was able to cover all but the last two sections in a 10-week course at UCLA, using the exercises provided within the notes to assess the students (who were generally second or third-year graduate students, having already taken a course or two in graduate real analysis).

Finally, I close this volume with a third (and largely unrelated) topic (Chapter 3), namely a series of lectures on recent developments in additive prime number theory, both by myself and my coauthors, and by others. These lectures are derived from a lecture I gave at the annual meeting of the AMS at San Diego in January of 2007, as well as a lecture series I gave at Penn State University in November 2007.

A remark on notation

For reasons of space, we will not be able to define every single mathematical term that we use in this book. If a term is italicised for reasons other than emphasis or definition, then it denotes a standard mathematical object, result, or concept, which can be easily looked up in any number of references.

---

[1] One also has to consider *extensions* of systems of one type by another, e.g. mixing extensions of periodic systems; see Section 2.15 for a precise statement.

(In the blog version of the book, many of these terms were linked to their Wikipedia pages, or other on-line reference pages.)

I will however mention a few notational conventions that I will use throughout. The cardinality of a finite set E will be denoted $|E|$. We will use the asymptotic notation $X = O(Y)$, $X \ll Y$, or $Y \gg X$ to denote the estimate $|X| \leq CY$ for some absolute constant $C > 0$. In some cases we will need this constant C to depend on a parameter (e.g. d), in which case we shall indicate this dependence by subscripts, e.g. $X = O_d(Y)$ or $X \ll_d Y$. We also sometimes use $X \sim Y$ as a synonym for $X \ll Y \ll X$.

In many situations there will be a large parameter n that goes off to infinity. When that occurs, we also use the notation $o_{n \to \infty}(X)$ or simply $o(X)$ to denote any quantity bounded in magnitude by $c(n)X$, where $c(n)$ is a function depending only on n that goes to zero as n goes to infinity. If we need $c(n)$ to depend on another parameter, e.g. d, we indicate this by further subscripts, e.g. $o_{n \to \infty;d}(X)$.

We will occasionally use the averaging notation

$$\mathbf{E}_{x \in X} f(x) := \frac{1}{|X|} \sum_{x \in X} f(x)$$

to denote the average value of a function $f : X \to \mathbf{C}$ on a non-empty finite set X.

Acknowledgments

The author is supported by a grant from the MacArthur Foundation, by NSF grant DMS-0649473, and by the NSF Waterman award.

Chapter 1

Expository Articles

1.1. The blue-eyed islanders puzzle

This is one of my favourite logic puzzles. It has a number of formulations, but I will use this one:

Problem 1.1.1. There is an island upon which a tribe resides. The tribe consists of 1000 people, with various eye colours. Yet, their religion forbids them to know their own eye color, or even to discuss the topic; thus, each resident can (and does) see the eye colors of all other residents, but has no way of discovering his or her own (there are no reflective surfaces). If a tribesperson does discover his or her own eye color, then their religion compels them to commit ritual suicide at noon the following day in the village square for all to witness. All the tribespeople are highly logical[1] and devout, and they all know that each other is also highly logical and devout (and they all know that they all know that each other is highly logical and devout, and so forth).

Of the 1000 islanders, it turns out that 100 of them have blue eyes and 900 of them have brown eyes, although the islanders are not initially aware of these statistics (each of them can of course only see 999 of the 1000 tribespeople).

One day, a blue-eyed foreigner visits the island and wins the complete trust of the tribe.

One evening, he addresses the entire tribe to thank them for their hospitality.

---

[1] For the purposes of this logic puzzle, "highly logical" means that any conclusion that can be logically deduced from the information and observations available to an islander, will automatically be known to that islander.

However, not knowing the customs, the foreigner makes the mistake of mentioning eye color in his address, remarking how unusual it is to see another blue-eyed person like myself in this region of the world.

What effect, if anything, does this *faux pas* have on the tribe?

The interesting thing about this puzzle is that there are two quite plausible arguments here, which give opposing conclusions:

Argument 1. The foreigner has no effect, because his comments do not tell the tribe anything that they do not already know (everyone in the tribe can already see that there are several blue-eyed people in their tribe).   □

Argument 2. 100 days after the address, all the blue eyed people commit suicide. This is proven as a special case of Proposition 1.1.2 below.   □

Proposition 1.1.2. *Suppose that the tribe had n blue-eyed people for some positive integer n. Then n days after the traveller's address, all n blue-eyed people commit suicide.*

Proof. We induct on n. When $n = 1$, the single blue-eyed person realizes that the traveler is referring to him or her, and thus commits suicide on the next day. Now suppose inductively that n is larger than 1. Each blue-eyed person will reason as follows: "If I am not blue-eyed, then there will only be $n - 1$ blue-eyed people on this island, and so they will all commit suicide $n - 1$ days after the traveler's address. But when $n - 1$ days pass, none of the blue-eyed people do so (because at that stage they have no evidence that they themselves are blue-eyed). After nobody commits suicide on the $(n-1)^{\text{st}}$ day, each of the blue eyed people then realizes that they themselves must have blue eyes, and will then commit suicide on the nth day."   □

Which argument is logically valid? Or are the hypotheses of the puzzle logically impossible to satisfy?[2]

Notes. I will not spoil the solution to this puzzle in this article; but one can find much discussion on this problem at the comments to the web page for this puzzle, at `terrytao.wordpress.com/2008/02/05`. See also `xkcd.com/blue_eyes.html` for some further discussion.

1.2. Kleiner's proof of Gromov's theorem

In this article, I would like to present the recent simplified proof by Kleiner [**Kl2007**] of the celebrated theorem of Gromov [**Gr1981**] on groups of polynomial growth.

---

[2]Note that this is not the same as the hypotheses being extremely implausible, which of course they are.

Let G be an at most countable group generated by a finite set S of generators, which we can take to be symmetric (i.e., $s^{-1} \in S$ whenever $s \in S$). Then we can form the *Cayley graph* Γ, whose vertices are the elements of G, with g and gs connected by an edge for every $g \in G$ and $s \in S$. This is a connected regular graph, with a transitive left action of G. For any vertex x and $R > 0$, one can define the ball $B(x, R)$ in Γ to be the set of all vertices connected to x by a path of length at most R. We say that G has *polynomial growth* if we have the bound $|B(x, R)| = O(R^{O(1)})$ as $R \to \infty$; one can easily show that the left-hand side is independent of x, and that the polynomial growth property does not depend on the choice of generating set S.

Examples of finitely generated groups of polynomial growth include

(1) Finite groups;

(2) Abelian groups (e.g. $\mathbf{Z}^d$);

(3) Nilpotent groups (a generalisation of (2));

(4) *Virtually nilpotent* groups, i.e., those that have a nilpotent subgroup of finite index (a combination of (1) and (3)).

In [**Gr1981**], Gromov proved that these are the only examples:

Theorem 1.2.1 (Gromov's theorem). *Let G be a finitely generated group of polynomial growth. Then G is virtually nilpotent.*

Gromov's original argument used a number of deep tools, including the Montgomery-Zippin-Yamabe [**MoZi1955**] structure theory of locally compact groups (related to *Hilbert's fifth problem*), as well as various earlier partial results on groups of polynomial growth. Several proofs have subsequently been found. Recently, Kleiner [**Kl2007**] obtained a proof which was significantly more elementary, although it still relies on some non-trivial partial versions of Gromov's theorem. Specifically, it needs the following result proven by Wolf [**Wo1968**] and by Milnor [**Mi1968**]:

Theorem 1.2.2 (Gromov's theorem for virtually solvable groups). *Let G be a finitely generated group of polynomial growth which is* virtually solvable *(i.e., it has a solvable subgroup of finite index). Then it is virtually nilpotent.*

The argument also needs a related result:

Theorem 1.2.3. *Let G be a finitely generated amenable[3] group which is linear, thus $G \subset GL_n(\mathbf{C})$ for some n. Then G is virtually solvable.*

---

[3]In this context, one definition of amenability is that G contains a *Følner sequence* $F_1, F_2, \ldots$ of finite sets, thus $\bigcup_{n=1}^{\infty} F_n = G$ and $\lim_{n \to \infty} |gF_n \Delta F_n|/|F_n| = 0$ for all $g \in G$.

This theorem is an immediate consequence of the Tits alternative [**Ti1972**], but also has a short elementary proof, due to Shalom [**Sh1998**]. An easy application of the pigeonhole principle to the sequence $|B(x, R)|$ for $R = 1, 2, \ldots$ shows that every group of polynomial growth is amenable. Thus Theorems 1.2.2 and 1.2.3 already give Gromov's theorem for linear groups.

Other than the use of Theorems 1.2.2 and 1.2.3, Kleiner's proof of Theorem 1.2.1 is essentially self-contained. The argument also extends to groups of *weakly polynomial growth*, which means that $|B(x, R)| = O(R^{O(1)})$ for some sequence of radii R going to infinity. (This extension of Gromov's theorem was first established in [**vdDrWi1984**].) But for simplicity we only discuss the polynomial growth case here.

1.2.1. Reductions. The first few reductions follow the lines of Gromov's original argument. The first observation is that it suffices to exhibit an infinite abelianisation of G, or more specifically to prove:

Proposition 1.2.4 (Existence of infinite abelian representation). *Let G be an infinite finitely generated group of polynomial growth. Then there exists a subgroup G' of finite index whose abelianisation $G'/[G', G']$ is infinite.*

Indeed, if G' has infinite abelianisation, then one can find a non-trivial homomorphism $\alpha : G' \to \mathbf{Z}$. The kernel K of this homomorphism is a normal subgroup of G'. Using the polynomial growth hypothesis, one can show that K is also finitely generated; furthermore, it is of polynomial growth of one lower order (i.e., the exponent in the $O(R^{O(1)})$ bound for $|B(x, R)|$ is reduced by 1). An induction hypothesis then gives that K is virtually nilpotent, which easily implies that G' (and thus G) is virtually solvable. Gromov's theorem for infinite G then follows from Theorem 1.2.2. (The theorem is of course trivial for finite G.)

Remark 1.2.5. The above argument not only shows that G is virtually solvable, but moreover that G' is the semidirect product $K \rtimes_\phi \mathbf{Z}$ of a virtually nilpotent group K and the integers, which acts on K by some automorphism ϕ. Thus one does not actually need the full strength of Theorem 1.2.2 here, but only the special case of semidirect products of the above form. In fact, most proofs of Theorem 1.2.2 proceed by reducing to this sort of case anyway.

To prove Proposition 1.2.4, it suffices to prove

Proposition 1.2.6 (Existence of infinite linear representation). *Let G be an infinite finitely generated group of polynomial growth. Then there exists a finite-dimensional representation $\rho : G \to GL_n(\mathbf{C})$ whose image $\rho(G)$ is infinite.*

Indeed, the image $\rho(G) \subset GL_n(\mathbf{C})$ is also finitely generated with polynomial growth, and hence, by Theorems 1.2.3 and 1.2.2, is virtually nilpotent (actually, for this argument we do not need Theorem 1.2.2 and would be content with virtual solvability). If the abelianisation of $\rho(G)$ is finite, one can easily pass to a subgroup G' of finite index and reduce the (virtual) step of $\rho(G')$ by 1, so one can quickly reduce to the case when the abelianisation is infinite, at which point Proposition 1.2.4 follows. So all we need to do now is to prove Proposition 1.2.6.

1.2.2. Harmonic functions on Cayley graphs. Kleiner's approach to Proposition 1.2.6 relies on the notion of a (possibly vector-valued) *harmonic function* on the Cayley graph Γ. This is a function $f : G \to H$ taking values in a Hilbert space H such that $f(g) = \frac{1}{|S|} \sum_{s \in S} f(gs)$ for all $g \in G$. Formally, harmonic functions are local minimisers of the energy functional

$$E(f) := \frac{1}{2} \sum_{g \in G} |\nabla f(g)|^2,$$

where

$$|\nabla f(g)|^2 := \frac{1}{|S|} \sum_{s \in S} \|f(gs) - f(g)\|_H^2,$$

though of course with the caveat that $E(f)$ is often infinite. (This property is also equivalent to a certain graph Laplacian of f vanishing.)

Of course, every constant function is harmonic. But there are other harmonic functions too: for instance, on $\mathbf{Z}^d$, any linear function is harmonic (regardless of the actual choice of generators). Kleiner's proof of Proposition 1.2.6 follows by combining the following two results:

Proposition 1.2.7. *Let G be an infinite finitely generated group of polynomial growth. Then there exists an (affine-) isometric (left) action of G on a Hilbert space H with no fixed points, and a harmonic map $f : G \to H$ which is G-equivariant (thus $f(gh) = gf(h)$ for all $g, h \in G$). (Note that in view of equivariance and the absence of fixed points, this harmonic map is necessarily non-constant.)*

Proposition 1.2.8. *Let G be a finitely generated group of polynomial growth, and let $d \geq 0$. Then the linear space of harmonic functions $u : G \to \mathbf{R}$ with growth of order at most d (thus $u(g) = O(R^d)$ on $B(\mathrm{id}, R)$) is finite-dimensional.*

Indeed, if f is the vector-valued map given by Proposition 1.2.7, then from the G-equivariance it is easy to see that f is of polynomial growth (indeed it is Lipschitz). But the linear projections $\{f \cdot v : v \in H\}$ of f to scalar-valued harmonic maps lie in a finite-dimensional space, by Proposition 1.2.8. This implies that the range $f(G)$ of f lies in a finite-dimensional space

V. On the other hand, the obvious action of G on V has no fixed points (being a restriction of the action of G on H), and so the image of G in $GL_n(V)$ must be infinite, and Proposition 1.2.6 follows.

It remains to prove Proposition 1.2.7 and Proposition 1.2.8. Proposition 1.2.7 follows from some more general results of Korevaar-Schoen [**KoSc1997**] and Mok [**Mo1995**], though Kleiner provided an elementary proof, which we sketch below. Proposition 1.2.8 was initially proven by Colding and Minicozzi [**CoMi1997**] (for finitely presented groups, at least) using Gromov's theorem; Kleiner's key new observation was that Proposition 1.2.8 can be proven directly by an elementary argument based on a Poincaré inequality.

1.2.3. A non-constant equivariant harmonic function. We now sketch the proof of Proposition 1.2.6. The first step is to just get the action on a Hilbert space with no fixed points:

Lemma 1.2.9. *Let G be a countably infinite amenable group. Then there exists an action of G on a Hilbert space H with no fixed points.*

This is essentially the well-known assertion that countably infinite amenable groups do not obey *Property (T)*, but we can give an explicit proof as follows. Using amenability, one can construct a nested *Følner sequence* $F_1 \subset F_2 \subset \ldots \subset \bigcup_n F_n = G$ of finite sets with the property that $|(F_{n-1} \cdot F_n) \Delta F_n| \leq 2^{-n}|F_n|$ (say). (In the case of groups of polynomial growth, one can take $F_n = B(\text{id}, R_n)$ for some rapidly growing, randomly chosen sequence of radii R_n.) We then look at $H := l^2(\mathbf{N}; l^2(G))$, the Hilbert space of sequences $f_1, f_2, \cdots \in l^2(G)$ with $\sum_n \|f_n\|_{l^2(G)}^2 < \infty$. This space has the obvious unitary action of G, defined as $g : (f_n(\cdot))_{n \in \mathbf{N}} \to (f_n(g \cdot))_{n \in \mathbf{N}}$. This action has a fixed point 0, but we can delete this fixed point by considering instead the affine-isometric action $f \mapsto gf + gh - h$, where h is the sequence $h = (\frac{1}{|F_n|^{1/2}} 1_{F_n})_{n \in \mathbf{N}}$. This sequence h does not directly lie in H, but observe that $gh - h$ lies in H for every g. One can then easily show that this action obeys the conclusions of Lemma 1.2.9.

Another way of asserting that an action of G on H has no fixed point is to say that the energy functional $E : H \to \mathbf{R}^+$ defined by $E(v) := \frac{1}{2} \sum_{s \in S} \|sv - v\|_H^2$ is always strictly positive. So Lemma 1.2.9 concludes that there exists an action of G on a Hilbert space on which E is strictly positive. It is possible to then conclude that there exists another action of G on another Hilbert space on which the energy E is not only strictly positive but actually attains its minimum at some vector v. This observation follows from more general results of Fisher and Margulis [**FiMa2005**], but one can also argue directly as follows. For every $0 < \lambda < 1$ and $A > 0$, there must exist a vector v which almost minimises E in the sense that $E(v') \geq \lambda E(v)$

whenever $\|v - v'\| \le AE(v)^{1/2}$, since otherwise one could iterate and sum a Neumann-type series to obtain a fixed point of v. But then by shifting v to the origin, and taking an ultrafilter limit (!) as $\lambda \to 1$ and $A \to \infty$, we obtain the claim.

Some elementary calculus of variations then shows that if v is the energy minimiser, the map $f : g \mapsto gv$ is a harmonic G-equivariant function from G to H, and Proposition 1.2.6 follows.

1.2.4. Poincaré's inequality, and the complexity of harmonic functions.

Now we turn to the proof of Proposition 1.2.8, which is the main new ingredient in Kleiner's argument. To simplify the exposition, let us cheat[4] and suppose that the polynomial growth condition $|B(x, R)| = O(R^{O(1)})$ is replaced by the slightly stronger doubling condition

$$|B(x, 2R)| = O(|B(x, R)|).$$

Similarly, to simplify the argument, let us pretend that the harmonic functions $u : G \to \mathbf{R}$ are not only of polynomial growth but also obey a doubling condition $\sum_{x \in B(\mathrm{id}, 2R)} u(x)^2 \ll \sum_{x \in B(\mathrm{id}, R)} u(x)^2$.

The key point is to exploit the fact that harmonic functions are fairly smooth. For instance, a simple "integration by parts" argument shows that if u is harmonic, then

$$(1.1) \qquad \sum_{x \in B(\mathrm{id}, R)} |\nabla u(x)|^2 \, dx \ll R^{-2} \sum_{x \in B(\mathrm{id}, 2R)} |u(x)|^2 \, dx;$$

this estimate can be established by the usual trick of replacing the summation over $B(\mathrm{id}, R)$ with a smoother cutoff function and then expanding out the gradient square.

To use this gradient control, Kleiner established the *Poincaré inequality*

$$(1.2) \qquad \sum_{x \in B(x_0, R)} \sum_{y \in B(x_0, R)} |u(x) - u(y)|^2 \ll R^2 |B(x_0, R)| \sum_{x \in B(x_0, 3R)} |\nabla u(x)|^2$$

assuming the doubling condition on balls. This inequality (a variant of a similar inequality of Coulhon and Saloff-Coste [**CoSC1993**]) is actually quite easy to prove. Observe that if $x, y \in B(x_0, R)$, then $x = yg$ for some $g \in B(0, 2R)$. Thus it suffices to show that

$$\sum_{x \in B(x_0, R)} |u(x) - u(xg)|^2 \ll R^2 \sum_{x \in B(x_0, 3R)} |\nabla u(x)|^2$$

for each such g. By expanding g as a product of at most $2R$ generators, splitting $u(x) - u(xg)$ as a telescoping series, and using the Cauchy-Schwarz inequality, the result follows easily.

---

[4]In practice, one can use pigeonholing arguments to show that polynomial growth implies doubling on large ranges of scales, which turns out to suffice.

Combining (1.1) and (1.2), one sees that a harmonic function which is controlled on a large ball $B(0, R)$, becomes nearly constant on small balls $B(x, \varepsilon R)$ (morally speaking, we have $|u(x) - u(y)| \ll \varepsilon |u(x)|$ "on the average" on such small balls). In particular, given any $0 < \varepsilon < 1$, one can now obtain an inequality of the form

$$\sum_{x \in B(\mathrm{id}, R)} |u(x)|^2 \ll \frac{1}{|B(\mathrm{id}, \varepsilon R)|} \sum_j \left| \sum_{x \in B_j} u(x) \right|^2 + \varepsilon^2 \sum_{x \in B(\mathrm{id}, 16R)} |u(x)|^2,$$

where B_j ranges over a cover of $B(\mathrm{id}, R)$ by balls B_j of radius εR (the number of such balls can be chosen to be polynomially bounded in $1/\varepsilon$, by the doubling condition). As a consequence, we see that if a harmonic function u obeys a doubling condition and has zero average on each ball B_j, then it vanishes identically. Morally speaking, this shows that the space of functions that obey the doubling condition has finite dimension (bounded by the number of such balls), yielding Proposition 1.2.8 in the doubling case. It requires a small amount of combinatorial trickery to obtain this conclusion in the case when u and the balls $B(\mathrm{id}, R)$ exhibit polynomial growth rather than doubling, but the general idea is still the same.

Notes. This lecture first appeared at

terrytao.wordpress.com/2008/02/14,

and was given as a talk in an IPAM workshop on expanders in pure and applied mathematics in February of 2008.

Yehuda Shalom pointed out that one does not need the full strength of Proposition 1.2.7 (constructing the vector-valued equivariant harmonic map) in order to deduce Proposition 1.2.6 from Proposition 1.2.8. Instead, all one needs is a single non-constant scalar harmonic map f of polynomial growth. Indeed, observe that G acts by left rotation on the space V of harmonic maps of a fixed polynomial growth, which is finite-dimensional by Proposition 1.2.8. If the image of f is infinite, then Proposition 1.2.6 is immediate, so suppose the image of f is finite. Then there is a normal subgroup N of G of finite index which stabilises f (and everything else in the image of f also). Because of this, the harmonic map f on G pushes down to a harmonic map on the quotient space $N \backslash G$ (which still has a right action of the generator set S). But it is easy to see that a harmonic map on a finite connected graph is constant, and so f is constant, a contradiction.

David Fisher pointed out that a simplified treatment of the Montgomery-Zippin-Yamabe theory of locally compact groups using non-standard analysis is given in [**Hi**].

1.3. The van der Corput lemma, and equidistribution on nilmanifolds

In this article I would like to record a version of van der Corput's lemma which is particularly applicable for equidistribution of orbits on nilmanifolds, and morally underlies my paper [**GrTa2009c**] with Ben Green on this topic. As an application, I reprove an old theorem of Leon Green (Theorem 2.16.18) that gives a necessary and sufficient condition as to whether a linear sequence $(g^n x)_{n=1}^{\infty}$ on a nilmanifold G/Γ is equidistributed, which generalises the famous theorem of Weyl on equidistribution of polynomials, Theorem 2.6.26.

1.3.1. The classical van der Corput trick. The classical van der Corput trick (first used implicitly by Weyl) gives a means to establish the equidistribution of a sequence $(x_n)_{n=1}^{\infty}$ in a torus $\mathbf{T}^d$ (e.g. a sequence $(P(n) \bmod 1)_{n=1}^{\infty}$ in the unit circle $\mathbf{T} = \mathbf{R}/\mathbf{Z}$ for some function P, such as a polynomial). Recall that such a sequence is said to be *equidistributed* if one has

$$(1.3) \qquad \frac{1}{N} \sum_{n=1}^{N} f(x_n) \to \int_{\mathbf{T}^d} f$$

as $N \to \infty$ for every continuous function $f : \mathbf{T}^d \to \mathbf{C}$; an equivalent[5] formulation of equidistribution is that

$$\frac{1}{N} |\{1 \le n \le N : x_n \in B\}| \to \mathrm{Vol}(B)$$

for every box B in the torus $\mathbf{T}^d$. Equidistribution is an important phenomenon to study in ergodic theory and number theory, but also arises in applications such as Monte Carlo integration and pseudorandom number generation.

A fundamental result in the subject is

Theorem 1.3.1 (Weyl equidistribution theorem). *A sequence $(x_n)_{n=1}^{\infty}$ in $\mathbf{T}^d$ is equidistributed if and only if the exponential sums*

$$(1.4) \qquad \frac{1}{N} \sum_{n=1}^{N} e^{2\pi i \chi(x_n)}$$

converge to zero for every non-trivial character $\chi : \mathbf{T}^d \to \mathbf{T}$, i.e., a non-zero continuous homomorphism to the unit cicle.

Proof. It is clear that (1.4) is a special case of (1.3). Conversely, (1.4) implies that (1.3) holds whenever f is a finite linear combination of characters $e^{2\pi i \chi}$. Applying the *Weierstrass approximation theorem*, we obtain the claim. $\qquad\square$

---

[5]The equivalence can be deduced easily from *Urysohn's lemma*.

The significance of the equidistribution theorem is that it reduces the study of equidistribution to the question of estimating exponential sums, which is a problem in analysis and number theory. For instance, from Theorem 1.3.1 and the geometric series formula we immediately obtain the following result:

Corollary 1.3.2 (Equidistribution of linear sequences in tori). *Let $\alpha \in \mathbf{T}^d$. Then the sequence $(\alpha n)_{n=1}^\infty$ is equidistributed in $\mathbf{T}^d$ if and only if α is totally irrational, which means that $\chi(\alpha) \neq 0$ for all non-zero characters χ.*

For instance, the linear sequence $(\sqrt{2}n \bmod 1, \sqrt{3}n \bmod 1)$ is equidistributed in the two-torus $\mathbf{T}^2$, since $(\sqrt{2}, \sqrt{3})$ is totally irrational, but the linear sequence $(\sqrt{2}n \bmod 1, \sqrt{8}n \bmod 1)$ is not[6] (the character $\chi : (x, y) \mapsto y - 2x$ annihilates $(\sqrt{2}, \sqrt{8})$ and thus obstructs equidistribution).

One elementary but very useful tool for estimating exponential sums is *Weyl's differencing trick*, that ultimately rests on the humble Cauchy-Schwarz inequality. One formulation of this trick can be phrased as the following inequality (cf. Lemma 2.12.7):

Lemma 1.3.3 (van der Corput inequality). *Let $a_1, a_2, \ldots$ be a sequence of complex numbers bounded in magnitude by 1. Then for any $1 \leq H \leq N$ we have*

$$(1.5) \qquad \left| \frac{1}{N} \sum_{n=1}^N a_n \right| \ll \left(\frac{1}{H} \sum_{h=0}^{H-1} \left| \frac{1}{N} \sum_{n=1}^N a_{n+h} \overline{a_n} \right| \right)^{1/2} + O\left(\frac{H}{N} \right).$$

Proof. Observe that

$$\frac{1}{N} \sum_{n=1}^N a_n = \frac{1}{N} \sum_{n=1}^N a_{n+h} + O\left(\frac{H}{N} \right)$$

for every $0 \leq h \leq H - 1$. Averaging this in h we obtain

$$\frac{1}{N} \sum_{n=1}^N a_n = \frac{1}{N} \sum_{n=1}^N \frac{1}{H} \sum_{h=0}^{H-1} a_{n+h} + O\left(\frac{H}{N} \right)$$

and hence by the Cauchy-Schwarz inequality

$$\left| \frac{1}{N} \sum_{n=1}^N a_n \right| \leq \left(\frac{1}{N} \sum_{n=1}^N \left| \frac{1}{H} \sum_{h=0}^{H-1} a_{n+h} \right|^2 \right)^{1/2} + O\left(\frac{H}{N} \right).$$

Expanding out the square and rearranging a bit, we soon obtain the upper bound (1.5) (in fact one can sharpen the constants slightly here, though this will not be important for this discussion). □

---

[6]Of course, in the latter case, the orbit is still equidistributed in a smaller torus, namely the kernel of the character χ mentioned above; this is an extremely simple case of *Ratner's theorem*. See Section 2.16 for further discussion.

The significance of this inequality is that it replaces the task of bounding a sum of coefficients a_n by that of bounding a sum of "differentiated" coefficients $a_{n+h}\overline{a_n}$. This trick is thus useful in "polynomial"-type situations when the differentiated coefficients are often simpler than the original coefficients. One particularly clean application of this inequality is as follows:

Corollary 1.3.4 (Van der Corput's difference theorem). *Let $(x_n)_{n=1}^{\infty}$ be a sequence in a torus $\mathbf{T}^d$ such that the difference sequences $(x_{n+h} - x_n)_{n=1}^{\infty}$ are equidistributed for every non-zero h. Then $(x_n)_{n=1}^{\infty}$ is itself equidistributed.*

Proof. By Theorem 1.3.1, it suffices to show that (1.4) holds for every non-trivial character χ. But by Lemma 1.3.3, we can bound the magnitude of the left-hand side of (1.4) by

$$(1.6) \qquad \ll \left(\frac{1}{H} \sum_{h=0}^{H-1} \left| \frac{1}{N} \sum_{n=1}^{N} e^{2\pi i \chi(x_{n+h})} e^{-2\pi i \chi(x_n)} \right| \right)^{1/2} + O\left(\frac{H}{N} \right)$$

for any fixed H.

Now we use the fact that χ is a character to simplify $e^{2\pi i \chi(x_{n+h})} e^{-2\pi i \chi(x_n)}$ as $e^{2\pi i \chi(x_{n+h} - x_n)}$. By hypothesis and the equidistribution theorem, the inner sum $\frac{1}{N} \sum_{n=1}^{N} e^{2\pi i \chi(x_{n+h})} e^{-2\pi i \chi(x_n)}$ goes to zero as $N \to \infty$ for any fixed non-zero h; when instead h is zero, this sum is of course just 1. We conclude that for fixed H, the expression (1.6) is bounded by $O(1/H)$ in the limit $N \to \infty$. Thus the limit (or limit superior) of the magnitude of (1.4) is bounded in magnitude by $O(1/H)$ for every H, and is thus zero. The claim follows. $\qquad\square$

By iterating this theorem, and using the observation that the difference sequence $(P(n + h) - P(n))_{n=1}^{\infty}$ of a polynomial sequence $(P(n))_{n=1}^{\infty}$ of degree d becomes a polynomial sequence of degree $d - 1$ for any non-zero h, we can conclude by induction the following famous result of Weyl, generalising Corollary 1.3.2 (see also Theorem 2.1.12, Corollary 2.4.4, and Theorem 2.6.26):

Theorem 1.3.5 (Equidistribution of polynomial sequences in tori). *Let $P : \mathbf{Z} \to \mathbf{T}^d$ be a polynomial sequence taking values in a torus. Then the sequence $(P(n))_{n=1}^{\infty}$ is equidistributed in $\mathbf{T}^d$ if and only if $\chi(P(\cdot))$ is non-constant for all non-zero characters χ.*

In the one-dimensional case $d = 1$, this theorem asserts that a polynomial $P : \mathbf{Z} \to \mathbf{R}$ with real coefficients is equidistributed modulo one if and only if it has at least one irrational non-constant coefficient; thus for instance the sequence $(\pi n^3 + \sqrt{2} n^2 + \frac{1}{4} n \bmod 1)_{n=1}^{\infty}$ is equidistributed.

1.3.2. A variant of the trick. It turns out that van der Corput's difference theorem (Corollary 1.3.4) can be generalised to deal not just on tori, but on more general measure spaces with a torus action. Given a topological probability space (X, μ) (which we will take to be a Polish space to avoid various technicalities) and a sequence $(x_n)_{n=1}^{\infty}$ in X, we say that such a sequence is *equidistributed with respect to μ* if we have

$$(1.7) \qquad\qquad \frac{1}{N} \sum_{n=1}^{N} f(x_n) \to \int_X f \, d\mu$$

for all continuous compactly supported functions $f : X \to \mathbf{C}$. This clearly generalises the previous notion of equidistribution, in which X was a torus and μ was a uniform probability measure.

To motivate our generalised version of Corollary 1.3.4, we observe that the hypothesis "the sequence $(x_{n+h} - x_n)_{n=1}^{\infty}$ is equidistributed in $\mathbf{T}^d$" can be phrased in a more dynamical fashion (eliminating the subtraction operation, which is algebraic) as the equivalent assertion that the sequence of pairs $((x_{n+h}, x_n))_{n=1}^{\infty}$ in $\mathbf{T}^d \times \mathbf{T}^d$, after quotienting out by the action of the diagonal subgroup $(\mathbf{T}^d)^{\Delta} := \{(y, y) : y \in \mathbf{T}^d\}$, becomes equidistributed on the quotient space $\mathbf{T}^d \times \mathbf{T}^d / (\mathbf{T}^d)^{\Delta}$. This convoluted reformulation is necessary for generalisations, in which we do not have a good notion of subtraction, but we still have a good notion of group action and quotient spaces.

We can now prove

Proposition 1.3.6 (Generalised van der Corput difference theorem)**.** *Let (X, μ) be a (Polish) probability space with a continuous (right) action of a torus $\mathbf{T}^d$, and let $\pi : X \to X/\mathbf{T}^d$ be the projection map onto the quotient space (which then has the pushforward measure $\pi_* \mu$). Let $(x_n)_{n=1}^{\infty}$ be a sequence in X obeying the following properties:*

(1) *(Horizontal equidistribution) The projected sequence $(\pi(x_n))_{n=1}^{\infty}$ in $X/\mathbf{T}^d$ is equidistributed with respect to $\pi_* \mu$.*

(2) *(Vertical differenced equidistribution) For every non-zero h, the sequence $((x_{n+h}, x_n)(\mathbf{T}^d)^{\Delta})_{n=1}^{\infty}$ in the quotiented product space $(X \times X)/(\mathbf{T}^d)^{\Delta}$ is equidistributed with respect to some measure ν_h which is invariant under the action of the torus $\mathbf{T}^d \times \mathbf{T}^d/(\mathbf{T}^d)^{\Delta}$.*

Then $(x_n)_{n=1}^{\infty}$ is equidistributed with respect to μ.

Note that Corollary 1.3.4 is the special case of Proposition 1.3.6 in which X is itself the torus $\mathbf{T}^d$ with the usual translation action and uniform measure (so that the quotient space is a point).

Proof. We need to verify the property (1.3). If the function f was invariant under the action of the torus $\mathbf{T}^d$, then we could push it down to the quotient space $X/\mathbf{T}^d$ and the claim would follow from hypothesis (1). We may therefore subtract the invariant component $\int_{\mathbf{T}^d} f(\cdot y)\, dy$ from our function and assume instead that f has zero vertical mean in the sense that $\int_{\mathbf{T}^d} f(xy)\, dy = 0$ for all x. A Fourier expansion in the vertical variable (or the Weierstrass approximation theorem) then allows us to reduce to the case when f has a *vertical frequency* given by some non-zero character $\chi : \mathbf{T}^d \to \mathbf{T}$ of the torus, in the sense that $f(xy) = f(x)e^{2\pi i \chi(y)}$ for all $x \in X$ and $y \in \mathbf{T}^d$.

Now we apply van der Corput's inequality as in the proof of Corollary 1.3.4. Using these arguments, we find that it suffices to show that

$$\frac{1}{N}\sum_{n=1}^{N} f(x_{n+h})\overline{f(x_n)} \to 0$$

for each non-zero h. But the summand here is just the tensor product function $f \otimes \overline{f} : X \times X \to \mathbf{C}$ applied to the pair (x_{n+h}, x_n). The fact that f has a vertical frequency implies that $f \otimes \overline{f}$ is invariant with respect to the diagonal action $(\mathbf{T}^d)^{\Delta}$, and thus this function descends to the quotient space $(X \times X)/(\mathbf{T}^d)^{\Delta}$. On the other hand, as the vertical frequency is non-trivial, the latter function also has zero mean on every orbit of $\mathbf{T}^d \times \mathbf{T}^d/(\mathbf{T}^d)^{\Delta}$ and thus vanishes when integrated against ν_h. The claim then follows from hypothesis (2). $\qquad\square$

As an application, let us prove the following result, first established in [**Gr1961**]:

Theorem 1.3.7 (Equidistribution of linear sequences in nilmanifolds). *Let G/Γ be a nilmanifold (where we take the nilpotent group G to be connected for simplicity, although this is not strictly necessary), and let $g \in G$ and $x \in G/\Gamma$. Then $(g^n x)_{n=1}^{\infty}$ is equidistributed with respect to Haar measure on G/Γ if and only if $\chi(g^n x)$ is non-constant in n for every non-trivial horizontal character $\chi : G/\Gamma \to \mathbf{T}$, where a horizontal character is any continuous homomorphism $\chi : G \to \mathbf{T}$ that vanishes on Γ (and thus descends to G/Γ).*

This statement happens to contain[7] Weyl's result (Theorem 1.3.5) as a special case, because polynomial sequences can be encoded as linear sequences in nilmanifolds; but it is actually stronger, allowing extensions to generalised polynomials that involve the floor function $\lfloor \cdot \rfloor$ or the fractional part function $\{\,\}$. For instance, if we take

$$G := \begin{pmatrix} 1 & \mathbf{R} & \mathbf{R} \\ 0 & 1 & \mathbf{R} \\ 0 & 0 & 1 \end{pmatrix}, \qquad \Gamma := \begin{pmatrix} 1 & \mathbf{Z} & \mathbf{Z} \\ 0 & 1 & \mathbf{Z} \\ 0 & 0 & 1 \end{pmatrix},$$

---

[7]It is also equivalent to Theorem 2.16.18.

and

$$g := \begin{pmatrix} 1 & \alpha & \beta \\ 0 & 1 & \gamma \\ 0 & 0 & 1 \end{pmatrix}, \qquad x = \Gamma$$

for some real numbers α, β, γ, then a computation shows that

$$g^n x = \begin{pmatrix} 1 & \{\alpha n\} & \{\beta n + \alpha \frac{n(n-1)}{2} - \{\alpha n\} \lfloor \gamma n \rfloor\} \\ 0 & 1 & \{\gamma n\} \\ 0 & 0 & 1 \end{pmatrix} \Gamma,$$

and then Theorem 1.3.7 asserts that the triple

$$\left(\{\alpha n\}, \left\{ \beta n + \alpha \frac{n(n-1)}{2} - \{\alpha n\} \lfloor \gamma n \rfloor \right\}, \{\gamma n\} \right)$$

is equidistributed in the unit cube $[0,1]^3$ if and only if the pair (α, γ) is totally irrational (the rationality of β turns out to be irrelevant). Even for concrete values such as $\alpha = \sqrt{2}$, $\beta = 0$, $\gamma = \sqrt{3}$, it is not obvious how to establish this fact directly; for instance a direct application of Corollary 1.3.4 does not obviously simplify the situation.

Proof of Theorem 1.3.7. (Sketch) It is clear that if $\chi(g^n x)$ is constant for some non-trivial character, then the orbit $g^n x$ is trapped on a level set of χ and thus cannot equidistribute. Conversely, suppose that $\chi(g^n x)$ is never constant. We induct on the step s of the nilmanifold. The case $s = 0$ is trivial, and the case $s = 1$ follows from Corollary 1.3.2, so suppose inductively that $s \geq 2$ and that the claim has already been proven for smaller s. We then look at the vertical torus $G_s/(\Gamma \cap G_s) \equiv \mathbf{T}^d$, where G_s is the last non-trivial group in the lower central series (and thus central). The quotient of the nilmanifold G/Γ by this torus action turns out to be a nilmanifold of one lower step (in which G is replaced by G/G_s), and so the projection of the orbit $(g^n x)_{n=1}^\infty$ is then equidistributed by induction hypothesis. Applying Proposition 1.3.6, it thus suffices to check that for each non-zero h, the sequence of pairs $(g^{n+h} x, g^n x)$ in $G/\Gamma \times G/\Gamma$, after quotienting out by the diagonal action of the torus, is equidistributed with respect to some measure which is invariant under the residual torus $\mathbf{T}^d \times \mathbf{T}^d / (\mathbf{T}^d)^\Delta$.

We first pass to the abelianisation (or *horizontal torus*) $G/G_2\Gamma$ of the nilmanifold, and observe that the projections $\pi(g^{n+h} x), \pi(g^n x)$ of the coefficients of the pair $(g^{n+h} x, g^n x)$ to this torus only differ by a constant $\pi(g^h)$. Thus the pair $(g^{n+h} x, g^n x)$ does not range freely in $G/\Gamma \times G/\Gamma$, but is instead constrained to a translate of a smaller nilmanifold $G/\Gamma \times_\pi G/\Gamma$, defined as the space of pairs (x, y) with $\pi(x) = \pi(y)$. After quotienting out also by the diagonal vertical torus, we obtain a nilmanifold coming from the group $(G \times_{G_2} G)/G_s^\Delta$, where $G \times_{G_2} G$ is the space of pairs (g, h) of group elements $g, h \in G$ whose projections to the abelianisation G/G_2 agree, and

$G_s^\Delta := \{(g_s, g_s) : g_s \in G_s\}$ is the vertical diagonal group. But a short computation shows that this new group is at most $s - 1$ step nilpotent. One can then apply the induction hypothesis to show the required equidistribution properties of (x_{n+h}, x_n), thus closing the induction by Proposition 1.3.6. $\square$

There are many further generalisations of these results, including a polynomial version of Theorem 1.3.7 in [**Le2005**], [**Le2005b**] that also permits G to be disconnected, and quantitative versions of all of these results in [**GrTa2009c**].

Notes. This article first appeared at

terrytao.wordpress.com/2008/06/14.

Thanks to an anonymous commenter for corrections.

1.4. The strong law of large numbers

Let X be a real-valued random variable, and let $X_1, X_2, X_3, \ldots$ be an infinite sequence of independent and identically distributed copies of X. Let $\overline{X}_n := \frac{1}{n}(X_1 + \ldots + X_n)$ be the empirical averages of this sequence. A fundamental theorem in probability theory is the *law of large numbers*, which comes in both a weak and a strong form:

Theorem 1.4.1 (Weak law of large numbers). *Suppose that the first moment* $\mathbf{E}|X|$ *of X is finite. Then $\overline{X}_n$ converges in probability to $\mathbf{E}X$, thus* $\lim_{n \to \infty} \mathbf{P}(|\overline{X}_n - \mathbf{E}X| \geq \varepsilon) = 0$ *for every $\varepsilon > 0$.*

Theorem 1.4.2 (Strong law of large numbers). *Suppose that the first moment* $\mathbf{E}|X|$ *of X is finite. Then $\overline{X}_n$ converges almost surely to $\mathbf{E}X$, thus* $\mathbf{P}(\lim_{n \to \infty} \overline{X}_n = \mathbf{E}X) = 1$.

Remark 1.4.3. The concepts of convergence in probability and almost sure convergence in probability theory are specialisations of the concepts of *convergence in measure* and *pointwise convergence almost everywhere* in measure theory.

Remark 1.4.4. If one strengthens the first moment assumption to that of finiteness of the second moment $\mathbf{E}|X|^2$, then we of course have a more precise statement than the (weak) law of large numbers, namely the *central limit theorem*, but I will not discuss that theorem here. With even more hypotheses on X, one similarly has more precise versions of the strong law of large numbers, such as the *Chernoff inequality*, which I will again not discuss here.

The weak law is easy to prove, but the strong law (which of course implies the weak law, by the dominated convergence theorem) is more subtle, and in fact the proof of this law (assuming just finiteness of the first moment) usually appears only in advanced graduate texts. So I thought I would present a proof here of both laws, which proceeds by the standard techniques of the moment method and truncation. The emphasis in this exposition will be on motivation and methods rather than brevity and strength of results; there do exist proofs of the strong law in the literature that have been compressed down to the size of one page or less, but this is not my goal here.

1.4.1. The moment method. The *moment method* seeks to control the tail probabilities of a random variable (i.e., the probability that it fluctuates far from its mean) by means of moments, and in particular the zeroth, first, or second moment. The reason that this method is so effective is because the first few moments can often be computed rather precisely. The *first moment method* usually employs *Markov's inequality*

$$(1.8) \qquad \mathbf{P}(|X| \geq \lambda) \leq \frac{1}{\lambda} \mathbf{E}|X|$$

(which follows by taking expectations of the pointwise inequality $\lambda I(|X| \geq \lambda) \leq |X|$), whereas the *second moment method* employs some version of *Chebyshev's inequality*, such as

$$(1.9) \qquad \mathbf{P}(|X| \geq \lambda) \leq \frac{1}{\lambda^2} \mathbf{E}|X|^2$$

(note that (1.9) is just (1.8) applied to the random variable $|X|^2$ and to the threshold λ^2).

Generally speaking, to compute the first moment one usually employs *linearity of expectation*,

$$\mathbf{E}(X_1 + \ldots + X_n) = \mathbf{E}X_1 + \ldots + \mathbf{E}X_n,$$

whereas to compute the second moment one also needs to understand *covariances* $\mathbf{Cov}(X_i, X_j) := \mathbf{E}(X_iX_j) - \mathbf{E}(X_i)\mathbf{E}(X_j)$ (which are particularly simple if one assumes pairwise independence), thanks to identities such as

$$\mathbf{E}(X_1 + \ldots + X_n)^2 = \mathbf{E}X_1^2 + \ldots + \mathbf{E}X_n^2 + 2 \sum_{1 \leq i < j \leq n} X_iX_j$$

or the normalised variant
$$(1.10)$$
$$\mathbf{Var}(X_1 + \ldots + X_n) = \mathbf{Var}(X_1) + \ldots + \mathbf{Var}(X_n) + 2 \sum_{1 \leq i < j \leq n} \mathbf{Cov}(X_i, X_j).$$

Higher moments can in principle give more precise information, but often require stronger assumptions on the objects being studied, such as joint independence.

Here is a basic application of the first moment method:

Lemma 1.4.5 (Borel-Cantelli lemma). *Let $E_1, E_2, E_3, \ldots$ be a sequence of events such that $\sum_{n=1}^{\infty} \mathbf{P}(E_n)$ is finite. Then almost surely, only finitely many of the events E_n are true.*

Proof. Let $I(E_n)$ denote the indicator function of the event E_n. Our task is to show that $\sum_{n=1}^{\infty} I(E_n)$ is almost surely finite. But by linearity of expectation, the expectation of this random variable is $\sum_{n=1}^{\infty} \mathbf{P}(E_n)$, which is finite by hypothesis. By Markov's inequality (1.8) we conclude that

$$\mathbf{P}\left(\sum_{n=1}^{\infty} I(E_n) \geq \lambda\right) \leq \frac{1}{\lambda} \sum_{n=1}^{\infty} \mathbf{P}(E_n).$$

Letting $\lambda \to \infty$ we obtain the claim. $\square$

Returning to the law of large numbers, the first moment method gives the following tail bound:

Lemma 1.4.6 (First moment tail bound). *If $\mathbf{E}|X|$ is finite, then*

$$\mathbf{P}(|\overline{X}_n| \geq \lambda) \leq \frac{\mathbf{E}|X|}{\lambda}.$$

Proof. By the triangle inequality, $|\overline{X}_n| \leq \overline{|X|}_n$. By linearity of expectation, the expectation of $\overline{|X|}_n$ is $\mathbf{E}|X|$. The claim now follows from Markov's inequality. $\square$

Lemma 1.4.6 is not strong enough by itself to prove the law of large numbers in either weak or strong form—in particular, it does not show any improvement as n gets large—but it will be useful to handle one of the error terms in those proofs.

We can get stronger bounds than Lemma 1.4.6—in particular, bounds which improve with n—at the expense of stronger assumptions on X.

Lemma 1.4.7 (Second moment tail bound). *If $\mathbf{E}|X|^2$ is finite, then*

$$\mathbf{P}(|\overline{X}_n - \mathbf{E}(X)| \geq \lambda) \leq \frac{\mathbf{E}|X - \mathbf{E}(X)|^2}{n\lambda^2}.$$

Proof. A standard computation, exploiting (1.10) and the pairwise independence of the X_i, shows that the variance $\mathbf{E}|\overline{X}_n - \mathbf{E}(X)|^2$ of the empirical averages $\overline{X}_n$ is equal to $\frac{1}{n}$ times the variance $\mathbf{E}|X - \mathbf{E}(X)|^2$ of the original variable X. The claim now follows from Chebyshev's inequality (1.9). $\square$

In the opposite direction, there is the *zeroth moment method*, more commonly known as the *union bound*

$$\mathbf{P}(E_1 \vee \ldots \vee E_n) \leq \sum_{j=1}^{n} \mathbf{P}(E_j)$$

or equivalently (to explain the terminology "zeroth moment")

$$\mathbf{E}(X_1 + \ldots + X_n)^0 \leq \mathbf{E}X_1^0 + \ldots + X_n^0$$

for any non-negative random variables $X_1, \ldots, X_n \geq 0$. Applying this to the empirical means, we obtain the *zeroth moment tail estimate*

$$(1.11) \qquad \mathbf{P}(\overline{X}_n \neq 0) \leq n\mathbf{P}(X \neq 0).$$

Just as the second moment bound (Lemma 1.4.7) is only useful when one has good control on the second moment (or variance) of X, the zeroth moment tail estimate (1.10) is only useful when we have good control on the zeroth moment $\mathbf{E}|X|^0 = \mathbf{P}(X \neq 0)$, i.e., when X is mostly zero.

1.4.2. Truncation. The second moment tail bound (Lemma 1.4.7) already gives the weak law of large numbers in the case when X has finite second moment (or equivalently, finite variance). In general, if all one knows about X is that it has finite first moment, then we cannot conclude that X has finite second moment. However, we can perform a truncation

$$(1.12) \qquad X = X_{\leq N} + X_{>N}$$

of X at any desired threshold N, where $X_{\leq N} := XI(|X| \leq N)$ and $X_{>N} := XI(|X| > N)$. The first term $X_{\leq N}$ has finite second moment; indeed we clearly have

$$\mathbf{E}|X_{\leq N}|^2 \leq N\mathbf{E}|X|$$

and hence also we have finite variance

$$(1.13) \qquad \mathbf{E}|X_{\leq N} - \mathbf{E}X_{\leq N}|^2 \leq N\mathbf{E}|X|.$$

The second term $X_{>N}$ may have infinite second moment, but its first moment is well controlled. Indeed, by the monotone convergence theorem, we have

$$(1.14) \qquad \mathbf{E}|X_{>N}| \to 0 \text{ as } N \to \infty.$$

By the triangle inequality, we conclude that the first term $X_{\leq N}$ has expectation close to $\mathbf{E}X$:

$$(1.15) \qquad \mathbf{E}X_{\leq N} \to \mathbf{E}(X) \text{ as } N \to \infty.$$

These are all the tools we need to prove the weak law of large numbers.

Proof of Theorem 1.4.1. Let $\varepsilon > 0$. It suffices to show that, whenever n is sufficiently large depending on ε, $\overline{X}_n = \mathbf{E}X + O(\varepsilon)$ with probability $1 - O(\varepsilon)$.

From (1.14), (1.15), we can find a threshold N (depending on ε) such that $\mathbf{E}|X_{\geq N}| = O(\varepsilon^2)$ and $\mathbf{E}X_{<N} = \mathbf{E}X + O(\varepsilon)$. Now we use (1.12) to split

$$\overline{X}_n = (\overline{X_{\geq N}})_n + (\overline{X_{<N}})_n.$$

From the first moment tail bound (Lemma 1.4.6), we know that $(\overline{X_{\geq N}})_n = O(\varepsilon)$ with probability $1 - O(\varepsilon)$. From the second moment tail bound (Lemma 1.4.7) and (1.13), we know that $(\overline{X_{<N}})_n = \mathbf{E}X_{<N} + O(\varepsilon) = \mathbf{E}X + O(\varepsilon)$ with probability $1 - O(\varepsilon)$ if n is sufficiently large depending on N and ε. The claim follows. $\qquad\square$

1.4.3. The strong law. The strong law can be proven by pushing the above methods a bit further, and using a few more tricks.

The first trick is to observe that to prove the strong law, it suffices to do so for non-negative random variables $X \geq 0$. Indeed, this follows immediately from the simple fact that any random variable X with finite first moment can be expressed as the difference of two non-negative random variables $\max(X, 0)$, $\max(-X, 0)$ of finite first moment.

Once X is non-negative, we see that the empirical averages $\overline{X}_n$ cannot decrease too quickly in n. In particular we observe that

(1.16) $\qquad \overline{X}_m \leq (1 + O(\varepsilon))\overline{X}_n$ whenever $(1 - \varepsilon)n \leq m \leq n$.

Because of this quasimonotonicity, we can *sparsify* the set of n for which we need to prove the strong law. More precisely, it suffices to prove

Theorem 1.4.8 (Strong law of large numbers, reduced version). *Let X be a non-negative random variable with $\mathbf{E}X < \infty$, and let $1 \leq n_1 \leq n_2 \leq n_3 \leq \dots$ be a sequence of integers which is* lacunary *in the sense that $n_{j+1}/n_j > c$ for some $c > 1$ and all sufficiently large j. Then $\overline{X}_{n_j}$ converges almost surely to $\mathbf{E}X$.*

Indeed, if we could prove the reduced version, then on applying that version to the lacunary sequence $n_j := \lfloor (1+\varepsilon)^j \rfloor$ and using (1.16), we would see that almost surely the empirical means $\overline{X}_n$ cannot deviate by more than a multiplicative error of $1 + O(\varepsilon)$ from the mean $\mathbf{E}X$. Setting $\varepsilon := 1/m$ for $m = 1, 2, 3, \dots$ (and using the fact that a countable intersection of almost sure events remains almost sure) we obtain the full strong law.

Remark 1.4.9. This sparsification trick is philosophically related to the *dyadic pigeonhole principle* philosophy; see [**Ta3**]. One could easily sparsify further, so that the lacunarity constant c is large instead of small, but this turns out not to help us too much in what follows.

Now that we have sparsified the sequence, it becomes economical to apply the Borel-Cantelli lemma (Lemma 1.4.5). Indeed, by many applications of that lemma we see that it suffices to show that[8]

$$(1.17) \qquad \sum_{j=1}^{\infty} \mathbf{P}(\overline{X}_{n_j} \neq \mathbf{E}(X) + O(\varepsilon)) < \infty$$

for non-negative X of finite first moment, any lacunary sequence $1 \leq n_1 \leq n_2 \leq \dots$, and any $\varepsilon > 0$.

Remark 1.4.10. If we did not first sparsify the sequence, the Borel-Cantelli lemma would have been too expensive to apply; see Remark 1.4.12 below. Generally speaking, the Borel-Cantelli lemma is only worth applying when one expects the events E_n to be fairly "disjoint" or "independent" of each other; in the non-lacunary case, the events E_n change very slowly in n, which makes the lemma very inefficient. We will not see how lacunarity is exploited until the punchline at the very end of the proof, but certainly there is no harm in taking advantage of this "free" reduction to the lacunary case now, even if it is not immediately clear how it will be exploited.

At this point we go back and apply the methods that already worked to give the weak law. Namely, to estimate each of the tail probabilities $\mathbf{P}(\overline{X}_{n_j} \neq \mathbf{E}(X) + O(\varepsilon))$, we perform a truncation (1.12) at some threshold N_j. It is not immediately obvious what truncation to perform, so we adopt the usual strategy of leaving N_j unspecified for now and optimising in this parameter later.

We should at least pick N_j large enough so that $\mathbf{E}X_{<N_j} = \mathbf{E}X + O(\varepsilon)$. From the second moment tail estimate (Lemma 1.4.7) we conclude that $(\overline{X}_{<N_j})_{n_j}$ is also equal to $\mathbf{E}X + O(\varepsilon)$ with probability $1 - O\left(\frac{1}{\varepsilon n_j}\mathbf{E}|X_{\leq N_j}|^2\right)$. One could attempt to simplify this expression using (1.13), but this turns out to be a little wasteful, so let us hold off on that for now. However, (1.13) does strongly suggest that we want to take N_j to be something like n_j, which is worth keeping in mind in what follows.

Now we look at the contribution of $X_{\geq N_j}$. One could use the first moment tail estimate (Lemma 1.4.6), but it turns out that the first moment $\mathbf{E}X_{>N_j}$ decays too slowly in j to be of much use (recall that we are expecting N_j to be like the lacunary sequence n_j); the root problem here is that the decay (1.14) coming from the monotone convergence theorem is *ineffective*.[9]

But there is one last card to play, which is the zeroth moment method tail estimate (1.11). As mentioned earlier, this bound is lousy in general—but is very good when X is mostly zero, which is precisely the situation

---

[8]This is a slight abuse of the $O(\ )$ notation, but it should be clear what is meant by this.

[9]One could effectivise this using the finite convergence principle (see Section 1.3 of *Structure and Randomness*), but this turns out to give very poor results here.

with $X_{>N_j}$, and in particular we see that $\overline{(X_{>N_j})}_{n_j}$ is zero with probability $1 - O(n_j \mathbf{P}(X > N_j))$.

Putting this all together, we see that

$$\mathbf{P}(\overline{X}_{n_j} \neq \mathbf{E}(X) + O(\varepsilon)) \ll \frac{1}{\varepsilon n_j} \mathbf{E}|X_{\leq N_j}|^2 + n_j \mathbf{P}(X > N_j).$$

Summing this in j, we see that we will be done as soon as we figure out how to choose N_j so that

(1.18)
$$\sum_{j=1}^{\infty} \frac{1}{n_j} \mathbf{E}|X_{\leq N_j}|^2$$

and

(1.19)
$$\sum_{j=1}^{\infty} n_j \mathbf{P}(X > N_j)$$

be both finite. As usual, we have a trade-off: making the N_j larger makes (1.19) easier to establish at the expense of (1.18), and vice versa when making N_j smaller.

Based on the discussion earlier, it is natural to try setting $N_j := n_j$. Happily, this choice works cleanly; the lacunary nature of n_j ensures (basically from the geometric series formula) that we have the pointwise estimates

$$\sum_{j=1}^{\infty} \frac{1}{n_j} X_{\leq n_j} = O(X)$$

and

$$\sum_{j=1}^{\infty} n_j I(X \geq n_j) = O(X)$$

(where the implied constant here depends on the sequence $n_1, n_2, \ldots$, and in particular on the lacunarity constant c). The claims (1.17), (1.18) then follow from one last application of linearity of expectation, giving the strong law of large numbers.

Remark 1.4.11. The above proof in fact shows that the strong law of large numbers holds even if one only assumes pairwise independence of the X_n, rather than joint independence.

Remark 1.4.12. It is essential that the random variables $X_1, X_2, \ldots$ are "recycled" from one empirical average $\overline{X}_n$ to the next, in order to get the crucial quasimonotonicity property (1.16). If instead we took completely independent averages $\overline{X}_n = \frac{1}{n}(X_{n,1} + \ldots + X_{n,n})$, where the $X_{i,j}$ are all iid, then the strong law of large numbers in fact breaks down[10] with just a first moment assumption. Of course, if one restricts attention to a lacunary sequence of n, then the above proof goes through in the independent case (since the Borel-Cantelli lemma is insensitive to this independence). By exploiting the joint independence further (e.g. by using *Chernoff's inequality*) one can also get the strong law for independent empirical means for the full sequence n under second moment bounds.

Remark 1.4.13. From the perspective of interpolation theory, one can view the above argument as an interpolation argument, establishing an L^1 estimate (1.17) by interpolating between an L^2 estimate (Lemma 1.4.7) and the L^0 estimate (1.11).

Remark 1.4.14. By viewing the sequence $X_1, X_2, \ldots$ as a stationary process, and thus as a special case of a measure-preserving system one can view the weak and strong law of large numbers as special cases of the mean and pointwise ergodic theorems respectively (see Exercise 2.8.9 and Theorem 2.9.4).

Notes. This article first appeared at

terrytao.wordpress.com/2008/06/18.

Thanks to toomuchcoffeeman and Joshua Batson for corrections.

Siva pointed out that for bounded random variables, a short proof of the strong law of large numbers (interpreting this law as an ergodic theorem for stationary processes) appeared in [**Ke1995**].

Giovanni Peccati noted that almost sure analogues of the central limit theorem exist; see e.g. [**Be1995**].

1.5. Tate's proof of the functional equation

The *Riemann zeta function* $\zeta(s)$, defined for $\mathrm{Re}(s) > 1$ by the formula

$$(1.20) \qquad \zeta(s) := \sum_{n \in \mathbf{N}} \frac{1}{n^s},$$

where $\mathbf{N} = \{1, 2, \ldots\}$ are the natural numbers, and extended meromorphically to other values of s by analytic continuation, obeys the remarkable *functional equation*

$$(1.21) \qquad \Xi(s) = \Xi(1 - s),$$

---

[10] For a counterexample, consider a random variable X which equals $2^m/m^2$ with probability 2^{-m} for $m = 1, 2, 3, \ldots$; this random variable (barely) has finite first moment, but for $n \sim 2^m/m^2$, we see that $\overline{X}_n$ deviates by at least absolute constant from its mean with probability $\gg 1/m^2$. As the empirical means $\overline{X}_n$ for $n \sim 2^m/m^2$ are now jointly independent, the probability that one of them deviates significantly is now extremely close to 1 (super-exponentially close in m, in fact), leading to the total failure of the strong law in this setting.

where

$$(1.22) \qquad \Xi(s) := \Gamma_\infty(s)\zeta(s)$$

is the *Riemann Xi function*,

$$(1.23) \qquad \Gamma_\infty(s) := \pi^{-s/2}\Gamma(s/2)$$

is the *Gamma factor at infinity*, and the *Gamma function* $\Gamma(s)$ is defined for $\mathrm{Re}(s) > 1$ by

$$(1.24) \qquad \Gamma(s) := \int_0^\infty e^{-t}t^s\,\frac{dt}{t}$$

and extended meromorphically to other values of s by analytic continuation.

There are many proofs known of the functional equation (1.21). One of them (dating back to [**Ri1859**][11]) relies on the *Poisson summation formula*

$$(1.25) \qquad \sum_{a\in\mathbf{Z}} f_\infty(at_\infty) = \frac{1}{|t|_\infty}\sum_{a\in\mathbf{Z}} \hat{f}_\infty(a/t_\infty)$$

for the reals[12] $k_\infty := \mathbf{R}$ and $t \in k_\infty^*$, where f is a *Schwartz function*, $|t|_\infty := |t|$ is the usual *Archimedean absolute value* on k_∞, and

$$(1.26) \qquad \hat{f}_\infty(\xi_\infty) := \int_{k_\infty} e_\infty(-x_\infty\xi_\infty)f_\infty(x_\infty)\,dx_\infty$$

is the Fourier transform on k_∞, with $e_\infty(x_\infty) := e^{2\pi i x_\infty}$ being the standard *character* $e_\infty : k_\infty \to S^1$ on k_∞. Applying this formula to the (Archimedean) Gaussian function

$$(1.27) \qquad g_\infty(x_\infty) := e^{-\pi|x_\infty|^2},$$

which is its own (additive) Fourier transform, and then applying the *multiplicative* Fourier transform (i.e., the *Mellin transform*), one soon obtains (1.21). One can "clean up" this proof a bit by replacing the Gaussian by a Dirac delta function, although one now has to work formally and "renormalise" by throwing away some infinite terms.[13] Note how this proof combines the additive Fourier transform with the multiplicative Fourier transform.[14]

In the famous 1950 thesis of Tate (see e.g. [**CaFr1967**, Chapter XV]), the above argument was reinterpreted using the language of the *adèle ring* $\mathbf{A}$,

---

[11] Riemann also had another proof of the functional equation relying primarily on contour integration, which I will not discuss here.

[12] The reason for this rather strange notation for the real line and its associated structures will be made clearer shortly.

[13] One can use the theory of distributions to make this approach rigorous, but I will not discuss this here.

[14] Continuing with this theme, the Gamma function (1.24) is an inner product between an additive character e^{-t} and a multiplicative character t^s, and the zeta function (1.20) can be viewed both additively, as a sum over n, or multiplicatively, as an *Euler product*.

with the Poisson summation formula (1.23) on k_∞ replaced by the Poisson summation formula

$$(1.28) \qquad \sum_{a \in k} f(at) = \sum_{a \in k} \hat{f}(t/a)$$

on $\mathbf{A}$, where $k = \mathbf{Q}$ is the rationals, $t \in \mathbf{A}$, and f is now a *Schwartz-Bruhat function* on $\mathbf{A}$. Applying this formula to the adelic (or global) Gaussian function $g(x) := g_\infty(x_\infty) \prod_p 1_{\mathbf{Z}_p}(x_p)$, which is its own Fourier transform, and then using the adelic Mellin transform, one again obtains (1.21). Again, the proof can be cleaned up by replacing the Gaussian with a Dirac mass, at the cost of making the computations formal (or requiring the theory of distributions).

In this section, I will write down both Riemann's proof and Tate's proof together (but omitting some technical details), to emphasise the fact that they are, in some sense, the same proof. However, Tate's proof gives a high-level clarity to the situation (in particular, explaining more adequately why the Gamma factor at infinity (1.23) fits seamlessly with the Riemann zeta function (1.20) to form the Xi function (1.21)), and allows one to generalise the functional equation relatively painlessly to other zeta-functions and L-functions, such as Dedekind zeta functions and Hecke L-functions.

1.5.1. Riemann's proof. Applying the Poisson summation formula (1.21) for k_∞ to the Schwartz function (1.27), we see that the *theta function*

$$(1.29) \qquad \Theta_\infty(x_\infty) := \sum_{n \in \mathbf{Z}} g_\infty(n x_\infty) = 1 + 2 \sum_{n=1}^{\infty} e^{-\pi n^2 |x_\infty|_\infty^2}$$

obeys the functional equation

$$(1.30) \qquad \Theta_\infty(x_\infty) = \frac{1}{|x_\infty|_\infty} \Theta_\infty\left(\frac{1}{x_\infty}\right)$$

for $x_\infty \in k_\infty^\times := k_\infty \backslash \{0\}$. In particular, since $\Theta_\infty(x_\infty) - 1$ is rapidly decreasing as $x_\infty \to \infty$, we see that $\Theta_\infty(x_\infty) - 1/x_\infty$ is rapidly decreasing as $x_\infty \to 0$.

Formally, we can take Mellin transforms of (1.30) and conclude that

$$(1.31) \qquad \int_{k_\infty^\times} \Theta_\infty(x_\infty) |x_\infty|_\infty^s d^\times x_\infty = \int_{k_\infty^\times} \Theta_\infty(x_\infty) |x_\infty|_\infty^{1-s} d^\times x_\infty$$

for any s, where $d^\times x_\infty := \frac{dx_\infty}{|x_\infty|_\infty}$ is the standard multiplicative Haar measure on $k_\infty^\times$. This does not make rigorous sense, because the integrands here diverge at 0 and at infinity (which is ultimately due to the poles of the Riemann Xi function at $s = 0$ and $s = 1$), but let us forge ahead regardless.

By making the change of variables $y := \pi n^2 t^2$ and using (1.22), (1.23), we see that

$$(1.32) \qquad \int_{k_\infty} e^{-\pi n^2 x_\infty^2} |x_\infty|_\infty^s d^\times x_\infty = \Gamma_\infty(s) n^{-s},$$

and so from (1.29) and (1.20) we formally have

$$(1.33) \qquad \int_{k_\infty} \Theta_\infty(x_\infty) |x_\infty|_\infty^s d^\times x_\infty = \int_{k_\infty} |x_\infty|^s d^\times x_\infty + 2\Gamma_\infty(s)\zeta(s).$$

If we casually discard the divergent integral $\int_{k_\infty} |x_\infty|^s d^\times x_\infty$ and apply (1.31), we formally obtain the functional equation (1.21).

Of course, the above computations were totally formal in nature. Nevertheless it is possible to make the argument rigorous. For instance, when $\mathrm{Re}(s) > 1$, we have a rigorous version of (1.33), namely

$$(1.34) \qquad \int_{k_\infty} (\Theta_\infty(x_\infty) - 1) |x_\infty|_\infty^s d^\times x_\infty = 2\Gamma_\infty(s)\zeta(s),$$

which can be deduced from (1.32) and Fubini's theorem (or by dominated convergence). Using (1.30) and a little undergraduate calculus, we can rewrite the left-hand side of (1.34) as

$$(1.35) \qquad 2\left(\int_1^\infty (\Theta_\infty(t) - 1)(t^s + t^{1-s}) \frac{dt}{t} - \frac{1}{s} - \frac{1}{1-s} \right).$$

Observe that this expression extends meromorphically to all of s and can thus be taken as a definition of $\Xi(s)$ for all $s \neq 0, 1$, and the functional equation (1.21) is then manifestly obvious.

Here is a slightly different way to view the above computations. Since the Gaussian (1.27) is its own Fourier transform, we see for every $t > 0$ that the Fourier transform of $e^{-\pi t^2 |x|_\infty^2}$ is $\frac{1}{t} e^{-\pi |x|_\infty^2 / t^2}$. Integrating this against $|t|^s d^\times t$ on $k_\infty^\times$ using (1.32), we obtain (formally at least) that the Fourier transform[15] of $\Gamma_\infty(s) |x|_\infty^{-s} d^\times x$ is $\Gamma_\infty(1-s) |x|_\infty^{1-s} d^\times x$. Formally applying the Poisson summation formula (1.23) to this, and casually discarding the singular terms at the origin, we obtain (1.21). One can make the above computations rigorous using the theory of distributions, and applying Gaussians to regularise the various integrals and summations appearing here, in which case the computations become essentially equivalent to the previous ones.

---

[15]Note that from scaling considerations it is formally clear that the Fourier transform of $|x|_\infty^{-s} d^\times x$ must be some sort of constant multiple of $|x|_\infty^{1-s} d^\times x$; the Gamma factors can thus be viewed as the normalisation of these multiplicative characters that is compatible with the Fourier transform.

1.5.2. p-adic analogues. The above "Archimedean" Fourier analysis on $k_\infty = \mathbf{R}$ has analogues in the p-adic completions $k_p = \mathbf{Q}_p$ of the rationals $k = \mathbf{Q}$. Recall that the reals $k_\infty = \mathbf{R}$ are the metric completion of the rationals $k = \mathbf{Q}$ with respect to the metric arising from the usual Archimedean absolute value $x \mapsto |x|_\infty$. This absolute value obeys the following basic properties:

(1) Positivity: $|x| \geq 0$ for all x, with equality if and only if $x = 0$.

(2) Multiplicativity: $|xy| = |x||y|$ for all x, y.

(3) Triangle inequality: $|x + y| \leq |x| + |y|$ for all x, y.

A function from k to $[0, +\infty)$ with the above three properties is known as an *absolute value* (or *valuation*) on k. In addition to the Archimedean absolute value, each prime p defines a p-adic absolute value $x \mapsto |x|_p$ on k, defined by the formula $|x|_p := p^{-n}$, where n is the number[16] of times p divides x. Equivalently, $|x|_p$ is the unique valuation such that $|p|_p = 1/p$ and $|n|_p = 1$ whenever n is an integer coprime to p. One easily verifies that $|x|_p$ is an absolute value; in fact, it not only obeys the triangle inequality but also the *ultra-triangle* inequality $|x + y| \leq \max(|x|, |y|)$, making the p-adic absolute value a non-Archimedean absolute value.

A classical theorem of Ostrowski asserts that the Archimedean absolute value $x \mapsto |x|_\infty$ and the p-adic absolute values $x \mapsto |x|_p$ are in fact the *only* absolute values on the rationals k, up to the renormalisation of replacing an absolute value $|x|$ with a power $|x|^\alpha$. If we define a *place* to be an absolute value up to renormalisation, we thus see that the rationals k have one Archimedean (or infinite) place ∞, together with one non-Archimedean (or finite) place p for every prime. One could have set $|p|_p$ to some other value between 0 and 1 than $1/p$ (thus replacing $|x|_p$ with some power $|x|_p^\alpha$) and still get an absolute value; but this normalisation is natural because it allows one to write the fundamental theorem of arithmetic in the appealing form

$$(1.36) \qquad\qquad \prod_\nu |x|_\nu = 1 \text{ for all } x \in k^\times,$$

where ν ranges over all places, and $k^\times := k \backslash \{0\}$ is the multiplicative group of k. If one takes the metric completion of the rationals $k = \mathbf{Q}$ using a p-adic (rather than the Archimedean) absolute value $|x|_p$, one obtains the p-adic field $k_p = \mathbf{Q}_p$. One can view this field as a kind of inverted version of the real field $\mathbf{R}$, in which p has been inverted to be small rather than large.

---

[16]This number could be negative if the denominator of the rational number x contains factors of p.

Some illustrations of this inversion:

(1) In k_∞, the sequence p^n goes to infinity as $n \to +\infty$ and goes to zero as $n \to -\infty$; in k_p, it is the other way around.

(2) Elements of k_∞ can be expressed in base p as strings of digits that need not terminate to the right of the decimal point, but must terminate to the left. In k_p, it is the other way around.[17]

(3) In k_∞, the set of integers $\mathbf{Z}$ is closed and forms a discrete cocompact additive subgroup. In k_p, the integers are not closed, but their closure $\mathcal{O}_p = \mathbf{Z}_p$ (the ring of p-adic integers) forms a compact codiscrete additive subgroup.

Despite this inversion, we can obtain analogues of most of the additive and multiplicative Fourier analytic computations of the previous section for the p-adics.

Let us begin with the additive Fourier structure. By the theory of Haar measures (Subsection 2.11.2), there is a unique translation-invariant measure dx_p on $k_p = \mathbf{Q}_p$ which assigns a unit mass to the compact codiscrete subgroup $\mathcal{O}_p = \mathbf{Z}_p$. One can check that this measure interacts with dilations in the expected manner, thus

$$(1.37) \qquad \int_{k_p} f(tx_p) \, dx_p = \frac{1}{|t|} \int_{k_p} f(x_p) \, dx_p$$

for all absolutely integrable f and all invertible $t \in k_p^\times := k_p \backslash \{0\}$.

Just as k_∞ has a standard character $e_\infty : x \mapsto e^{2\pi i x}$, we can define a standard character $e_p : k_p \to S^1$ as the unique character (i.e., continuous homomorphism from k_p to S^1) such that $e_p(p^n) = e^{2\pi i p^n}$ for all integers n (in particular, e_p is trivial on the integers, just as e_∞ is). One easily verifies that this is indeed a character. From this and the additive Haar measure dx_p, we can now define the p-adic Fourier transform

$$(1.38) \qquad \hat{f}(\xi_p) := \int_{k_p} e_p(-x_p \xi_p) f(x_p) \, dx_p$$

for reasonable (e.g. absolutely integrable) f, and it is a routine matter to verify all the usual Fourier-analytic identities for this transform (or one can appeal to the general theory of Fourier analysis on locally compact abelian groups).

In k_∞, we have the Gaussian function (1.27), which is its own Fourier transform. In k_p, the analogous Gaussian function $g_p : k_p \to \mathbf{C}$ is given by

---

[17]The famous ambiguity $0.999\ldots = 1.000\ldots$ in k_∞ does not occur in the p-adic field k_p, because the latter has the topology of a Cantor space rather than a continuum. See also Section 1.6 of *Structure and Randomness*.

the formula

(1.39) $g_p := 1_{\mathcal{O}_p},$

i.e., the p-adic Gaussian is just the indicator function of the p-adic integers. One easily verifies that this function is also its own Fourier transform.

Now we turn to the multiplicative Fourier theory for k_p. The natural multiplicative Haar measure $d^\times x_p$ on $k_p^\times$ is given by the formula $d^\times x_p := \frac{p}{p-1}\frac{dx_p}{|x_p|_p}$; the normalisation factor $\frac{p}{p-1}$ is natural in order for the group of units $\mathcal{O}_p^\times = \mathbf{Z}_p \backslash p\mathbf{Z}_p$ to have unit mass.

In k_∞, we see from (1.32) that the Gamma factor at infinity can be expressed (for $\mathrm{Re}(s) > 1$) as the Mellin transform of the Gaussian:

(1.40) $\Gamma_\infty(s) = \int_{k_\infty^\times} g_\infty(x_\infty)|x_\infty|_\infty^s d^\times x_\infty.$

In analogy with this, we can define the Gamma factor at p by the formula

(1.41) $\Gamma_p(s) = \int_{k_p^\times} g_p(x_p)|x_p|_p^s d^\times x_p.$

Due to the simple and explicit nature of all the expressions on the right-hand side, it is a straightforward matter to compute this factor explicitly; it becomes

(1.42) $\Gamma_p(s) = (1 - p^{-s})^{-1}$

for $\mathrm{Re}(s) > 1$, at least; of course, one can then extend Γ_p meromorphically in the obvious manner.

In k_∞, we showed (formally, at least) that $\Gamma_\infty(s)|x|_\infty^s$ and $\Gamma_\infty(1-s)|x|_\infty^{1-s}$ were Fourier transforms of each other. One can similarly show that $\Gamma_p(s)|x|_p^s$ and $\Gamma_p(1-s)|x|_p^{1-s}$ are Fourier transforms of each other in k_p. On the other hand, there is no obvious analogue of the Poisson summation formula manipulations for k_p, because (unlike k_∞), k_p lacks a discrete cocompact subgroup.

1.5.3. Tate's proof. We have just performed some *local*[18] additive and multiplicative Fourier analysis at a single place. In his famous thesis, Tate observed that all these local Fourier-analytic computations could be unified into a single global Fourier-analytic computation, using the langue of the

---

[18]This use of "local" may seem unrelated to the topological or analytical notion of "local", as in "in the vicinity of a single point", but it is actually much the same concept; compare for instance the formal power series in p for a p-adic number with the Taylor series expansion in $t-t_0$ of a function $f(t)$ around a point t_0. Indeed one can view local analysis at a place p as being the analysis of the integers or rationals when p is "close to zero"; one can make this precise using the language of schemes, but we will not do so here.

adèle ring $\mathbf{A}$. This ring is the set[19] of all tuples $x = (x_\nu)_\nu$, where ν ranges over places and $x_\nu \in k_\nu$, and furthermore all but finitely many of the x_ν lie in their associated ring of integers $\mathcal{O}_\nu$. This restriction that each x_ν consists mostly of integers in k_ν is important for a large variety of analytic and algebraic reasons; for instance, it keeps the adèle ring σ-compact.

Many of the structures and objects on the local fields k_ν can be multiplied together to form corresponding global structures on the adèle ring. For instance:

(1) The commutative ring structures on the k_ν multiply together to give a commutative ring structure on $\mathbf{A}$.

(2) The locally compact Hausdorff structures on the k_ν multiply together to give a locally compact Hausdorff structure on $\mathbf{A}$.

(3) The local additive Haar measures dx_ν on the k_ν multiply together to give a global additive Haar measure dx on $\mathbf{A}$.

(4) The local characters $e_\nu : k_\nu \to S^1$ on the k_ν multiply together to give a global character $e : \mathbf{A} \to S^1$ (here it is essential that most components of an adèle are integers, and so are trivial with respect to their local character).

(5) The local additive Fourier transforms on the k_ν then multiply to form a global additive Fourier transform on $\mathbf{A}$, defined as $\hat{f}(\xi) := \int_{\mathbf{A}} e(-x\xi) f(x) \, dx$ for reasonable f.

(6) The local absolute values $x_\nu \mapsto |x_\nu|_\nu$ on k_ν multiply to form a global absolute value $x \mapsto |x|$ on $\mathbf{A}$, though with the important caveat that $|x|$ can vanish for non-zero x (indeed, a simple calculation using Euler's observation $\prod_p (1 - \frac{1}{p}) = 0$ shows that almost every x does this, with respect to additive Haar measure). The x for which $|x|$ is non-zero are invertible and known as idèles; they form a multiplicative group $\mathbf{A}^\times$; the idèles have measure zero inside the adèles.

(7) The local Gaussians $g_\nu : k_\nu \to \mathbf{C}$ multiply together to form a global Gaussian $g : \mathbf{A} \to \mathbf{C}$, which is its own Fourier transform.

(8) The embeddings $k \subset k_\nu$ at each place ν multiply together to form a diagonal embedding $k \subset \mathbf{A}$. This embedding is both discrete (by the fundamental theorem of arithmetic) and cocompact (this is basically because the integers are cocompact in the adelic integers).

(9) The local multiplicative Haar measures $d^\times x_\nu$ multiply together to form a global multiplicative Haar measure $d^\times x$, though one should

---

[19]Equivalently, the adèle ring is the tensor product of the rationals $k = \mathbf{Q}$ with the ring of integral adèles $\mathbf{R} \times \prod_p \mathbf{Z}_p$.

caution that this measure is supported on the idèles $\mathbf{A}^\times$ rather than the adèles $\mathbf{A}$.

(10) The local Gamma factors $\Gamma_\nu(s)$ for each place ν multiply together to form the Riemann Xi function (1.22) (for $\text{Re}(s) > 1$ at least), thanks to the Euler product formula $\zeta(s) = \prod_p (1 - p^{-s})^{-1}$.

Recall that the local Gamma factors were the local Mellin transforms of the local Gaussians. Multiplying this together, we see that the Riemann Xi function is the global Mellin transform of the global Gaussian:

$$(1.43) \qquad \Xi(s) = \int_{\mathbf{A}^\times} g(x)|x|^s d^\times x.$$

Our derivation of (1.43) used the Euler product formula. Another way to establish (1.43) using the original form (1.20) of the zeta function is to observe (thanks to the fundamental theorem of arithmetic) that the set $J := \mathbf{R}^+ \times \prod_p \mathcal{O}_p^\times$ is a fundamental domain for the action of $k^\times$ on $\mathbf{A}^\times$, thus

$$(1.44) \qquad \mathbf{A}^\times = \biguplus_{a \in k^\times} a \cdot J.$$

Partitioning (1.43) using (1.44) and then using (1.32) and (1.20) one can give an alternative derivation of (1.43).[20]

In his thesis, Tate established the Poisson summation formula (1.28) for the adèles for all sufficiently nice f (e.g. any f in the *Schwartz-Bruhat class* would do). Applying this to the global Gaussian g, we conclude that the global Theta function $\Theta(x) := \sum_{a \in k} g(ax)$ obeys the functional equation

$$(1.45) \qquad \Theta(x) = \frac{1}{|x|}\Theta\left(\frac{1}{x}\right)$$

for all idèles t. This formally implies that

$$(1.46) \qquad \int_J \Theta(x)|x|^s d^\times x = \int_J \Theta(x)|x|^{1-s} d^\times x,$$

which on applying (1.44) and (1.43) (and casually discarding the singular contributions of $a = 0$) yields the functional equation (1.21). One can make this formal computation rigorous in exactly the same way that Riemann's proof was made rigorous in previous sections.

Recall that Riemann's proof could also be established by inspecting the Fourier transforms of $\Gamma_\infty(s)|x_\infty|_\infty^s d^\times x_\infty$. A similar approach can work here.

---

[20]The two derivations are ultimately the same, of course, since the Euler product formula is itself essentially a restatement of the fundamental theorem of arithmetic.

If we (very formally!) apply the Poisson summation formula (1.28) to the measure $1_J(x)|x|^s \, d^\times x$, we obtain

$$(1.47) \qquad 1 = \sum_{a \in k} \int_J e(-ax)|x|^s \, d^\times x.$$

Unpacking this summation using (1.44) (and (1.36)), and casually discarding the $a = 0$ term, we formally conclude that

$$(1.48) \qquad 1 = \int_{\mathbb{A}^\times} e(-x)|x|^s \, d^\times x.$$

Rescaling this, we formally conclude that the Fourier transform of $|x|^s \, d^\times x$ is $|x|^{1-s} \, d^\times x$. Inserting this into (1.43), the functional equation (1.21) formally follows from *Parseval's theorem*; alternatively, one can derive it by multiplying together all the local facts that the Fourier transform of $\Gamma_\nu(s)|x_\nu|_\nu^s \, d^\times x_\nu$ in k_ν is $\Gamma_\nu(1-s)|x_\nu|_\nu^{1-s} \, d^\times x_\nu$. These arguments can be made rigorous using the theory of distributions (and much care), but we will not do so here.

Notes. This article first appeared at

terrytao.wordpress.com/2008/07/27.

Thanks to Chandan Singh Dalawat for corrections.

Richard Borcherds recalled Andre Weil's characterisation of the gamma factors as the unique constant (up to some standard normalisations) that one could place in front of the obvious homogeneous distributions $|x|^s$ to make the resultant distribution holomorphic for *all* complex s; this uniqueness can then be used to easily derive the functional equation. In higher rank groups, the analogue of Weil's description of the gamma factor gives the local L-factor of an automorphic form.

There was some discussion as to when special values (or residues) of zeta-type functions could be recovered from the same sort of adelic analysis discussed here; for instance, the class number formula could be obtained by these methods, but many other special values seem to require deeper tools to establish.

1.6. The divisor bound

Given a positive integer n, let $d(n)$ denote the *number of divisors* of n (including 1 and n), thus for instance $d(6) = 4$, and more generally, if n has a prime factorisation

$$(1.49) \qquad n = p_1^{a_1} \dots p_k^{a_k},$$

then (by the fundamental theorem of arithmetic)

$$(1.50) \qquad d(n) = (a_1 + 1) \dots (a_k + 1).$$

Clearly, $d(n) \leq n$. The *divisor bound* asserts that, as n gets large, one can improve this trivial bound to

$$(1.51) \qquad d(n) \leq C_\varepsilon n^\varepsilon$$

for any $\varepsilon > 0$, where C_ε depends only on ε; equivalently, in asymptotic notation, one has $d(n) = n^{o(1)}$. In fact one has a more precise bound

$$(1.52) \qquad d(n) \leq n^{O(1/\log\log n)} = \exp\left(O\left(\frac{\log n}{\log\log n} \right) \right).$$

The divisor bound is useful in many applications in number theory, harmonic analysis, and even PDE (on periodic domains); it asserts that for any large number n, only a "logarithmically small" set of numbers less than n will actually divide n exactly, even in the worst-case scenario when n is *smooth* (i.e., has many small prime factors).[21]

The divisor bound is elementary to prove (and not particularly difficult), and I was asked about it recently, so I thought I would provide the proof here, as it serves as a case study in how to establish worst-case estimates in elementary multiplicative number theory.

1.6.1. Proof of (1.51). We first prove the weak form of the divisor bound (1.51), which is already good enough for many applications (because a loss of $n^{o(1)}$ is often negligible if, say, the final goal is to extract some polynomial factor in n in one's eventual estimates). By rearranging a bit, our task is to show that for any $\varepsilon > 0$, the expression

$$(1.53) \qquad \frac{d(n)}{n^\varepsilon}$$

is bounded uniformly in n by a constant depending on ε. Using (1.49) and (1.50), we can express (1.53) as a product

$$(1.54) \qquad \prod_{j=1}^{k} \frac{a_j + 1}{p_j^{\varepsilon a_j}},$$

where each term involves a different prime p_j, and the a_j are at least 1. We have thus "localised" the problem to studying the effect of each individual prime independently. (We are able to do this because $d(n)$ is a multiplicative function.)

---

[21] The average value of $d(n)$ is much smaller, being about $\log n$ on the average, as can be easily seen from the *double counting identity*

$$\sum_{n \leq N} d(n) = \#\{(m, l) \in \mathbf{N} \times \mathbf{N} : ml \leq N\} = \sum_{m=1}^{N} \left\lfloor \frac{N}{m} \right\rfloor \sim N \log N,$$

or from the heuristic that a randomly chosen number m less than n has a probability about $1/m$ of dividing n, and $\sum_{m<n} \frac{1}{m} \sim \log n$. However, (1.52) is the correct "worst-case" bound, as I discuss below.

Let us fix a prime p_j and look at a single term $\frac{a_j+1}{p_j^{\varepsilon a_j}}$. The numerator is linear in a_j, while the denominator is exponential. Thus, as per Malthus, we expect the denominator to dominate, at least when a_j is large. However, because of the ε, the numerator might exceed the denominator when a_j is small—but only if p_j is also small.

Following these heuristics, we now divide our argument into cases. Suppose that p_j is large, say $p_j \geq \exp(1/\varepsilon)$. Then $p_j^{\varepsilon a_j} \geq \exp(a_j) \geq 1 + a_j$ (by Taylor expansion), and so the contribution of p_j to the product (1.54) is at most 1. Thus all the large primes give a net contribution of at most 1 here.

What about the small primes, in which $p_j < \exp(1/\varepsilon)$? Well, due to Malthus, we know that the sequence $\frac{a+1}{p_j^{\varepsilon a}}$ goes to zero as $a \to \infty$. Since convergent sequences are bounded, we therefore have a bound of the form $\frac{a_j+1}{p_j^{\varepsilon a_j}} \leq C_{p_j,\varepsilon}$ for some $C_{p_j,\varepsilon}$ depending only on p_j and ε, but not on a_j. So, each small prime gives a bounded contribution to (1.54) (uniformly in n). But the number of small primes is itself bounded (uniformly in n). Thus the total product in (1.54) is also bounded uniformly in n, and the claim follows.

1.6.2. Proof of (1.52). One can refine the above analysis to get a more explicit value of C_ε, which will let us get (1.52), as follows.

Again consider the product (1.54) for some $\varepsilon > 0$. As discussed previously, each prime larger than $\exp(1/\varepsilon)$ gives a contribution of at most 1. What about the small primes? Here we can estimate the denominator from below by Taylor expansion:

$$p_j^{\varepsilon a_j} = \exp(\varepsilon a_j \log p_j) \geq 1 + \varepsilon a_j \log p_j$$

and hence

$$\frac{a_j + 1}{p_j^{\varepsilon a_j}} \leq \frac{a_j + 1}{1 + \varepsilon a_j \log p_j} \ll \frac{1}{\varepsilon \log p_j}$$

(the point here being that our bound is uniform in a_j). One can of course use undergraduate calculus to try to sharpen the bound here, but this does not improve it by too much, basically because the Taylor approximation $\exp(x) \approx 1 + x$ is quite accurate when x is small, which is the important case here.

Anyway, inserting this bound into (1.54), we see that (1.54) is in fact bounded by

$$\prod_{p < \exp(1/\varepsilon)} O\left(\frac{1}{\varepsilon \log p}\right).$$

Now let us be very crude[22] and bound $\log p$ from below by $\log 2$, and bound the number of primes less than $\exp(1/\varepsilon)$ by $\exp(1/\varepsilon)$. We thus conclude that

$$(1.55) \qquad (1.54) \leq O\left(\frac{1}{\varepsilon}\right)^{\exp(1/\varepsilon)} = \exp(\exp(O(1/\varepsilon)));$$

unwinding what this means for (1.51), we obtain

$$d(n) \leq \exp(\exp(O(1/\varepsilon)))n^{\varepsilon}$$

for all $n \geq 1$ and $\varepsilon > 0$. If we now set $\varepsilon = C/\log\log n$ for a sufficiently large C, then the second term of the right-hand side dominates the first (as can be seen by taking logarithms), and the claim (1.52) follows.

The above argument also suggests the counterexample that will demonstrate that (1.52) is basically sharp. Pick $\varepsilon > 0$, and let n be the product of all the primes up to $\exp(1/\varepsilon)$. The prime number theorem[23] tells us that $\log n \sim \exp(1/\varepsilon)$. On the other hand, the prime number theorem also tells us that the number of primes dividing n is $\sim \varepsilon \exp(1/\varepsilon)$, so by (1.50), $\log d(n) \sim \varepsilon \exp(1/\varepsilon)$. Using these numbers we see that (1.52) is tight up to constants.

1.6.3. Why is the divisor bound useful? One principal application of the divisor bound (and some generalisations of that bound) is to control the number of solutions to a Diophantine equation. For instance, (1.51) immediately implies that for any fixed positive n, the number of solutions to the equation

$$xy = n$$

with x, y integer,[24] is only $n^{o(1)}$ at most. This can be leveraged to some other Diophantine equations by high school algebra. For instance, thanks to the identity $x^2 - y^2 = (x + y)(x - y)$, we conclude that the number of integer solutions to

$$x^2 - y^2 = n$$

---

[22]One can of course be more efficient about this, but again it does not improve the final bounds too much. A general principle is that when one estimates an expression such as A^B, or more generally the product of B terms, each of size about A, then it is far more important to get a good bound for B than to get a good bound for A, except in those cases when A is very close to 1.

[23]If one does not care about the constants, then one does not need the full strength of the prime number theorem to show that (1.52) is sharp; the more elementary bounds of Chebyshev, which say that the number of primes up to N is comparable to $N/\log N$ up to constants, would suffice.

[24]For x and y real, the number of solutions is of course infinite.

is also at most $n^{o(1)}$; similarly, the identity $x^3 + y^3 = (x+y)(x^2 - xy + y^2)$ implies[25] that the number of integer solutions to

$$x^3 + y^3 = n$$

is at most $n^{o(1)}$.

Now consider the number of solutions to the equation

$$x^2 + y^2 = n.$$

In this case, $x^2 + y^2$ does not split over the rationals $\mathbf{Q}$, and so one cannot directly exploit the divisor bound for the rational integers $\mathbf{Z}$. However, we can factor $x^2 + y^2 = (x+iy)(x-iy)$ over the Gaussian rationals $\mathbf{Q}[\sqrt{-1}]$. Happily, the Gaussian integers $\mathbf{Z}[\sqrt{-1}]$ enjoy essentially the same divisor bound as the rational integers $\mathbf{Z}$; the Gaussian integers have unique factorisation, but perhaps more importantly, they only have a finite set of units ($\{-1, +1, -i, +i\}$ to be precise). Because of this, one can easily check that $x^2 + y^2 = n$ also has at most $n^{o(1)}$ solutions.

One can similarly exploit the divisor bound on other number fields; for instance the divisor bound for $\mathbf{Z}[\sqrt{-3}]$ allows one to count solutions to $x^2 + xy + y^2 = n$ or $x^2 - xy + y^2 = n$. On the other hand, not all number fields have the divisor bound. For instance, $\mathbf{Z}[\sqrt{2}]$ has an infinite number of units, which means that the number of solutions to *Pell's equation*

$$x^2 - 2y^2 = 1$$

is infinite.

Another application of the divisor bound comes up in *sieve theory*. Here, one often deals with functions of the form $\nu(n) := \sum_{d|n} a_d$, where the sieve weights a_d typically have size $O(n^{o(1)})$, and the sum is over all d that divide n. The divisor bound (1.51) then implies that the sieve function $\nu(n)$ also has size $O(n^{o(1)})$. This bound is too crude to deal with the most delicate components of a sieve theory estimate, but it is often very useful for dealing with error terms (especially those that have gained a factor of n^{-c} relative to the main terms for some $c > 0$).

Notes. This article first appeared at

terrytao.wordpress.com/2008/09/23.

Marius Overholt and Emmanuel Kowalski noted that the implied constant in (1.52) is known to be $\log 2 + o(1)$, a classical result of Wigert from 1906

---

[25]Note from Bezout's theorem (or direct calculation) that $x + y$ and $x^2 - xy + y^2$ determine x, y up to at most a finite ambiguity.

(with a simpler proof given by Ramanujan in 1914). Explicit constants are also known for the moments of the divisor function; for instance

$$\frac{1}{x} \sum_{n \leq x} d(n)^2 = \left(\frac{1}{\pi^2} + o(1) \right) \log^3 x$$

and more generally

$$\frac{1}{x} \sum_{n \leq x} d(n)^k = (c_k + o(1)) \log^{2^k - 1} x$$

for any fixed $k \geq 1$ and some explicit constant c_k.

Ben Green noted that there appeared to be no bound approaching the strength of the divisor bound if one perturbs the integers in a manner that destroys the number-theoretic structure. For instance, for a fixed shift θ, it does not seem possible to obtain a bound of $O(n^{o(1)})$ to the number of solutions to the equation $(a + \theta)(b + \theta) = n + O(1)$ for integer a, b.

1.7. The Lucas-Lehmer test for Mersenne primes

Recently, the *Great Internet Mersenne Prime Search* (GIMPS) announced the discovery of two new Mersenne primes, both over ten million digits in length, including one discovered by the computing team at UCLA.

The GIMPS approach to finding Mersenne primes relies of course on modern computing power, parallelisation, and efficient programming, but the number-theoretic heart of it—aside from some basic optimisation tricks such as *fast multiplication* and preliminary sieving to eliminate some obviously non-prime Mersenne number candidates—is the *Lucas-Lehmer primality test for Mersenne numbers*, which is much faster for this special type of numbers than any known general-purpose (deterministic) primality test (such as, say, the *AKS primality test* [**AgKaSa2004**]). This test is easy enough to describe, and I will do so later in this section, and it also has some short elementary proofs of correctness; but the proofs are sometimes presented in a way that involves pulling a lot of rabbits out of hats, giving the argument a magical feel rather than a natural one. In this article, I will try to explain the basic ideas that make the primality test work, seeking a proof which might be less elementary and a little longer than some of the proofs in the literature, but is perhaps a little better motivated.

1.7.1. Order. We begin with a general discussion of how to tell when a given number n (which is not necessarily of the Mersenne form $n = 2^m - 1$) is prime or not. One should think of n as being moderately large, e.g. $n \sim 10^{10^7}$ (which is broadly the size of the Mersenne primes discovered recently).

Our starting point will be *Lagrange's theorem*, which asserts that

$$(1.56) \qquad\qquad a^{|G|} = 1$$

for any finite group G and any $a \in G$, thus the *order* $\mathrm{ord}_G(a)$ of a in G divides $|G|$. Specialising this to the multiplicative group $\mathbf{F}_p^\times$ of a finite field of prime order p, we obtain *Fermat's little theorem*

$$(1.57) \qquad\qquad a^{p-1} = 1 \bmod p$$

for p prime and a coprime to p; applying it instead to the multiplicative group $(\mathbf{Z}/n\mathbf{Z})^\times$ of a cyclic group of order n, we obtain *Euler's theorem*

$$(1.58) \qquad\qquad a^{\phi(n)} = 1 \bmod n$$

whenever a is coprime to n, where $\phi(n) := |(\mathbf{Z}/n\mathbf{Z})^\times|$ is the *Euler totient function* of n.

Fermat's little theorem (1.57) already gives a necessary condition for the primality of a candidate prime n: take any a coprime to n (typically one picks a small number such as $a = 2$ or $a = 3$), and compute a^{n-1} modulo n. If it is not equal to 1, then n cannot be prime. This is a (barely) feasible test to execute for n as large as 10^{10^7}, because one can compute exponents such as a^{n-1} relatively quickly, by the trick of repeatedly squaring a modulo n to obtain $a, a^2, a^{2^2}, a^{2^3}, \dots$ mod n, and then decomposing $n-1$ into binary[26] to compute a^{n-1}. Unfortunately, while this test is necessary for primality, it is not sufficient, due to the existence of *pseudoprimes*.[27] So Fermat's little theorem alone does not provide the answer, at least if one wants a *deterministic* certificate of primality rather than a *probabilistic* one.

Nevertheless, the above facts do provide some important information about the order $\mathrm{ord}_n(a) := \mathrm{ord}_{(\mathbf{Z}/n\mathbf{Z})^*}(a)$ of a modulo n. If n is prime, Fermat's little theorem (1.57) tells us that $\mathrm{ord}_n(a)$ is at most $n-1$ (in fact, it divides $n-1$). On the other hand, if n is not prime, then Euler's theorem (1.58) tells us that $\mathrm{ord}_n(a)$ cannot be as large as $n-1$, but is instead at most $\phi(n)$, which is now strictly less than $n-1$. Thus, if we can find a number a coprime to n such that $\mathrm{ord}_n(a)$ is exactly $n-1$, we have certified that n is prime.

Testing whether a given number a is coprime to n is very easy and fast (thanks to the *Euclidean algorithm*). Unfortunately, it is difficult in general to compute[28] the order $\mathrm{ord}_n(a)$ of a number a if the base n is large; a brute-force approach would require one to compute up to $n-1$ powers of a, which

---

[26]If n is a Mersenne number, then there are some pretty obvious shortcuts one can take for this last step, using division instead of multiplication.

[27]For instance, the number $n = 561 = 3 * 11 * 17$ is not prime, but $a^{n-1} = 1 \bmod n$ is true for all a coprime to n (in other words, 561 is a *Carmichael number*).

[28]More generally, the problem of computing order exactly is in general closely related to the *discrete logarithm problem*, which is notoriously difficult.

is prohibitively expensive if n has size comparable to 10^{10^7}. However, there are a few cases in which the order can be found very quickly. Suppose that we somehow find positive integers a, k such that

(1.59) $$a^{2^k} = -1 \bmod n$$

(which in particular implies that a is coprime to n). Squaring this, we obtain

$$a^{2^{k+1}} = 1 \bmod n$$

and so we see that $\operatorname{ord}_n(a)$ divides 2^{k+1} but not 2^k, and thus must be exactly equal to 2^{k+1}. Conversely, if $\operatorname{ord}_n(a) = 2^{k+1}$, then we must have (1.59). So in the special case when the order of a is a power of 2, we can compute the order using only one exponentiation, which is computationally feasible for the orders of magnitude we are considering.

Unfortunately, this is not quite what we need for the Mersenne prime problem, because if n is a Mersenne number, then it is n *plus* 1, which is[29] a power of 2, rather than n *minus* 1.

So this is a frustrating near miss: if n is a Mersenne number, we can easily check if a number has order $n + 1$ modulo n, but we needed a test for when a number has order $n - 1$ instead. And indeed, even when n is prime, Fermat's little theorem (1.57) shows that it is impossible for a number to have order $n + 1$ modulo n, since the order needs to divide $n - 1$. So we seem to be a bit stuck.

But while $n + 1$ clearly does not divide $n - 1$, it *does* divide $n^2 - 1$. Looking at Lagrange's theorem (1.56), we then see that it could be possible to find elements of order $n + 1$ in a multiplicative group of order $n^2 - 1$ rather than $n - 1$. Recall that if n was prime, then the multiplicative group $\mathbf{F}_n^\times$ of the finite field $\mathbf{F}_n$ had order $n - 1$. But $\mathbf{F}_{n^2}$ is also a finite field, and its multiplicative group $\mathbf{F}_{n^2}^\times$ has order $n^2 - 1$. Aha!

So the plan (assuming for the sake of argument that n is prime) is to somehow work in the finite field $\mathbf{F}_{n^2}$ instead of $\mathbf{F}_n$, in order to find elements of order $n+1$. We can get our hands on this larger finite field more concretely by viewing it as a *quadratic extension* $\mathbf{F}_n[\sqrt{a}]$, where a is a *quadratic non-residue* of n.

Let us now take n to be a Mersenne prime. What numbers are quadratic non-residues? A quick appeal to *quadratic reciprocity* and some elementary number theory soon reveals that 2 is a quadratic residue of n, but that 3 is not. Thus we can[30] take $\mathbf{F}_{n^2} \equiv \mathbf{F}_n[\sqrt{3}]$. Henceforth all calculations will be in this field $\mathbf{F}_n[\sqrt{3}]$, which of course contains $\mathbf{F}_n$ as a subfield.

---

[29]The above method would be ideal for finding *Fermat primes* rather than Mersenne primes, but it is likely that in fact there are no more such primes to be found.

[30]One could work with other numbers than 3 here, but being the smallest quadratic non-residue available, it is the simplest one to use, and the one which is most likely to be able to take

Now we need to look for a field element a of order $n+1$, which is a power of 2. Thus (by adapting (1.59) to $\mathbf{F}_n[\sqrt{3}]$) we need to find a solution to the equation

(1.60) $$a^{(n+1)/2} = -1$$

in this field.

Let us compute some expressions of the form $a^{(n+1)/2}$. From Fermat's little theorem (1.57) we have

$$3^{n-1} = 1;$$

because 3 is not a quadratic residue, we see (from taking discrete logarithms) that

(1.61) $$3^{(n-1)/2} = -1$$

and thus

$$3^{(n+1)/2} = -3.$$

Similarly we have

(1.62) $$2^{(n+1)/2} = +2.$$

These are pretty close to (1.60), but not quite right. To go further, it is convenient to work with n^{th} powers rather than $((n+1)/2)^{\text{th}}$ powers; i.e., we work with the *Frobenius endomorphism* $x \mapsto x^n$. Indeed, since $\mathbf{F}_n[\sqrt{3}]$ has characteristic n, we have the endomorphism properties

(1.63) $$(x+y)^n = x^n + y^n; \quad (xy)^n = x^n y^n.$$

From (1.61) we have $(\sqrt{3})^n = -\sqrt{3}$, while from (1.57) we have $a^n = a$ for $a \in \mathbf{F}_n$. From (1.63) we thus see that

$$(a + b\sqrt{3})^n = a - b\sqrt{3}$$

for $a, b \in \mathbf{F}_n$; thus the Frobenius automorphism is nothing more than Galois conjugation.[31]

Now we go back from n^{th} powers to $((n+1)/2)^{\text{th}}$ powers. Multiplying both sides of the preceding equation by $a + b\sqrt{3}$, we obtain

$$(a + b\sqrt{3})^{n+1} = a^2 - 3b^2.$$

Squaring $a + b\sqrt{3}$, we conclude

$$(a^2 + 3b^2 + 2ab\sqrt{3})^{(n+1)/2} = a^2 - 3b^2.$$

---

advantage of the *strong law of small numbers* [**Gu1988**], which is the informal assertion that numerical coincidences are most likely to occur amongst small numbers than amongst large ones.

[31] Actually, this can be deduced quite readily from standard Galois theory.

Now we are in a good position to solve the equation (1.60). We cannot make $a^2 - 3b^2$ equal to -1, since -1 is not a quadratic residue modulo 3, but we can make it equal to, say, -2, by setting $a = 1$ and $b = -1$ (say):

$$(4 + 2\sqrt{3})^{(n+1)/2} = -2.$$

Dividing this by (1.62) we obtain the desired solution

$$\text{(1.64)} \qquad\qquad\qquad \omega^{(n+1)/2} = -1$$

to (1.60), where[32] $\omega := 2 + \sqrt{3}$.

To summarise, we have shown that

Proposition 1.7.1. *If n is a Mersenne prime, then (1.64) holds.*

Based on our previous discussion, we expect to be able to reverse this implication. Indeed, we have the following converse:

Lemma 1.7.2. *Let n be a Mersenne number. If (1.64) holds (in the ring $(\mathbf{Z}/n\mathbf{Z})[\sqrt{3}]$), then n is prime.*

Proof. We use an argument of Bruce [**Br1993**]. Let q be a prime divisor of n. Then $\omega^{(n+1)/2} = -1$ in the field $\mathbf{F}_q[\sqrt{3}]$ (which we define as $\mathbf{F}_q$ if 3 is a quadratic residue there), thus ω has order exactly $n + 1$ (cf. (1.59)). By Lagrange's theorem (1.56), this means that $n + 1$ divides the multiplicative order of $\mathbf{F}_q[\sqrt{3}]^\times$, which is $q^2 - 1$ (if 3 is a non-residue modulo q) or $q - 1$ (if 3 is a residue modulo q). In particular, q has to exceed $\sqrt{n}$. Thus the only prime divisors of n exceed $\sqrt{n}$, and so by the *sieve of Eratosthenes*, n is prime. $\qquad\square$

We have thus shown

Corollary 1.7.3 (Lucas-Lehmer test, preliminary version). *Let $n = 2^m - 1$ with m odd. Then n is prime if and only if (1.64) holds in $\mathbf{Z}/n\mathbf{Z}[\sqrt{3}]$.*

This is already a reasonable criterion, but it is a little non-elementary (and also a little unpleasant numerically) due to the presence of the quadratic extension by $\sqrt{3}$. One can get rid of this extension by the Galois theory trick of taking *traces*. Indeed, observe that $\omega^{-1} = 2 - \sqrt{3}$ is the *Galois conjugate* of ω. Basic Galois theory tells us that $\omega^{(n+1)/4} + \omega^{-(n+1)/4}$ lies in $\mathbf{Z}/n\mathbf{Z}$, and vanishes precisely when $\omega^{(n+1)/2}$ is equal to -1. So it suffices to show that

$$\omega^{(n+1)/4} + \omega^{-(n+1)/4} = \omega^{2^{m-2}} + \omega^{-2^{m-2}} =: S_{m-2}$$

vanishes in $\mathbf{Z}/n\mathbf{Z}$. The quantity $\omega^{(n+1)/4} = \omega^{2^{m-2}}$ could be computed by repeated squaring in $\mathbf{Z}/n\mathbf{Z}[\sqrt{3}]$. The quantity S_{m-2} can be computed by a

---

[32]Note that one could also use $\omega^{-1} = 2 - \sqrt{3}$ here; indeed, Galois theory tells us that $+\sqrt{3}$ and $-\sqrt{3}$ are interchangeable in these computations.

similar device in $\mathbf{Z}/n\mathbf{Z}$. Indeed, the sequence $S_j := \omega^{2^j} + \omega^{-2^j}$ is easily seen to obey the recursion

$$(1.65) \qquad\qquad S_j = S_{j-1}^2 - 2; \quad S_0 = 4,$$

and so we have

Theorem 1.7.4 (Lucas-Lehmer test, final version). *Let $n = 2^m - 1$ with m odd. Then n is prime if and only if S_{m-2} vanishes modulo n, where S_{m-2} is given by the recursion* (1.65).

To apply this test, one needs to perform about m squaring operations modulo n. Doing everything as efficiently as possible (in particular, using *fast multiplication*), the total cost of testing a single Mersenne number $n = 2^m - 1$ for primality is about $O(m^2)$ (modulo some $\log m$ terms). This turns out to barely be within reach[33] of modern computers for $m \sim 10^7$, especially since the algorithm is somewhat parallelisable. There are general-purpose probabilistic tests (such as the *Miller-Rabin test*) which have run-time comparable to the Lucas-Lehmer test, but as mentioned at the beginning, we are only interested here in deterministic (and unconditional, in particular not relying on the generalised Riemann hypothesis) certificates of primality.

Notes. This article first appeared at

`terrytao.wordpress.com/2008/10/02`.

Thanks to René Schoof, Jernej, and several anonymous commenters for corrections.

More information on the recently found Mersenne primes can be found at `www.math.ucla.edu/~edson/prime`. (I was not involved in this computing effort.) As for the question "Why do we want to find such big primes anyway?", see `primes.utm.edu/notes/faq/why.html`.

An anonymous commenter pointed out that an application of large Mersenne primes to coding theory appeared recently in [**Ye2007**].

1.8. Finite subsets of groups with no finite models

Additive combinatorics is largely focused on the additive properties of finite subsets A of an additive group $G = (G, +)$. This group can be finite or infinite, but there is a very convenient trick, the *Ruzsa projection trick*, which allows one to reduce the latter case to the former. For instance, consider the set $A = \{1, \ldots, N\}$ inside the integers $\mathbf{Z}$. The integers of

---

[33] In contrast, the best known general-purpose deterministic primality testing algorithm, the AKS algorithm [**AgKaSa2004**], has a run time of about $O(m^6)$ (with a sizable implicit constant), which is not feasible for $m \sim 10^7$.

course form an infinite group, but if we are only interested in sums of at most two elements of A at a time, we can embed A inside the finite cyclic group $\mathbf{Z}/2N\mathbf{Z}$ without losing any combinatorial information. More precisely, there is a *Freiman isomorphism of order* 2 between the set $\{1, \ldots, N\}$ in $\mathbf{Z}$ and the set $\{1, \ldots, N\}$ in $\mathbf{Z}/2N\mathbf{Z}$. One can view the latter version of $\{1, \ldots, N\}$ as a *model* for the former version of $\{1, \ldots, N\}$. More generally, it turns out that any finite set A in an additive group can be modeled in the above set by an equivalent set in a finite group, and in fact one can ensure that this ambient modeling group is not much larger than A itself if A has some additive structure; see [**Ru1994**] or [**TaVu2006**, Lemma 5.26] for a precise statement. This projection trick has a number of important uses in additive combinatorics, most notably in Ruzsa's simplified proof [**Ru1994**] of Freiman's theorem [**Fr1973**].

Given the interest in non-commutative analogues of Freiman's theorem (see Section 3.2 of *Structure and Randomness*), it is natural to ask whether one can similarly model finite sets A in multiplicative (and non-commutative) groups $G = (G, \times)$ using finite models. Unfortunately (as I learned recently from Akshay Venkatesh, via Ben Green), this turns out to be impossible in general, due to an old example of Higman [**Hi1951**]. More precisely, Higman proved

Theorem 1.8.1. *There exists an infinite group G generated by four distinct elements a, b, c, d that obey the relations*

$$(1.66) \qquad ab = ba^2, \quad bc = cb^2, \quad cd = dc^2, \quad da = ad^2;$$

in fact, a and c generate the free non-abelian group in G. On the other hand, if G' is a finite group that contains four elements a, b, c, d obeying (1.66), then a, b, c, d are all trivial.

As a consequence, the finite set $A := \{1, a, b, c, d, ab, bc, cd, da\}$ in G has no model (in the sense of *Freiman isomorphisms*) in a finite group. Theorem 1.8.1 is proven by a small amount of elementary group theory and number theory, and it was neat enough that I thought I would reproduce it here.

1.8.1. No non-trivial finite models. We first show the second part of Theorem 1.8.1. The key point is that in a finite group G', all elements have finite *order*, thanks to *Lagrange's theorem*. From (1.66) we have

$$b^{-1}ab = a^2$$

and hence by induction

$$(1.67) \qquad b^{-n}ab^n = a^{2^n}$$

for any positive n. One consequence of (1.67) is that if $b^n = 1$, then $a = a^{2^n}$, and thus $a^{2^n - 1} = 1$. Applying this with n equal to the order $\operatorname{ord}(b)$ of b, we conclude that

$$\mathrm{ord}(a)|2^{\mathrm{ord}(b)} - 1.$$

As a consequence, if $\mathrm{ord}(a)$ is divisible by some prime p, then $2^{\mathrm{ord}(b)} - 1$ is divisible by p, which forces p to be odd and $\mathrm{ord}(b)$ to be divisible by the multiplicative order of 2 modulo p. This is at most $p - 1$ (by *Fermat's little theorem*), and so $\mathrm{ord}(b)$ is divisible by a prime strictly smaller than the prime dividing $\mathrm{ord}(a)$. But we can cyclically permute this argument and conclude that $\mathrm{ord}(c)$ is divisible by an even smaller prime than the prime dividing $\mathrm{ord}(b)$, and so forth, creating an *infinite descent*, which is absurd. Thus none of $\mathrm{ord}(a)$, $\mathrm{ord}(b)$, $\mathrm{ord}(c)$, $\mathrm{ord}(d)$ can be divisible by any prime, and so a, b, c, d are trivial as claimed.

Remark 1.8.2. There is nothing special here about using four generators; the above arguments work with any number of generators (adapting (1.66) appropriately). But we will need four generators in order to establish the infinite model below.

Remark 1.8.3. The above argument also shows that the group G has no non-trivial finite-dimensional linear representation. Indeed, let a, b, c, d be matrices obeying (1.66); then b is conjugate to $b^2 = c^{-1}bc$, which by the spectral theorem forces the eigenvalues of b to be roots of unity, which implies in particular that b^n grows at most polynomially in n; similarly for a^n. Applying (1.67) we see that a^{2^n} grows at most polynomially in n, which by the *Jordan normal form* (see Section 1.13 of *Structure and Randomness*) for a implies that a is diagonalisable; since its eigenvalues are roots of unity, it thus has finite order. Similarly for b, c, d. Now apply the previous argument to conclude that a, b, c, d are trivial.

1.8.2. Existence of an infinite model. To build the infinite group G that obeys the relations (1.66), we need the notion of an *amalgamated free product* of groups. Recall that the *free product* $G_1 * G_2$ of two groups G_1 and G_2 can be defined (up to group isomorphism) in one of three equivalent ways:

(1) (Relations-based definition) $G_1 * G_2$ is the group *generated* by the disjoint union $G_1 \uplus G_2$ of G_1 and G_2, with no further relations between these elements beyond those already present in G_1 and G_2 separately.

(2) (Category-theoretic definition) $G_1 * G_2$ is a group with *homomorphisms* from G_1 and G_2 into $G_1 * G_2$, which is *universal* in the sense that any other group G' with homomorphisms from G_1, G_2 will have these homomorphisms factor uniquely through $G_1 * G_2$.

(3) (Word-based definition) $G_1 * G_2$ is the collection of all *words* $g_1 g_2 \ldots g_n$, where each g_i lies in either G_1 or G_2, with no two adjacent g_i, g_{i+1} lying in the same G_j (let us label G_1, G_2 here to be disjoint to avoid notational confusion), with the obvious group operations.

It is not hard to see that all three definitions are equivalent, and that the free product exists and is unique up to group isomorphism.

Example 1.8.4. The free product of the free cyclic group $\langle a \rangle$ with one generator a, and the free cyclic group $\langle b \rangle$ with one generator b, is the free (non-abelian) group $\langle a, b \rangle$ on two generators.

We will need a "relative" generalisation of the free product concept, in which the groups G_1, G_2 are not totally disjoint, but instead share a common subgroup H (or if one wants to proceed more category-theoretically, with a group H that embeds into both G_1 and G_2). In this situation, we define the amalgamated free product $G_1 *_H G_2$ by one of the following two equivalent definitions:

(1) (Relations-based definition) $G_1 *_H G_2$ is the group *generated* by the relative disjoint union $G_1 \uplus_H G_2$ of G_1 and G_2 (which is the same as the disjoint union but with the common subgroup H identified), with no further relations between these elements beyond those already present in G_1 and G_2 separately.

(2) (Category-theoretic definition) $G_1 *_H G_2$ is a group with *homomorphisms* from G_1 and G_2 into $G_1 *_H G_2$ that agree on H, which is *universal* in the sense that any other group G' with homomorphisms from G_1, G_2 that agree on H will have these homomorphisms factor uniquely through $G_1 *_H G_2$.

Example 1.8.5. Let $G_1 := \langle a, b | ab = ba^2 \rangle$ be the group generated by two elements a, b with one relation $ab = ba^2$. It is not hard to see that all elements of G_1 can be expressed uniquely as $b^n a^m$ for some integer n and dyadic rational $m = q/2^k$, and in particular that $H := \langle b \rangle$ is a free cyclic group. Let $G_2 := \langle b, c | bc = cb^2 \rangle$ be the group generated by two elements b, c with one relation $bc = cb^2$; then again $H := \langle b \rangle$ is a free cyclic group, and isomorphic to the previous copy of H. The amalgamated free product $G_1 *_H G_2 = \langle a, b, c | ab = ba^2, bc = cb^2 \rangle$ is then generated by three elements a, b, c with two relations $ab = ba^2, bc = cb^2$.

It is not hard to see that the above two definitions are equivalent, and that $G_1 *_H G_2$ exists and is unique up to group isomorphism. But note that I did not give the word-based definition of the amalgamated free product yet. We will need to do so now; I will use the arguments from [**Ne1954**], though the basic result I need here (namely, Corollary 1.8.8) dates all the way back to the work of Schreier in 1927.

In order to analyse these groups, we will need to study how they act on various spaces. If G is a group, we define a *G-space* to be a set X together with an *action* $(g, x) \mapsto gx$ of G on X (or equivalently, a homomorphism from G to the permutation group $\mathrm{Sym}(X)$ of X). Thus for instance G is itself a G-space. A G-space X is *transitive* if for every $x, y \in X$, there exists $g \in G$ such that $gx = y$. A *morphism* from one G-space X to another G-space Y is a map $\phi : X \to Y$ such that $\phi(gx) = g\phi(x)$ for all $g \in G$ and $x \in X$. If a morphism has an inverse that is also a morphism, we say that it is an *isomorphism*.

The first observation is that a G-space with certain properties will necessarily be isomorphic to G itself.

Lemma 1.8.6 (Criterion for isomorphism with G). *Let G be a group, let X be a non-empty transitive G-space, and suppose there is a morphism $\pi : X \to G$ from the G-space X to the G-space G. Then π is in fact an isomorphism of G-spaces.*

Proof. It suffices to show that π is both injective and surjective. To show surjectivity, observe that the image $\pi(X)$ is G-invariant and non-empty. But the action of G on G is transitive, and so $\pi(X) = G$ as desired. To show injectivity, observe from transitivity that if x, x' are distinct elements of X, then $x' = gx$ for some non-identity $g \in G$, thus $\pi(x') = g\pi(x)$, thus $\pi(x') \neq \pi(x)$, establishing injectivity. $\qquad\square$

Now we can give the word formulation of the amalgamated free product.

Lemma 1.8.7 (Word-based description of amalgamated free product). *Let G_1, G_2 be two groups with common subgroup H, and let $G := G_1 *_H G_2$ be the amalgamated free product. Let $G_1 = \bigcup_{s_1 \in S_1} H \cdot s_1$, $G_2 = \bigcup_{s_2 \in S_2} H \cdot s_2$ be some partitions of G_1, G_2 into right cosets of H. Let X be the space of all formal words of the form $hs_1s_2 \ldots s_n$, where $h \in H$, each s_i lies in either S_1 or S_2, and no two adjacent s_i, s_{i+1} lie in the same S_j. Let $\pi : X \to G$ be the obvious evaluation map. Then there is an action of G on X for which π becomes an isomorphism of G-spaces.*

Proof. It is easy to verify that G_1 and G_2 act separately on X in a manner consistent (via π) with their action on G, and these actions agree on H. Hence the amalgamated free product G also acts on this space and turns π into a morphism of G-spaces. From construction of X we see that the G-action is transitive, and the claim now follows from Lemma 1.8.6. $\quad\square$

Corollary 1.8.8. *Let G_1, G_2 be two groups with common subgroup H, and let $G := G_1 *_H G_2$ be the amalgamated free product. Let $g_1 \in G_1$ and $g_2 \in G_2$*

be such that the cyclic groups $\langle g_1 \rangle, \langle g_2 \rangle$ are infinite and have no intersection with H. Then g_1, g_2 generate a free subgroup in G.

Proof. By hypothesis (and the axiom of choice), we can find a partition $G_1 = \bigcup_{s_1 \in S_1} H \cdot s_1$, where S_1 contains the infinite cyclic group $\langle g_1 \rangle$, and similarly we can find a partition $G_2 = \bigcup_{s_2 \in S_2} H \cdot s_2$. Let X be the space in Lemma 2. Each reduced word formed by g_1, g_2 then generates a distinct element of X, and thus (by Lemma 1.8.7) a distinct element of G. The claim follows. $\square$

Remark 1.8.9. The above corollary can also be established by the *ping-pong lemma* (which is not surprising, since the proof of that lemma uses many of the same ideas, and in particular exploits an action of G on a space X in order to distinguish various words in G from each other). Indeed, observe that g_1, g_1^{-1} map those words $hs_1 s_2 \ldots s_n$ in X with $s_1 \notin S_1$ into words $hs_0 s_1 \ldots s_n$ with $s_0 \in S_1$, and similarly for g_2, g_2^{-1}, which is the type of hypothesis needed to apply the ping-pong lemma. [Thanks to Ben Green for this observation.]

Now we can finish the proof of Theorem 1.8.1. As discussed in Example 1.8.5, the group $G_1 := \langle a, b, c | ab = ba^2, bc = cb^2 \rangle$ is the amalgamated free product of $\langle a, b | ab = ba^2 \rangle$ and $\langle b, c | bc = cb^2 \rangle$ relative to $\langle b \rangle$. By Corollary 1.8.8, a and c generate the free group here, thus G_1 contains $H = \langle a, c \rangle$ as a subgroup. Similarly, the group $G_2 := \langle c, d, a | cd = dc^2; da = ad^2 \rangle$ also contains $H = \langle a, c \rangle$ as a subgroup. We may then form the amalgamated free product

$$G := G_1 *_H G_2 = \langle a, b, c, d | ab = ba^2, bc = cb^2, cd = dc^2, da = ad^2 \rangle,$$

and another application of Corollary 1.8.8 shows that b, d generate the free group (and are in particular distinct); similarly a, c are distinct. Finally, the group $\langle a, b | ab = ba^2 \rangle$ embeds into G_1, which embeds into G, and so a, b, are also distinct; cyclically permuting this we conclude that all of the a, b, c, d are distinct as claimed.

Notes. This article first appeared at

terrytao.wordpress.com/2008/10/06.

Thanks to an anonymous commenter for corrections.

Ben Green pointed out that for the specific "non-commutative Freiman theorem" application of trying to characterise finite sets A of small doubling (thus $|A \cdot A| \leq K|A|$), it may still be possible that some large *subset* A' of A has a finite model, even if A itself need not be. Currently, all known examples of finite sets of small doubling have this property.

David Fisher pointed out that Remark 1.8.3 can be deduced from Theorem 1.8.1 using the general fact that all linear groups are residually finite (i.e., have finite quotients that separate any finite set). The proof of this latter fact is non-trivial, however.

1.9. Small samples, and the margin of error

In view of this year's U.S. presidential election, I would like to talk about some of the basic mathematics underlying electoral polling, and specifically to explain the fact, which can be highly unintuitive to those not well versed in statistics, that polls can be accurate even when sampling only a tiny fraction of the entire population.

Take for instance a nationwide poll of U.S. voters on which presidential candidate they intend to vote for. A typical poll will ask a number n of randomly selected voters for their opinion; a typical value here is $n = 1000$. In contrast, the total voting-eligible population of the U.S.—let us call this set X—is about 200 million.[34] Thus, such a poll would sample about 0.0005% of the total population X—an incredibly tiny fraction. Nevertheless, the margin of error (at the 95% *confidence level*) for such a poll, if conducted under idealised conditions (see below), is about 3%. In other words, if we let p denote the proportion of the entire population X that will vote for a given candidate A, and let $\overline{p}$ denote the proportion of the polled voters that will vote for A, then the event $\overline{p} - 0.03 \leq p \leq \overline{p} + 0.03$ will occur with probability at least 0.95. Thus, for instance (and oversimplifying somewhat by ignoring the probability-altering effects of conditional expectation—see below), if the poll reports that 55% of respondents would vote for A, then the true percentage of the electorate that would vote for A has at least a 95% chance of lying between 52% and 58%. Larger polls will of course give a smaller margin of error; for instance the margin of error for an (idealised) poll of 2,000 voters is about 2%.

I will give a rigorous proof of a weaker version of the above statement (giving a margin of error of about 7%, rather than 3%) in an appendix at the end of this section. But the main point here is a little different, namely to address the common misconception that the accuracy of a poll is a function of the *relative* sample size rather than the *absolute* sample size, which would suggest that a poll involving only 0.0005% of the population could not possibly have a margin of error as low as 3%. I also want to point out some limitations of the mathematical analysis; depending on the methodology and the context, some polls involving 1000 respondents may have a much higher margin of error than the idealised rate of 3%.

---

[34]The actual *turnout* for the 2008 election ended up being approximately 130 million, but we ignore this fact for the sake of discussion.

1.9.1. Assumptions and conclusion. Not all polls are created equal; there are a certain number of hypotheses on the methodology and effectiveness of the poll that we have to assume in order to make our mathematical conclusions valid. We will make the following idealised assumptions:

(1) **Simple question.** Voters polled can only offer one of two responses, which I will call A and not-A; thus we ignore the effect of third-party candidates, undecided voters, or refusals to respond. In particular, we do not try to combine this data with other questions about the polled voters, such as demographic data. We also assume that the question is unambiguous and cannot be misinterpreted by respondents (see hypothesis (3) below).

(2) **Perfect response rate.** All voters polled offer a response; there are no refusals to respond to the poll, or failures to make contact with the voter being polled. (This is a special case of hypothesis (1), but it deserves to be emphasised.) In particular, this excludes polls that are self-selected, such as internet polls (since in most cases, a large fraction of viewers of a web page with a poll will refuse to respond to that poll).

(3) **Honest responses.** The response given by a voter to the poll is an accurate representation whether that voter intends to vote for A or not; thus we ignore response-distorting effects such as the *Bradley effect*, *push-polling*, *tactical voting*, frivolous responses, misunderstanding of the question, or attempts to "game" a poll by the respondents.

(4) **Fixed poll size.** The number n of polled voters is fixed in advance; in particular, one cannot keep polling until one has achieved some desired outcome, and then stop.

(5) **Simple random sampling (without replacement).** Each one of the n voters polled is selected uniformly at random among the entire population X, thus each voter is equally likely to be selected by the poll, and no non-voter can be selected by the poll. (In particular, we make the important assumption that there is no *selection bias*.) Furthermore, each polled voter is chosen *independently* of all the others, except for the one condition that we do not poll any given voter more than once. (Thus, once a voter is polled, that voter is "crossed off the list" of the pool X of voters from which one randomly selects the next voter to be polled.) In particular, we assume that the poll is not *clustered*.

(6) **Honest reporting.** The results of the poll are always reported, with no inaccuracies; one cannot cancel, modify, or ignore a poll

once it has begun. In particular, one cannot conduct multiple polls and only report the "best" results (thus running the risk of *confirmation bias*).

Polls which deviate significantly from these hypotheses (e.g. due to complex questions, self-selection or other selection bias, confirmation bias, inaccurate responses, a high refusal rate, variable poll size, or clustering) will generally be less accurate than an idealised poll with the same sample size. Of course, there is a substantial literature in statistics (and polling methodology) devoted to measuring, mitigating, avoiding, or compensating for these less ideal situations, but we will not discuss those (important) issues here. We will remark though that in practice it is difficult to make the poll selection truly uniform. For instance, if one is conducting a telephone poll, then the sample will of course be heavily biased towards those voters who actually own phones; a little more subtly, it will also be biased toward those voters who are near their phones at the time the poll was conducted, and have the time and inclination to answer phone calls. As long as these factors are not strongly correlated with the poll question (i.e., whether the voter will vote for A), this is not a major concern, but in some cases, the poll methodology will need to be adjusted (e.g. by reweighting the sample) to compensate for the non-uniformity.

As stated in the introduction, we let p be the proportion of the entire population X that will vote for A, and $\bar{p}$ be the proportion of the polled voters that will vote for A (which, by hypotheses (2) and (3), is exactly equal to the proportion of polled voters that say that they will vote for A). Under the above idealised conditions, if the number n of polled voters is $1,000$, and the size of the population X is 200 million, then the margin of error is about 3%, thus $\mathbf{P}(\bar{p} - 0.03 \le p \le \bar{p}+0.03) \ge 0.95$.

There is an important subtlety here: it is only the *unconditional* probability of the event $\bar{p} - 0.03 \le p \le \bar{p} + 0.03$ that is guaranteed to be greater than 0.95. If one has additional *prior information* about p and $\bar{p}$, then the *conditional* probability of this event, relative to this information, may be very different. For instance, if one had, prior to the poll, a very good reason to believe that p is almost certainly between 0.4 and 0.6, and then the poll reports $\bar{p}$ to be 0.1, then the conditional probability that $\bar{p}-0.03 \le p \le \bar{p}+0.03$ occurs should be lower[35] than the unconditional probability. The question of how to account for prior information is a very delicate one in *Bayesian probability*, and will not be discussed here.

One special case of the above point is worth emphasising: the statement that $\bar{p}-0.03 \le p \le \bar{p}+0.03$ is true with at least 95% probability is only valid

---

[35]Note though that having prior information just about p, and not $\bar{p}$, will not cause the probability to drop below 95%, as this bound on the confidence level is uniform in p.

before one actually conducts the poll and finds out the value of $\overline{p}$. Once $\overline{p}$ is computed, the statement $\overline{p} - 0.03 \leq p \leq \overline{p} + 0.03$ is either true or false, i.e., occurs with probability[36] 1 or 0 (unless one takes a Bayesian approach, as mentioned above).

1.9.2. Nobody asked for my opinion! One intuitive argument against a poll of small relative size being accurate goes something like this: a poll of just $1,000$ people among a population of $200,000,000$ is almost certainly not going to poll myself, or any of my friends or acquaintances. If the opinions of myself, and everyone that I know, is not being considered at all in this poll, how could this poll possibly be accurate?

It is true that if you know, say, $5,000$ voting-eligible people, then chances are that none of them (or maybe one of them, at best) will be contacted by the above poll. However, even though the opinions of all these people are not being directly polled, there will be many other people with *equivalent* opinions that *will* be contacted by the poll. Through those people, the views of yourself and your friends are being represented. (This may seem like a very weak form of representation, but recall that you and your $5,000$ friends and acquaintances still only represent 0.0025% of the total electorate.)

Now one may argue that no two voters are identical, and that each voter arrives at a decision for whom to vote, following his/her own unique reasons. True enough—but recall that this poll is asking only a *simple* question: whether one is going to vote for A or not. Once one narrowly focuses on this question alone, any two voters who both decide to vote for A, or to not vote for A, are considered equivalent, even if they arrive at this decision for totally different reasons. So, for the purposes of this poll, there are only two types of voters in the world—A-voters, and not-A-voters—with all voters in one of these two types considered equivalent. In particular, any given voter is going to have millions of other equivalent voters distributed throughout the population X, and a representative fraction of those equivalent voters is likely to be picked up by the poll.

As mentioned before, polls which offer complex questions (for instance, trying to discern the motivation behind one's voting choices) will inherently be less accurate; there are now fewer equivalent voters for each individual, and it is harder for a poll to pick up each equivalence class in a representative manner.[37]

---

[36]This phenomenon of course occurs all the time in probability. For instance, if x denotes the outcome of rolling a fair six-sided die, then before one performs this roll, the probability that x equals 1 will be $1/6$, but after one has seen what the value of this die is, the probability that x equals 1 will be either 1 or 0.

[37]In particular, the more questions asked, the more likely the responses to at least one of these questions will be inaccurate by an amount exceeding its margin of error. This provides a

1.9.3. Is there enough information? Another common objection to the accuracy of polls argues that there is not enough information (or "degrees of freedom") present in the poll sample to accurately describe the much larger amount of data present in the full population; $1,000$ bits of data cannot possibly contain $200,000,000$ bits of information. However, we are not asking to find out so much information; the purpose of the poll is to estimate just a *single* piece of information, namely the number p. If one is willing to accept an error of up to 3%, then one can represent this piece of information in about five bits rather than $200,000,000$. So, in principle at least, there is more than enough information present in the poll to recover this information; one does not need to sample the entire population to get a good reading.[38]

As before, the accuracy degrades as one asks more and more complicated questions. For instance, if one were to poll $1,000$ voters for their opinions on two unrelated questions A and B, each of the answers to A and B would be accurate to within 3% with probability 95%, but the probability that the answers to A and B were simultaneously accurate to within 3% would be lower (around 90% or so), and so any data analysis that relies on the responses to both A and B may not have as high a confidence level as data analysis that relies on A and B separately. This is consistent with the information-theoretic perspective: we are demanding more and more bits of information on our population, and it is harder for our fixed data set to supply so much information accurately and confidently.

1.9.4. Swings. One intuitive way to gauge the margin of error of a poll is to see how likely such a poll is to accurately detect a swing in the electorate. Suppose for instance that over the course of a given time period (e.g. a week), 7% of the voters switch their vote from not-A to A, while another 2% of the voters switch their vote from A to not-A, leading to a net increase of 5% in the proportion p of voters voting for A. How would this swing in the vote affect the proportion $\overline{p}$ of the voters being polled, if one imagines the same voters being polled at both the start of the week and the end of the week? (Recall that we are assuming that voters will honestly report their change of mind from one poll to the next.)

If the poll was conducted by simple random sampling, then each of the $1,000$ voters polled would have a 7% probability of switching from not-A to A, and a 2% probability of switching from A to not-A. Thus, one would expect about 70 of the $1,000$ voters polled to switch to A, and about 20 to

---

limit as to how much information one can confidently extract from data mining any given data set.

[38]The same general philosophy underlies *compressed sensing* (see Section 1.2 of *Structure and Randomness*), but that is another story.

switch to not-A, leading to a net swing of 50 voters, that would increase $\bar{p}$ by 5%, thus matching the increase in p. Now, in practice, there will be some variability here; due to the luck of the draw, the poll may pick up more or less than 70 of the voters switching to A, and more or less than 20 of the voters switching to not-A. But having $1,000$ voters to sample is just about large enough for the law of large numbers[39] (Section 1.4) to kick in and ensure that the number of voters switching to A picked up by the poll will be significantly larger than the number of voters switching to not-A. Thus, this poll will have a good chance of detecting a swing of size 5% or more, which is consistent with the assertion of a margin of error of about 3%.

It is worth noting that this swing of 5% in an electorate of $200,000,000$ voters represents quite a large shift in absolute terms: fourteen million voters switching to A and four million switching away from A. Quite a few of these shifting voters will be picked up by the poll (in contrast to one's sphere of friends and acquaintances, which is likely to be missed completely).

1.9.5. Irregularity. Another intuitive objection to polling accuracy is that the voting population is far from homogeneous. For instance, it is clear that voting preferences for the U.S. presidential election vary widely among the 50 states—shouldn't one need to multiply the poll size by 50 just to accomodate this fact? Similarly for distinctions in voting patterns based on gender, race, party affiliation, etc.

Again, these irregularities in voter distribution do not affect the final accuracy of the poll, for two reasons. Firstly, we are asking only the simple question of whether a voter votes for A or not-A, and are not breaking down the answers to this question by state, gender, race, or any other factor; as stated before, two voters are considered equivalent as long as they have the same preference for A, even if they are in different states, have different genders, etc. Secondly, while it is conceivable that the poll will cluster its sample in one particular state (or one particular gender, etc.), thus potentially skewing the poll, the fact that the voters are selected uniformly *and* independently of each other prevents this from happening very often.[40]

The independence hypothesis is rather important. If for instance one were to poll by picking one particular location[41] in the U.S. at random, and polling $1,000$ people from that location, then the responses would be highly

---

[39] In appealing to the law of large numbers, we are implicitly exploiting the uniformity and independence assumptions in hypothesis (5).

[40] And in any event, clustering in a demographic or geographic category is not what is of direct importance to the accuracy of the poll; the only thing that really matters in the end is whether there is clustering in the category of A-voters or not-A-voters.

[41] Incidentally, in the specific case of the U.S. presidential election, statewide polls are in fact more relevant to the outcome of the election than nationwide polls, due to the mechanics of the *U.S. Electoral College*, but this does not detract from the above points.

correlated (as one could have picked a location which happens to highly favour A, or highly favour not-A) and would have a much larger margin of error than if one polled $1,000$ people at random across the U.S.

1.9.6. Analogies. Some analogies may help explain why the relative size of a sample is largely irrelevant to the accuracy of a poll.

Suppose one is in front of a large body of water (e.g. a sea or ocean), and wants to determine whether it is a freshwater or saltwater body. This can be done very easily: dip one's finger into the body of water and taste a single drop. This gives an extremely accurate result, even though the relative proportion of the sample size to the population size is, literally, a drop in the ocean; the quintillions of water molecues and salt molecues present in that drop are more than sufficient to give a good reading of the salinity[42] of the water body.

Another analogy comes from digital imaging. As we all know, a digital camera takes a picture of a real-world object (e.g. a human face) and converts it into an array of pixels; an image with a larger number of pixels will generally lead to a more accurate image than one with fewer. But even with just a handful of pixels, say 1,000 pixels, one is already able to make crude distinctions between different images, for instance to distinguish a light-skinned face from a dark-skinned face (despite the fact that skin colour is determined by millions of cells and quintillions of pigment molecues). For instance, there is a well-known image in [**Ha1973**] of a portrait of Abraham Lincoln, which is clearly recognisable despite consisting of only a few hundred pixels.

1.9.7. Appendix: Mathematical justification. One can compute the margin of error for this simple sampling problem very precisely using the *binomial distribution*; however I would like to present here a cruder but more robust estimate, based on the *second moment method*, that works in much greater generality than the setting discussed here. (It is closely related to the arguments in Section 1.4.) The main mathematical result we need is

Theorem 1.9.1. *Let X be a finite set, let A be a subset of X, and let $p := |A|/|X|$ be the proportion of elements of X that lie in A. Let $x_1, \ldots, x_n$ be sampled independently and uniformly at random from X (in particular, we allow repetitions). Let $\overline{p} := |\{1 \le i \le n : x_i \in A\}|/n$ be the proportion of*

---

[42]To be fair, in order for this reading to be accurate, one needs to assume that the salinity is uniformly distributed across the body of water; if for instance the body happened to be nearly fresh on one side and much saltier on the other, then dipping one's finger in just one of these two sides would lead to an inaccurate measurement of average salinity. But if one were to stir the body of water vigorously, this irregularity of distribution disappears. The procedure of taking a random sample, with each sample point being independent of all the others, is analogous to this stirring procedure.

the $x_1, \ldots, x_n$ *(counting repetition) that lie in A. Then for any $r > 0$, one has*

(1.68)
$$\mathbf{P}(|\bar{p} - p| \le r) \ge 1 - \frac{1}{4nr^2}.$$

Proof. We use the second moment method. For each $1 \le i \le n$, let I_i be the *indicator* of the event $x_i \in A$; thus $I_i := 1$ when $x_i \in A$, and $I_i = 0$ otherwise. Observe that each I_i has a probability of p of equaling 1, thus

$$p = \mathbf{E}I_i.$$

On the other hand, we have

$$\bar{p} = \frac{1}{n} \sum_{i=1}^{n} I_i.$$

Thus

$$\bar{p} - p = \frac{1}{n} \sum_{i=1}^{n} I_i - \mathbf{E}(I_i);$$

squaring this and taking expectations, we obtain

$$\mathbf{E}|\bar{p} - p|^2 = \frac{1}{n^2} \sum_{i=1}^{n} \mathbf{Var}(I_i) + \frac{2}{n} \sum_{1 \le i < j \le n} \mathbf{Cov}(I_i, I_j),$$

where $\mathbf{Var}(I_i) := \mathbf{E}(I_i - \mathbf{E}I_i)^2$ is the *variance* of I_i, and $\mathbf{Cov}(I_i, I_j) := \mathbf{E}((I_i - p)(I_j - p))$ is the *covariance* of I_i, I_j.

By assumption, the random variables I_i, I_j for $i \ne j$ are independent, and so the covariances $\mathbf{Cov}(I_i, I_j)$ vanish. On the other hand, a direct computation shows that

$$\mathbf{Var}(I_i) = p - p^2 = \frac{1}{4} - \left(p - \frac{1}{2}\right)^2 \le \frac{1}{4}$$

for each i. Putting all this together we conclude that

$$\mathbf{E}|\bar{p} - p|^2 \le \frac{1}{4n}$$

and the claim (1.68) follows from *Markov's inequality*. $\square$

Applying this theorem with $n = 1000$ and $r = 1/\sqrt{200} \approx 0.07$, we conclude that p and $\bar{p}$ lie within about 7% of each other with probability at least 95%, regardless of how large the population X is. In the context of an election poll, this means that if one samples 1000 voters independently at random (with replacement) whether they would vote for A, the margin of error for the answer would be at most 7% at the 95% confidence level.

Remark 1.9.2. Observe that the proof of the above theorem did not really need the x_i to be fully independent of each other; the key thing was that each x_i was close to uniformly distributed, and that the covariances between the indicators I_i, I_j were small. (In particular, one only needs *pairwise independence* rather than joint independence for the theorem to hold.) Because of this, one can also obtain variants of the above theorem when one selects $x_1, \ldots, x_n$ for random sampling without replacement (known as *simple random sampling*); now there is a slight correlation between I_i, I_j, but it turns out to be negligible when X is large, for instance[43] when $n = 1000$ and $|X| \sim 10^8$.

Remark 1.9.3. If one assumes joint independence instead of pairwise independence, one can obtain slightly sharper inequalities than (1.68) (e.g. by using the *Chernoff inequality*), but at the 95% confidence level, this gives a relatively modest improvement only in the margin of error (in our specific example, the optimal margin of error is about 3% rather than 7%).

Remark 1.9.4. An inspection of the argument shows that if p is known to be very small or very large, then the margin of error is better than what (1.68) predicts. (In the most extreme case, if $p = 0$ or $p = 1$, then it is easy to see that the margin of error is zero.) But in the case of election polls, p is generally expected to be close to $1/2$, and so one does not expect to be able to improve the margin of error much from this effect. And in any case, we do not know the value of p exactly in practice (otherwise why would we be doing the poll in the first place?).

Remark 1.9.5. In real world situations, it can be difficult or impractical to get the x_i to be close to uniformly distributed (because of *sampling bias*), and to keep the correlations low (because of effects such as *clustering*). For this reason, one often needs to perform a more complicated sampling procedure than simple random sampling, which requires more sophisticated statistical analysis than given by the above theorem. This is beyond the scope of this article, though.

Notes. This article first appeared at

terrytao.wordpress.com/2008/10/10.

Thanks to jonm and Kieran for corrections.

A calculator to compute margins of error for various sample sizes and population sizes can be found at

www.americanresearchgroup.com/moe.html.

---

[43] For this range of parameters, there is a non-trivial probability of a *birthday paradox* occurring, so the two sampling methods are genuinely different from each other; but they turn out to have almost the same margin of error anyway.

1.10. Non-measurable sets via non-standard analysis

In Section 1.4 of *Structure and Randomness*, I sketched out a non-rigorous probabilistic argument justifying the following well-known theorem:

Theorem 1.10.1 (Non-measurable sets exist). *There exists a subset E of the unit interval $[0,1]$ which is not Lebesgue-measurable.*

The idea was to let E be a "random" subset of $[0,1]$. If one (non-rigorously) applies the law of large numbers (Section 1.4), one expects E to have "density" $1/2$ with respect to every subinterval of $[0,1]$, which would contradict the *Lebesgue differentiation theorem*.

I was recently asked whether I could in fact make the above argument rigorous. This turned out to be more difficult than I had anticipated, due to some technicalities in trying to make the concept of a random subset of $[0,1]$ (which requires an uncountable number of "coin flips" to generate) both rigorous and useful. However, there is a simpler variant of the above argument which can be made rigorous. Instead of letting E be a "random" subset of $[0,1]$, one takes E to be an "alternating" set that contains "every other" real number in $[0,1]$; this again should have density $1/2$ in every subinterval and thus again contradict the Lebesgue differentiation theorem.

Of course, in the standard model of the real numbers, it makes no sense to talk about "every other" or "every second" real number, as the real numbers are not discrete. If however one employs the language of *non-standard analysis*, then it is possible to make the above argument rigorous, and this is the purpose of this article. I will assume some basic familiarity with non-standard analysis, for instance as discussed in Section 1.5 of *Structure and Randomness*.

We begin by selecting a *non-principal ultrafilter* $p \in \beta\mathbf{N}\backslash\mathbf{N}$ and use it to construct non-standard models $*\mathbf{N}$, $*[0,1]$ of the natural numbers $\mathbf{N}$ and the unit interval $[0,1]$ by the usual *ultrapower* construction. We then let $N \in *\mathbf{N}$ be an unlimited non-standard number, i.e., a non-standard natural number larger than any standard natural number.[44]

We can partition the non-standard unit interval $*[0,1]$ into 2^N (non-standard) intervals $J_j := [j/2^N, (j+1)/2^N]$ for $j = 0, \ldots, 2^N - 1$ (the overlap of these intervals will have a negligible impact in our analysis). We then define the non-standard set $*E \subset *[0,1]$ to be the union of those J_j with j odd; this is the formalisation of the idea of "every other real number" in the introduction. The key feature of $*E$ that we need here is the following symmetry property: If $I = [a/2^n, (a+1)/2^n]$ is any standard *dyadic interval*, then the (non-standard counterpart to the) interval I can

---

[44]For instance, one could take N to be the equivalence class in $*\mathbf{N}$ of the sequence $1, 2, \ldots$.

be partitioned (modulo (non-standard) rationals) as the set $*E \cap I$ and its reflection $(2a+1)/2^n - (*E \cap I)$. This demonstrates in particular that $*E$ has (non-standard) density $1/2$ inside I, but we will need to use the symmetry property directly, rather than the non-standard density property, because the former is much easier to transfer to the standard setting than the latter.

We now return to the standard world, and introduce the *standard* set $E \subset [0,1]$ defined as the collection of all standard $x \in [0,1]$ whose non-standard representative $*x$ lies in $*E$. This is certainly a standard subset of $[0,1]$; informally, it is the set of all standard numbers whose N^{th} digit is 1. We claim that it is not Lebesgue measurable, thus establishing Theorem 1.10.1. To see this, recall that for any standard dyadic interval $I = [a/2^n, (a+1)/2^n]$, every non-standard irrational element of I lies in exactly one of $*E$ or $(2a+1)/2^n - *E$. Applying the transfer principle, we conclude that every standard irrational element of I lies in exactly one of E or $(2a+1)/2^n - E$. Thus, if E were measurable, the density of E in the dyadic interval I would be exactly $1/2$. But this contradicts the *Lebesgue differentiation theorem* (see e.g. Section 1.4 of *Structure and Randomness*), and we are done.

Remark 1.10.2. One can eliminate the non-standard analysis from this argument and rely directly on the non-principal ultrafilter p. Indeed, if one inspects the ultrapower construction carefully, one sees that (outside of the terminating binary rationals, which do not have a unique binary expansion), E consists of those numbers in $[0,1]$ whose 1s in the binary expansion lie in p. The symmetry property of E then reflects the non-principal ultrafilter nature of p, in particular the fact that membership of a set $A \subset \mathbf{N}$ in p is insensitive to any finite modification of A, and reversed by replacing A with its complement. On the other hand, one cannot eliminate the non-principal ultrafilter entirely; once one drops the axiom of choice, there exist models of the real line in which every set is Lebesgue measurable, and so it is necessary to have *some* step in the proof of Theorem 1.10.1 that involves this axiom. In the above argument, choice is used to find the non-principal ultrafilter p.

Remark 1.10.3. It is tempting to also make the original argument, based on randomness, work, but I was unable to push it through completely. Certainly, if one sets $*E$ to be a (non-standardly) random collection of the J_j, then with probability infinitesimally close to 1, the (non-standard) density of $*E$ in any standard dyadic interval (or indeed, any standard interval) is infinitesimally close to $1/2$, thanks to the law of large numbers. However, I was not able to transfer this fact to tell me anything about E; indeed, I could not even show that E was non-empty. (Question: Does there exist a non-standard subset of $*[0,1]$ of positive (non-infinitesimal) measure which avoids all standard real numbers? I do not know the answer to this.)

Notes. This article first appeared at

`terrytao.wordpress.com/2008/10/14`.

Thanks to Timothy Gowers, Wonghang, and an anonymous commenter for corrections.

K. P. Hart and Kevin O'Bryant noted that this construction is quite classical, going back to [**Ul1929**], [**Ta**]. K. P. Hart also noted that in the product space $2^{\mathbf{R}}$, the set of measurable and the set of non-measurable sets both have outer measure 1 with respect to product measure, so a naive attempt to formalise the statement that "a randomly chosen set is non-measurable" does not work.

1.11. A counterexample to a strong polynomial Freiman-Ruzsa conjecture

One of my favourite open problems in additive combinatorics is the *polynomial Freiman-Ruzsa conjecture* [**Gr2005**]. It has many equivalent formulations (which is always a healthy sign when considering a conjecture), but here is one involving "approximate homomorphisms":

Conjecture 1.11.1 (Polynomial Freiman-Ruzsa conjecture). *Let* $f : F_2^n \to F_2^m$ *be a function which is an approximate homomorphism in the sense that* $f(x + y) - f(x) - f(y) \in S$ *for all* $x, y \in F_2^n$ *and some set* $S \subset F_2^m$. *Then there exists a genuine homomorphism* $g : F_2^n \to F_2^m$ *such that* $f - g$ *takes at most* $O(|S|^{O(1)})$ *values.*

Remark 1.11.2. The key point here is that the bound on the range of $f - g$ is at most polynomial in $|S|$. An exponential bound of $2^{|S|}$ can be trivially established by splitting F_2^m into the subspace spanned by S (which has size at most $2^{|S|}$) and some complementary subspace, and then letting g be the projection of f to that complementary subspace.

In a forthcoming paper with Ben Green, we showed that this conjecture is equivalent to a certain polynomially quantitative strengthening of the inverse conjecture for the Gowers norm $U^3(F_2^n)$. For this (somewhat technical) section, I want to comment on a possible further strengthening of this conjecture, namely

Conjecture 1.11.3 (Strong polynomial Freiman-Ruzsa conjecture). *Let* $f : F_2^n \to F_2^m$ *be a function which is an approximate homomorphism in the sense that* $f(x + y) - f(x) - f(y) \in S$ *for all* $x, y \in F_2^n$ *and some set* $S \subset F_2^m$. *Then there exists a genuine homomorphism* $g : F_2^n \to F_2^m$ *such that* $f - g$ *takes values in the sumset* $CS := S + \ldots + S$ *for some fixed* $C = O(1)$.

This conjecture is known to be true for certain types of set S (e.g. for Hamming balls, this was shown in [**Fa2000**]). Unfortunately, it is false in general; the purpose of this article is to describe one counterexample (related to the failure of the inverse conjecture for the Gowers norm for $U^4(F_2^n)$ for classical polynomials; in particular, the arguments here have several features in common with those in [**LoMeSa2008**], [**GrTa2007**]; a somewhat different counterexample also appears in [**Fa2000**]. The verification of the counterexample is surprisingly involved, ultimately relying on the multidimensional Szemerédi theorem [**FuKa1979**].

1.11.1. Description of counterexample. We let n be a large number, and replace F_2^m by the $\frac{n(n+1)}{2}$-dimensional vector space V of quadratic forms $Q : F_2^n \to F_2$ (with a basis given by the monomials $x_i x_j$ with $1 \leq i \leq j \leq n$). We let $f : F_2^n \to V$ be defined by the formula

$$f(h_1, \ldots, h_n)(x_1, \ldots, x_n) := \sum_{1 \leq i < j \leq n} h_i x_i h_j x_j.$$

A brief computation shows that for any $h, k \in F_2^n$, the quadratic form $f(h + k) - f(h) - f(k)$ is of rank at most three, by which we mean that it is a function of at most three linear forms. More specifically, we have

$$(1.69) \qquad f(h + k) - f(h) - f(k) = a_{h,k}b_{h,k} + b_{h,k}c_{h,k} + c_{h,k}a_{h,k},$$

where

$$a_{h,k}(x) := \sum_{i=1}^{n} h_i(1 - k_i)x_i,$$

$$b_{h,k}(x) := \sum_{i=1}^{n}(1 - h_i)k_i x_i,$$

$$c_{h,k}(x) := \sum_{i=1}^{n} h_i k_i x_i.$$

Thus, if we let S be the space of quadratic forms of rank at most 3, the hypotheses of Conjecture 1.11.3 hold.

1.11.2. Verification of counterexample. To establish the counterexample, we assume for contradiction that there exists a linear function $g : F_2^n \to V$ such that $f(h) - g(h)$ has bounded rank for all h, and deduce a contradiction (for n sufficiently large).

By hypothesis, we have linear forms $L_{h,1}, \ldots, L_{h,d}$ for all $h \in F_2^n$ and some $d = O(1)$ and coefficients $c_{h,i,j} \in F_2$ for all $1 \leq i \leq j \leq d$ such that

$$f(h) - g(h) = \sum_{1 \leq i \leq j \leq n} c_{h,i,j} L_{h,i} L_{h,j}.$$

and in particular (by (1.69) and linearity of g)

$$a_{h,k}b_{h,k} + b_{h,k}c_{h,k} + c_{h,k}a_{h,k}$$

$$(1.70) \qquad = \sum_{1\le i\le j\le n} c_{h+k,i,j}L_{h+k,i}L_{h+k,j} - c_{h,i,j}L_{h,i}L_{h,j} - c_{k,i,j}L_{k,i}L_{k,j}.$$

The key point is that the linear forms $a_{h,k}, b_{h,k}, c_{h,k}$ are usually "independent" of the linear forms $L_{h,i}, L_{k,i}, L_{h+k,i}$. The crucial lemma in this regard is

Lemma 1.11.4. *If h, k are selected uniformly and independently at random, then with probability $1 - o(1)$, $a_{h,k}$ is not a linear combination of the $L_{h,i}, L_{k,i}, L_{h+k,i}$. Similarly for $b_{h,k}, c_{h,k}$.*

Proof. By cyclically permuting h, k, $h + k$ it suffices to show this for $c_{h,k}$. Since there are at most $O(1)$ possible linear combinations amongst the $L_{h,i}, L_{k,i}, L_{h+k,i}$, it suffices to show that for any given assignments $h \mapsto L'_h, k \mapsto L''_k, h + k \mapsto L'''_{h+k}$ of linear forms, the probability of the event

$$(1.71) \qquad\qquad c_{h,k} = L'_h + L''_k + L'''_{h+k}$$

is $o(1)$. Suppose for contradiction that the event (1.71) holds for a set E of pairs (h, k) in $F_2^n \times F_2^n$ of positive density. Applying the multidimensional Szemerédi theorem [**FuKa1979**] we can find (for n large enough) a square $(h, k), (h + r, k), (h, k + r), (h + r, k + r)$ in E with r non-zero. Applying (1.71) for all four pairs and summing, we obtain

$$c_{h,k} + c_{h+r,k} + c_{h,k+r} + c_{h+r,k+r} = 0$$

(recall that we are in characteristic 2). But the left-hand side is equal to the linear form $\sum_i r_i x_i$, which is non-zero, a contradiction. $\qquad\square$

Now we can obtain the desired contradiction. For a generic choice of h, k, we now know that none of the $a_{h,k}, b_{h,k}, c_{h,k}$ are linear combinations of the $L_{h,i}, L_{k,i}, L_{h+k,i}$. Thus, on a given level set of the $L_{h,i}, L_{k,i}, L_{h+k,i}$ (which form a subspace of F_2^n), the linear functions $a_{h,k}, b_{h,k}, c_{h,k}$ are non-constant, and so the range of the triplet $(a_{h,k}, b_{h,k}, c_{h,k})$ must be an affine subspace of F_2^3 which is not contained in any coordinate plane. But then the function $(a, b, c) \mapsto ab + bc + ca$ cannot be constant on this space (as can be seen by an axhaustive check of all possible cases), contradicting (1.70), and so Conjecture 1.11.3 fails.

Remark 1.11.5. The function f appearing in the above example is closely related to the symmetric polynomial

$$S_4(x) := \sum_{1\le i<j<k<l\le n} x_i x_j x_k x_l.$$

Indeed, one can show that the derivative $S_4(x + h) - S_4(x)$ of S_4 is equal to $f(h)$, plus some additional terms which involve only a finite number of linear forms, and the quadratic polynomial $S_2(x) := \sum_{1 \leq i < j \leq n} x_i x_j$. If it was the case that f could be approximated by a linear map g modulo low rank errors, then one could use this to eventually show that S_4 is correlated with a cubic polynomial; but it is known [**LoMeSa2008**], [**GrTa2007**] that this is not the case. Thus there is an alternative way to verify that the above example is indeed a counterexample to the strong polynomial Freiman-Ruzsa conjecture.

Notes. This article first appeared at

terrytao.wordpress.com/2008/11/09,

and is derived from forthcoming joint work with Ben Green.

1.12. Some notes on "non-classical" polynomials in finite characteristic

Let $k \geq 0$ be an integer. The concept of a polynomial $P : \mathbf{R} \to \mathbf{R}$ of one variable of degree $< k$ (or $\leq k - 1$) can be defined in one of two equivalent ways:

- (Global definition) $P : \mathbf{R} \to \mathbf{R}$ is a polynomial of degree $< k$ iff it can be written in the form $P(x) = \sum_{0 \leq j < k} c_j x^j$ for some coefficients $c_j \in \mathbf{R}$.
- (Local definition) $P : \mathbf{R} \to \mathbf{R}$ is a polynomial of degree $< k$ if it is k-times continuously differentiable and $\frac{d^k}{dx^k} P \equiv 0$.

From single variable calculus we know that if P is a polynomial in the global sense, then it is a polynomial in the local sense; conversely, if P is a polynomial in the local sense, then from the Taylor series expansion

$$P(x) = \sum_{0 \leq j < k} \frac{P^{(j)}(0)}{j!} x^j$$

we see that P is a polynomial in the global sense. We make the trivial remark that we have no difficulty dividing by $j!$ here, because the field $\mathbf{R}$ is of characteristic zero.

The above equivalence carries over to higher dimensions:

- (Global definition) $P : \mathbf{R}^n \to \mathbf{R}$ is a polynomial of degree $< k$ iff it can be written in the form

$$P(x_1, \ldots, x_n) = \sum_{0 \leq j_1, \ldots, j_n; \, j_1 + \ldots + j_n < k} c_{j_1, \ldots, j_n} x_1^{j_1} \ldots x_n^{j_n}$$

  for some coefficients $c_{j_1, \ldots, j_n} \in \mathbf{R}$.

- (Local definition) $P : \mathbf{R}^n \to \mathbf{R}$ is a polynomial of degree $< k$ if it is k-times continuously differentiable and $(h_1 \cdot \nabla) \ldots (h_k \cdot \nabla)P \equiv 0$ for all $h_1, \ldots, h_k \in \mathbf{R}^n$.

Again, it is not difficult to use several variable calculus to show that these two definitions of a polynomial are equivalent.

The purpose of this (somewhat technical) section is to record some basic analogues of the above facts in finite characteristic, in which the underlying domain of the polynomial P is F or F^n for some finite field F. In the "classical" case when the range of P is also the field F, it is a well-known fact (which we reproduce here) that the local and global definitions of polynomial are equivalent. But in the "non-classical" case, when P ranges in a more general group (and in particular in the unit circle $\mathbf{R}/\mathbf{Z}$), the global definition needs to be corrected somewhat by adding some new monomials to the classical ones $x_1^{j_1} \ldots x_n^{j_n}$. Once one does this, one can recover the equivalence between the local and global definitions.

1.12.1. General theory. One can extend the local definition of a polynomial to cover maps $P : G \to H$ for any additive groups G, H.[45] Given any such map, and any $h \in G$, define the shift $T^h P : G \to H$ and the (discrete) derivative $\Delta_h P : G \to H$ by the formulae

$$T^h P(x) := P(x + h), \qquad \Delta_h P = T^h P - P,$$

thus schematically we have

$$(1.72) \qquad\qquad \Delta_h = T^h - 1.$$

We say that P is an *(additive) polynomial of degree $< k$* (or degree $\leq k-1$) if $\Delta_{h_1} \ldots \Delta_{h_k} P = 0$ for all $h_1, \ldots, h_k \in G$. Note that this corresponds to the definition of a classical polynomial from $\mathbf{R}^n$ to $\mathbf{R}$ once one adds some regularity conditions, such as k-times differentiability (actually, measurability will already suffice).

Examples 1.12.1. The zero function has degree < 0. A constant function has degree ≤ 0. A homomorphism has degree ≤ 1. Composing a polynomial of degree $\leq k$ with a homomorphism (either on the left or right) will give another polynomial of degree $\leq k$. The sum of two polynomials of degree $\leq k$ is again of degree $\leq k$. The derivative $\Delta_h P$ of a polynomial of degree $< k$ is of degree $< k-1$. Since $T^h P = P + \Delta_h P$, we conclude that the shift of any polynomial of degree $< k$ is also of degree $< k$.

Now we show that the product and composition of polynomials are again polynomials.

---

[45] There is also an important generalisation of this concept to the case of nilpotent groups; we will not concern ourselves with this generalisation here, but see [**Le1998**], [**Le2002**].

Lemma 1.12.2 (Product of polynomials is polynomial). *Let $P : G \to H$, $Q : G \to K$ be polynomials of degree $\leq h$, $\leq k$, respectively, for some $h, k \geq 0$, and let $B : H \times K \to L$ be a bilinear map. Then $B(P, Q)$ is a polynomial of degree $\leq h + k$.*

Proof. We induct on $h + k$. The claim is easy when h or k is zero, so suppose that $h, k > 0$ and the claim has already been proven for smaller values of $h + k$. From the discrete product rule

$$\Delta_g B(P, Q) = B(\Delta_g P, Q) + B(T^g P, \Delta_g Q)$$

and induction we see that $\Delta_g B(P, Q)$ is of degree $\leq h + k - 1$, and thus $B(P, Q)$ has degree $\leq h + k$ as desired. $\square$

Corollary 1.12.3. *If H is a ring, then the product of two polynomials from G to H of degree $\leq h$, $\leq k$, respectively, is of degree $\leq h + k$.*

Lemma 1.12.4 (Composition of polynomials is polynomial). *Let $P : G \to H$, $Q : H \to K$ be polynomials of degree $\leq h$, $\leq k$, respectively, for some $h, k \geq 0$. Then $Q \circ P : G \to K$ is a polynomial of degree $\leq hk$.*

Proof. For inductive reasons it is convenient to prove the following more general statement: If $P : G \to H$, $Q : H \to K$ are polynomials of degree $\leq h + m$, $\leq k$, respectively, for some $m, h, k \geq 0$, and $R_1, \ldots, R_m : G \to H$ are polynomials of degree $\leq r_1, \ldots, \leq r_m$, respectively, where $0 \leq r_j \leq k$ for all j, then the function $S : G \to K$ defined by

$$S(x) := [\Delta_{R_1(x)} \ldots \Delta_{R_m(x)} P](Q(x))$$

is a polynomial of degree $hk + r_1 + \ldots + r_m$. Clearly Lemma 1.12.4 follows from the $m \geq 0$ case of this claim.

We prove this claim by induction on h; then, for fixed h, by induction on m; and then, for fixed h and m, by induction on $r_1 + \ldots + r_m$. Thus, assume that the claim has already been shown for all smaller values of h, or for the same value of h and all smaller values of m, or for the same values of h, m and all smaller values of $r_1 + \ldots + r_m$.

If $r_m = 0$ then R_m is constant, and by replacing P with $\Delta_{R_m} P$ and decrementing m, we see that the claim follows from the induction hypothesis. Similarly, if any other of the r_j vanish (since the derivative operators commute with each other). So we may assume that $r_j > 0$ for all j.

Let $g \in G$. By considering the successive differences between the quantities

$$S(x) = [\Delta_{R_1(x)} \Delta_{R_2(x)} \cdots \Delta_{R_m(x)}](Q(x)),$$
$$[\Delta_{R_1(x)} \Delta_{R_2(x)} \cdots \Delta_{R_m(x)}](T_g Q(x)),$$
$$[\Delta_{T_g R_1(x)} \Delta_{R_2(x)} \cdots \Delta_{R_m(x)}](T_g Q(x)),$$
$$[\Delta_{T_g R_1(x)} \Delta_{T_g R_2(x)} \cdots \Delta_{R_m(x)}](T_g Q(x)),$$
$$\vdots$$
$$T_g S(x) = [\Delta_{T_g R_1(x)} \Delta_{T_g R_2(x)} \cdots \Delta_{T_g R_m(x)}](T_g Q(x)),$$

we see that $\Delta_g S(x)$ is the sum of

$$[\Delta_{R_1(x)} \Delta_{R_2(x)} \cdots \Delta_{R_m(x)} \Delta_{\Delta_g Q(x)} P](Q(x)),$$
$$[\Delta_{\Delta_g R_1(x)} \Delta_{R_2(x)} \cdots \Delta_{R_m(x)}](R_1(x) + T_g Q(x)),$$
$$[\Delta_{T_g R_1(x)} \Delta_{\Delta_g R_2(x)} \cdots \Delta_{R_m(x)} \Delta_g](R_2(x) + T_g Q(x))$$
$$\vdots$$
$$[\Delta_{T_g R_1(x)} \Delta_{T_g R_2(x)} \cdots \Delta_{\Delta_g R_m(x)} \Delta_g](R_m(x) + T_g Q(x)).$$

By the induction hypothesis, each of these terms is a polynomial of degree $\leq hk + r_1 + \ldots + r_m - 1$. The claim follows. $\square$

1.12.2. The classical case. Now we consider polynomials taking values in a finite field F.

Lemma 1.12.5 (Global description of classical one-dimensional polynomials). *Let F be a field of prime order p. For any $k \geq 0$, a function $P : F \to F$ is of degree $< k$ if and only if we can expand $P(x) = \sum_{0 \leq j < k} c_j x^j$ for some coefficients $c_j \in F$; this expansion is unique for $k \leq p$. Also, every function $P : F \to F$ is a polynomial of degree $< p$.*

Proof. The "if" portion of the lemma follows from Corollary 1.12.3 (since the identity function $x \mapsto x$ is clearly of degree ≤ 1). For the "only if" part, observe from the binomial identity

$$T^h = (1 + \Delta_1)^h = \sum_{j=0}^{h} \binom{h}{j} \Delta_1^j$$

that for any non-negative integer h,

$$f(h) = \sum_{j=0}^{h} \binom{h}{j} \Delta_1^j f(0).$$

Since $h \mapsto \binom{h}{j} = \frac{h(h-1)\dots(h-j+1)}{j!}$ can be meaningfully defined on F for $0 \leq j < p$, we conclude in particular that

$$f(h) = \sum_{j=0}^{p-1} \binom{h}{j} \Delta_1^j f(0).$$

Since $\binom{h}{j}$ can be expanded as a linear combination over F of $1, h, \dots, h^j$, we obtain the remaining claims in Lemma 1.12.5. (Note that as the space of functions from F to F is p-dimensional, and generated by $1, x, \dots, x^{p-1}$, these functions must be linearly independent.) □

Corollary 1.12.6 (Integration lemma). *Let $f : F \to F$ be a polynomial of degree $\leq k$ for some $0 \leq k \leq p - 2$, and let $h \in F \backslash 0$. Then there exists a polynomial $P : F \to F$ of degree $\leq k+1$ such that $f = \Delta_h P$. (In particular, this implies the mean zero condition $\sum_{x \in F} f(x) = 0$.) Conversely, any function $f : F \to F$ with $\sum_{x \in F} f(x) = 0$ is a polynomial of degree $\leq p - 2$.*

Proof. From Lemma 1.12.5, the space of polynomials of degree $\leq k$ and $\leq k + 1$ is a vector space over F of dimension $k + 1$ and $k + 2$ respectively. The derivative operator Δ_h is a linear transformation from the latter to the former with a one-dimensional kernel (the space of constants), and must therefore be surjective. The first claim follows. The second claim follows by a similar dimension counting argument. □

We can iterate Lemma 1.12.5 to describe polynomials in higher dimensions:

Lemma 1.12.7 (Global description of classical multidimensional polynomials). *Let F be a field of prime order p, and let $n \geq 1$. For any $k \geq 0$, a function $P : F^n \to F$ is of degree $< k$ if and only if we can expand $P(x_1, \dots, x_n) = \sum_{0 \leq j_1, \dots, j_n : j_1 + \dots + j_n < k} c_{j_1, \dots, j_n} x_1^{j_1} \dots x_n^{j_n}$ for some coefficients $c_{j_1, \dots, j_n} \in F$.*

Proof. As before, the "if" portion follows from Corollary 1.12.3, so it suffices to show the "only if" portion. But this follows by a multidimensional version of the analogous argument used to prove Lemma 1.12.5, starting with the identity

$$T^{(h_1, \dots, h_n)} = (1 + \Delta_{e_1})^{h_1} \dots (1 + \Delta_{e_n})^{h_n}$$

for non-negative integers $h_1, \dots, h_n$, where $e_1, \dots, e_n$ is the standard basis of F^n; we leave the details to the reader. □

Remark 1.12.8. The above discussion was for fields $F = F_p$ of prime order, but we can use these results to describe classical polynomials for fields $F = F_{p^m}$ of prime power order, by viewing any vector space over F_{p^m}

as a vector space over F_p. Of course, the resulting polynomials one obtains are merely polynomials over F_p, rather than over F_p^m.

1.12.3. The non-classical case. Now we consider polynomials from F or F^n into other additive groups, where $F = F_p$ is as before a field of prime order p. Thanks to Pontryagin duality, it suffices (in principle, at least) to consider polynomials taking values in the unit circle $\mathbf{R}/\mathbf{Z}$. The first basic lemma is the following:

Lemma 1.12.9 (Multiplication by p reduces degree). *Let $f : F^n \to \mathbf{R}/\mathbf{Z}$ be of degree $\leq k + p - 1$ for some $k \geq 0$. Then pf is of degree $\leq k$.*

Proof. Since $\Delta_h(pf) = p\Delta_h f$ for any h, we see by induction that it suffices to show this lemma when $k = 0$. Let $h \in F^n$. Raising (1.72) to the pth power we have

$$T^{ph} = 1 + p\Delta_h + \frac{p(p-1)}{2}\Delta_h^2 + \ldots + p\Delta_h^{p-1} + \Delta_h^p.$$

Of course, $T^{ph} = 1$. Applying this identity to f and noting that $\Delta_h^p f = 0$ by hypothesis, we conclude that

$$\left(1 + \frac{p-1}{2}\Delta_h + \ldots + \Delta_h^{p-2}\right)\Delta_h(pf) = 0.$$

Inverting $(1 + \frac{p-1}{2}\Delta_h + \ldots + \Delta_h^{p-2})$ by using Neumann series (and the finite degree of f) we conclude that $\Delta_h(pf) = 0$ for all h, thus pf has degree ≤ 0 as required.   □

Corollary 1.12.10 (Polynomials are discretely valued). *If $f : F^n \to \mathbf{R}/\mathbf{Z}$ is of degree $\leq k$, then after subtracting a constant from f, we see that f takes values in the $(p^{\lfloor (k-1)/(p-1) \rfloor + 1})^{\text{th}}$ roots of unity.*

In one dimension, there is a converse to Lemma 1.12.9:

Lemma 1.12.11. *Let $f : F^n \to \mathbf{R}/\mathbf{Z}$ be such that pf has degree $\leq k$. Then f has degree $\leq k + p - 1$.*

Proof. As in Lemma 1.12.9, it suffices to establish the case $k = 0$. But this then follows from the last part of Lemma 1.12.5.   □

As a corollary we can classify all non-classical polynomials:

Theorem 1.12.12 (Global description of non-classical multidimensional polynomials). *A function $f : F^n \to \mathbf{R}/\mathbf{Z}$ is a polynomial of degree $< k$ if and only if it has the form*

$$f(x) = c_0 + \sum_{\substack{0 \leq j_1,\ldots,j_n \leq p-1; \\ m \geq 1: j_1 + \ldots + j_n + (p-1)(m-1) < k}} c_{j_1,\ldots,j_m,m} |x_1|^{j_1} \ldots |x_n|^{j_n} / p^m$$

for some $c_0 \in \mathbf{R}/\mathbf{Z}$ and $c_{j_1,\ldots,j_m,m} \in \{0,\ldots,p-1\}$, where $x \mapsto |x|$ is the obvious map from F to $\{0,\ldots,p-1\}$.

Proof. The "if" part follows easily from Lemma 1.12.7 in the case $k \leq p$, and then from Lemma 1.12.11 and induction in the general case. The "only if" part follows from Corollary 1.12.10 and Lemma 1.12.7 in the case $k \leq p$. Now suppose inductively that $k > p$ and the claim has already been proven for smaller values of k. By Lemma 1.12.9 and the induction hypothesis, pf takes the form

$$pf(x) = c_0' + \sum_{\substack{0 \leq j_1,\ldots,j_n \leq p-1; \\ m \geq 1: j_1 + \ldots + j_n + (p-1)(m-1) < k-p+1}} c_{j_1,\ldots,j_n,m}' |x_1|^{j_1} \ldots |x_n|^{j_n}/p^m$$

and thus

$$f(x) = c_0 + \sum_{\substack{0 \leq j_1,\ldots,j_n \leq p-1; \\ m \geq 1: j_1 + \ldots + j_n + (p-1)(m-1) < k-p+1}} c_{j_1,\ldots,j_n,m}' |x_1|^{j_1} \ldots |x_n|^{j_n}/p^{m+1}$$
$$+ g(x),$$

where c_0 is a pth root of c_0', and g takes values in p^{th} roots of unity. Applying Lemma 1.12.7 to expand g in monomials, we obtain the claim. $\square$

As a corollary to this theorem we obtain a converse to Lemma 1.12.9:

Corollary 1.12.13 (p^{th} roots of minimal degree)**.** *Let $f : F^n \to \mathbf{R}/\mathbf{Z}$ be of degree $\leq k$ for some $k \geq 0$. Then there exists $g : F^n \to \mathbf{R}/\mathbf{Z}$ of degree $\leq k + p - 1$ such that $pg = f$.*

Interestingly, there does not seem to be a way to establish this theorem without going through a global classification theorem such as Theorem 1.12.12.

Another corollary to Theorem 1.12.12 is that any function from a finite dimensional vector space F^n to a p^m-torsion group for some m will be a polynomial of finite degree.

Notes. This article first appeared at

`terrytao.wordpress.com/2008/11/13`,

and is derived from [**BeTaZi2009**]. Thanks to James Cranch for corrections.

1.13. Cohomology for dynamical systems

Recall from Section 2.1 that a dynamical system is a space X, together with an action $(g,x) \mapsto gx$ of some group $G = (G,\cdot)$.[46] A useful notion in the

---

[46]In practice, one often places topological or measure-theoretic structure on X or G (see Section 2.2), but this will not be relevant for the current discussion. In most applications, G is an

subject is that of an (abelian) *cocycle*; this is a function $\rho : G \times X \to U$ taking values in an abelian group $U = (U, +)$ that obeys the *cocycle equation*

$$(1.73) \qquad\qquad \rho(gh, x) = \rho(h, x) + \rho(g, hx)$$

for all $g, h \in G$ and $x \in X$.[47] The significance of cocycles in the subject is that they allow one to construct (abelian) *extensions* or *skew products* $X \times_\rho U$ of the original dynamical system X, defined as the Cartesian product $\{(x, u) : x \in X, u \in U\}$ with the group action $g(x, u) := (gx, u + \rho(g, x))$. (The cocycle equation (1.73) is needed to ensure that one indeed has a group action, and in particular that $(gh)(x, u) = g(h(x, u))$.) This turns out to be a useful means to build complex dynamical systems out of simpler ones.[48]

A special type of cocycle is a *coboundary*; this is a cocycle $\rho : G \times X \to U$ that takes the form $\rho(g, x) := F(gx) - F(x)$ for some function $F : X \to U$. (Note that the cocycle equation (1.73) is automaticaly satisfied if ρ is of this form.) An extension $X \times_\rho U$ of a dynamical system by a coboundary $\rho(g, x) := F(gx) - F(x)$ can be conjugated to the trivial extension $X \times_0 U$ by the change of variables $(x, u) \mapsto (x, u - F(x))$.

While every coboundary is a cocycle, the converse is not always true.[49] One can measure the extent to which this converse fails by introducing the *first cohomology group* $H^1(G, X, U) := Z^1(G, X, U)/B^1(G, X, U)$, where $Z^1(G, X, U)$ is the space of cocycles $\rho : G \times X \to U$ and $B^1(G, X, U)$ is the space of coboundaries (note that both spaces are abelian groups). In [**BeTaZi2009**], we make substantial use of some basic facts about this cohomology group (in the category of measure-preserving systems), which were established in a [**HoKr2005**].

The terminology of cocycles, coboundaries, and cohomology groups of course comes from the theory of *cohomology* in algebraic topology. Comparing the formal definitions of cohomology groups in that theory with those given above, we certainly see quite a bit of similarity, but in the dynamical systems literature the precise connection does not seem to be heavily emphasised. The purpose of this article is to record the precise fashion in which dynamical systems cohomology is a special case of cochain complex cohomology from algebraic topology, and more specifically, is analogous to singular cohomology (and can also be viewed as the group cohomology of

---

abelian (additive) group such as the integers $\mathbf{Z}$ or the reals $\mathbf{R}$, but I prefer to use multiplicative notation here.

[47] Again, if one is placing topological or measure-theoretic structure on the system, one would want ρ to be continuous or measurable, but we will ignore these issues.

[48] For instance, one can build nilsystems by starting with a point and taking a finite number of abelian extensions of that point by a certain type of cocycle; see Section 2.16.

[49] For instance, if X is a point, the only coboundary is the zero function, whereas a cocycle is essentially the same thing as a homomorphism from G to U, so in many cases there will be more cocycles than coboundaries. For a contrasting example, if X and G are finite (for simplicity) and G acts *freely* on X, it is not difficult to see that every cocycle is a coboundary.

the space of scalar-valued functions on X, when viewed as a G-module); this is not particularly difficult, but I found it an instructive exercise (especially given that my algebraic topology is extremely rusty), though perhaps this article is more for my own benefit that for anyone else's.

1.13.1. Chains. Throughout this discussion, the dynamical system X, the group G, and the group U will be fixed.

For any $n \geq 0$, we define an n-*chain* to be a formal integer linear combination of $n+1$-tuples $(g_1, \ldots, g_n, x)$, where $x \in X$ and $g_1, \ldots, g_n \in G$. One may wish to think of every such tuple as an "oriented simplex" connecting the $n+1$ points $x, g_n x, g_{n-1} g_n x, \ldots, g_1 \ldots g_n x$. Thus, a 0-chain is a formal combination $\sum_{i=1}^m c_i x_i$ of points, a 1-chain is a formal combination $\sum_{i=1}^m c_i (g_i, x_i)$ of "line segments" from x_i to $g_i x_i$, and so forth. Let $C_n(G, X)$ be the space of n-chains; this is an abelian group. We also adopt the convention that $C_n(G, X)$ is trivial for $n < 0$.

For each $n > 0$, we define the *boundary map* $\partial : C_n(G, X) \to C_{n-1}(G, X)$ to be the unique homomorphism such that

$$\partial(g_1, \ldots, g_n, x) = (g_1, \ldots, g_{n-1}, g_n x)$$
$$+ \sum_{i=1}^{n-1} (-1)^{n-i} (g_1, \ldots, g_{i-1}, g_i g_{i+1}, g_{i+2}, \ldots, g_n, x)$$
$$+ (-1)^n (g_2, \ldots, g_n, x);$$

thus for instance

$$\partial(g, x) = gx - x,$$
$$\partial(g, h, x) = (g, hx) - (gh, x) + (h, x),$$
$$\partial(g, h, k, x) = (g, h, kx) - (g, hk, x) + (gh, k, x) - (h, k, x),$$

and so forth. Note that this is analogous to the boundary map in singular homology, if one views the $n+1$-tuple $(x, g_1, \ldots, g_n)$ as a simplex, as discussed earlier. We also define the boundary maps $\partial : C_n(G, X) \to C_{n-1}(G, X)$ for $n \leq 0$ to be the trivial map, thus for instance $\partial x = 0$. It is not hard to verify the fundamental relation

$$\partial^2 = 0$$

thus turning the sequence of groups $C_n(G, X)$ into a *chain complex*.

An n-chain with vanishing boundary is called an n-*cycle*, while an n-chain which is the boundary of an $(n-1)$-chain is called an n-*boundary*; the spaces of n-cycles and n-boundaries are denoted $Z_n(G, X)$ and $B_n(G, X)$, respectively. Thus for instance $(gh, x) - (h, x) - (g, hx)$ is both a 1-cycle and a 1-boundary. However, if g is a non-trivial group element that fixes x and G is abelian, one can show that (g, x) is a 1-cycle but not a 1-boundary.

We define the *homology groups* $H_n(G, X) := Z_n(G, X)/B_n(G, X)$ for all n. It is a nice exercise to compute these groups in some simple cases, e.g.

- If G acts transitively on X, then $H_0(G, X) \equiv \mathbf{Z}$.
- If G acts freely on X, then $H_n(G, X)$ is trivial for $n > 0$.
- If X is a point, then $H_1(G, X) \equiv G/[G, G]$ is the abelianisation of G. [Question: Is there a nice description of the higher homology groups $H_n(G, X)$, $n > 1$, in this case?]

However, I do not know of any application of these homology groups to the theory of dynamical systems.

1.13.2. Cochains. An *n-cochain* is a homomorphism from the space $C_n(G, X)$ of n-chains to U. Since $C_n(G, X)$ is a free abelian group generated by the simplices $(x, g_1, \ldots, g_n)$, we can view an n-cochain as a function $F : (x, g_1, \ldots, g_n) \to F(x, g_1, \ldots, g_n)$ from $G \times \ldots \times G \times X$ to U. (Again, we ignore all measure-theoretic or topological considerations here.) The space of all n-cochains is denoted $C^n(G, X, U) := \mathrm{Hom}(C_n(G, X), U)$; this is an abelian group.

The boundary map $\partial : C_n(G, X) \to C_{n-1}(G, X)$ defines by duality a coboundary map $\delta : C^{n-1}(G, X, U) \to C^n(G, X, U)$, defined by the formula

$$\delta F(c) := F(\partial c)$$

for all $F \in C^{n-1}(G, X, U)$ and $c \in C_n(G, X)$; viewing F as a function on simplices, we thus have

$$\delta F(g_1, \ldots, g_n, x) = F(g_1, \ldots, g_{n-1}, g_n x)$$
$$+ \sum_{i=1}^{n-1} (-1)^{n-i} F(g_1, \ldots, g_{i-1}, g_i g_{i+1}, \ldots, g_n, x)$$
$$+ (-1)^n F(g_2, \ldots, g_n, x).$$

Thus for instance

$$\delta F(g, x) = F(gx) - F(x)$$

for 0-cochains $F : X \to U$,

$$\delta \rho(g, h, x) = \rho(g, hx) - \rho(gh, x) + \rho(g, x)$$

for 1-cochains $\rho : G \times X \to U$, and so forth.

Because $\partial^2 = 0$, we have $\delta^2 = 0$, and so $C^n(G, X, U)$ becomes a *cochain complex*; n-cochains whose coboundary vanishes are known as *n-cocycles*, and n-cochains which are the coboundary of an $(n-1)$-cochain are known as *n-coboundaries*. The spaces of n-cocycles and n-cochains are denoted $Z^n(G, X, U)$ and $B^n(G, X, U)$, respectively, allowing us to define the nth *cohomology group* $H^n(G, X, U) := Z^n(G, X, U)/B^n(G, X, U)$.

When $n = 0$, and if the action of G is *transitive* (in the discrete category), *minimal* (in the topological category; see Section 2.2), or *ergodic* (in the measure-theoretic category; see Section 2.9), the only 0-cocycles are the constants, and the only 0-coboundary is the zero function, so $H^0(G, X, U) \equiv U$. When $n = 1$, it is not hard to see that the notion of 1-cocycle and 1-coboundary correspond to the notion of cocycle and coboundary discussed at the beginning of this section.

This whole theory raises the obvious question as to whether the higher cocycles, coboundaries, and cohomology groups have any relevance in dynamical systems. For instance, a 2-cocycle becomes (after minor notational changes) a function $\psi : G \times G \times X \to U$ that obeys the 2-cocycle equation

$$\psi(g, h, kx) - \psi(g, hk, x) + \psi(gh, k, x) - \psi(h, k, x) = 0,$$

while a 2-coboundary is a function of the form

$$\psi(g, h, x) := \rho(gh, x) - \rho(h, x) - \rho(g, hx)$$

for some $\rho : G \times X \to U$. Is there a dynamical system interpretation of these objects, much as 1-cocycles and 1-coboundaries can be interpreted as describing abelian extensions and essentially trivial abelian extensions, respectively? (See Subsection 1.13.3 for a partial answer.) In [**BeTaZi2009**], we do briefly encounter 2-coboundaries (we have to deal with various "quasi-cocycles": 1-chains ρ whose 2-coboundary $\delta\rho$ does not vanish completely, as with 1-cocycles, but is still of a relatively simple form, such as a constant or a polynomial) but we do not make systematic use of this concept. (We also rely heavily in our paper on the cubic complexes $X^{[k]}$ of Host and Kra, which have some superficial resemblance to the simplex structures appearing here, but I do not know if there is a substantive connection in this regard.)

Another oddity is that homology and cohomology, as it is classically defined, requires the spaces of chains, cochains, etc. to all be abelian groups; but for dynamical systems one can certainly talk about cocycles and coboundaries taking values in a non-abelian group U by modifying the definitions slightly, leading to the concept of a *group extension* of a dynamical system. (In this context, the first cohomology $H^1(G, X, U)$ becomes a quotient space rather than a group; see also Section 1.4 of Part II.) It seems to me that in this case, the dynamical system concept of a cocycle or coboundary cannot be interpreted in terms of classical cohomology theory (but presumably can be handled by *non-abelian group cohomology*).

1.13.3. Epilogue: An interpretation of the second cohomology group. Minhyong Kim has provided a nice answer to my question about the relevance of higher order cohomology, such as $H^2(G, X, U)$, to the problem of extending dynamical systems. Suppose one has a short exact sequence

$$0 \to V \to \tilde{U} \to U \to 0$$

of abelian groups, thus one can view $\tilde{U}$ as the space of pairs (u, v) with $u \in U$, $v \in V$, with a group addition law

(1.74) $(u, v) + (u', v') := (u + u', v + v' + B(u, u'))$

for some function $B : U \times U \to V$, that needs to obey a certain set of axioms to make $\tilde{U}$ an abelian group, which we will not write down here. We then claim that we have a long exact sequence

(1.75) $\to H^1(G, X, \tilde{U}) \to H^1(G, X, U) \to H^2(G, X, V) \to,$

thus $H^2(G, X, V)$ is capable of detecting whether a U-extension of a G-system X can be lifted to a $\tilde{U}$-extension.

The first map in (1.75) is obvious: the projection from $\tilde{U}$ to U induces a projection from 1-cocycles $\tilde{\rho} : G \times X \to \tilde{U}$ to 1-cocycles $\rho : G \times X \to U$ which maps 1-coboundaries to 1-coboundaries, and thus maps $H^1(G, X, \tilde{U})$ to $H^1(G, X, U)$. The second map requires a bit more thought. Suppose one is given a 1-cocycle $\rho : G \times X \to U$ and asks whether it can be lifted to a 1-cocycle $\tilde{\rho} : G \times X \to \tilde{U}$ by the above projection. Writing $\tilde{\rho} = (\rho, \sigma)$ for some $\sigma : G \times X \to V$ and using (1.73), (1.74), we see that the question is equivalent to finding a σ that obeys the equation

$$\sigma(gh, x) = \sigma(h, x) + \sigma(g, hx) + B(\rho(h, x), \rho(g, hx)),$$

or, in other words, to showing that the map

$$\Phi(\rho) : (g, h, x) \mapsto B(\rho(h, x), \rho(g, hx))$$

is a V-valued 2-coboundary. The same observation (now setting $\sigma = 0$) shows that the map $(g, h, x) \mapsto (0, \Phi(\rho))$ is a $\tilde{U}$-valued 2-coboundary (indeed, it is the coboundary of $(\rho, 0)$), hence a $\tilde{U}$-valued 2-cocycle, and thus $\Phi(\rho)$ is a V-valued 2-cocycle, and so the map $\rho \mapsto \Phi(\rho)$ is a map from 1-cocycles $\rho : G \times X \to U$ to 2-cocycles $\Phi(\rho) : G \times G \times X \to V$. Similarly, given two 1-cocycles $\rho, \rho' : G \times X \to U$, we see that $(\rho + \rho', 0)$ differs from $(\rho, 0) + (\rho', 0)$ by some V-valued 1-cochain, so on taking derivatives we see that $\Phi(\rho + \rho')$ differs from $\Phi(\rho) + \Phi(\rho')$ by some 2-coboundary, thus Φ is linear modulo 2-coboundaries. Finally, if ρ is a U-valued 1-coboundary, then $(\rho, 0)$ is the sum of a $\tilde{U}$-valued 1-coboundary and a V-valued 1-cochain, and so on taking derivatives we see that Φ maps 1-coboundaries to 2-coboundaries.[50] Hence it induces a map from $H^1(G, X, U)$ to $H^2(G, X, V)$, and then (1.75) is exact by the preceding discussion.

---

[50]Presumably the above arguments are a special case of one of the standard diagram chasing lemmas in homological algebra, but I do not know which one it is. One could also verify these facts from the axioms of B induced from (1.74) and the abelian group structure on $\tilde{U}$, but this turns out to be remarkably tedious.

Notes. This article first appeared at

`terrytao.wordpress.com/2008/12/21`.

Thanks to AA for corrections.

Mikael Vejdemo Johansson pointed out that the group cohomology formalism developed above also extends to bimodules over G, though it is not clear what the dynamical interpretation of such bimodules would be.

Marlowe noted more generally that, as a general rule of thumb, if a certain cohomology group helps to classify extensions up to conjugation, the next cohomology group helps you find out if a certain candidate for an extension can be extended to a full extension; the discussion in Section 1.13.3 of course supports this rule.

Peter Samuelson also pointed out that this homology is a special case of Hochschild homology.

Further discussion on this topic can also be found at

`http` : `//golem.ph.utexas.edu/category/2008/12/bridge_building.html`.

Chapter 2

Ergodic Theory

2.1. Overview

In this lecture, I define the basic notion of a *dynamical system* (as well as the more structured notions of a *topological dynamical system* and a *measure-preserving system*) and describe the main topics we will cover in this course.

We'll begin abstractly. Suppose that X is a non-empty set (whose elements will be referred to as *points*), and $T : X \to X$ is a transformation. Later on, we shall put some structures on X (such as a topology, a σ-algebra, or a probability measure) and some assumptions on T, but let us work in total generality for now.[1]

One can think of X as a state space for some system, and T as the evolution of some discrete deterministic (autonomous) dynamics on X: if x is a point in X, denoting the current state of a system, then Tx can be interpreted as the state of the same system after one unit of time has elapsed.[2] More geometrically, one can think of T as some sort of shift operation (e.g. a rotation) on the space X.

Given X and T, we can define the iterates $T^n : X \to X$ for every non-negative integer n; if T is also invertible, then we can define T^n for a negative integer n as well. In the language of representation theory, T induces a representation[3] of either the additive semigroup $\mathbf{Z}^+$ or the additive group $\mathbf{Z}$.

---

[1] Indeed, a guiding philosophy in the first half of the course will be to try studying dynamical systems in as much generality as possible; later on, though, when we turn to more algebraic dynamical systems such as nilsystems, we shall exploit the specific structure of such systems more thoroughly.

[2] In particular, evolution equations that are well posed can be viewed as a continuous dynamical system.

[3] From the dynamical perspective, this representation is the mathematical manifestation of *time*.

More generally, one can consider representations of other groups, such as the real line $\mathbf{R}$ (corresponding to the dynamics $t \mapsto T^t$ of a continuous time evolution) or a lattice $\mathbf{Z}^d$ (which corresponds to the dynamics of d commuting shift operators $T_1, \ldots, T_d : X \to X$), or of many other semigroups or groups (not necessarily commutative). However, for simplicity we shall mostly restrict our attention to $\mathbf{Z}$-actions in this course, though many of the results here can be generalised to other actions (under suitable hypotheses on the underlying semigroup or group, of course).

Henceforth we assume T to be invertible, in which case we refer to the pair (X, T) as a *cyclic dynamical system*, or *dynamical system* for short. Here are some simple examples of such systems:

Example 2.1.1 (Finite systems). X is a finite set, and $T : X \to X$ is a permutation on X.

Example 2.1.2 (Group actions). Let G be a group, and let X be a homogeneous space for G, i.e., a non-empty space with a transitive G-action; thus X is isomorphic to G/Γ, where $\Gamma := \mathrm{Stab}(x)$ is the stabiliser of one of the points x in X. Then every group element $g \in G$ defines a dynamical system (X, T_g) given by $T_g x := gx$.

Example 2.1.3 (Circle rotations). As a special case of Example 2.1.2 (or Example 2.1.1), every real number $\alpha \in \mathbf{R}$ induces a dynamical system $(\mathbf{R}/\mathbf{Z}, T_\alpha)$ given by the rotation $T_\alpha x := x + \alpha$. This is the prototypical example of a very *structured* system, with plenty of algebraic structure (e.g. the shift map T_α is an isometry on the circle, thus two points always stay the same distance apart under shifts).

Example 2.1.4 (Cyclic groups). Another special case of Example 2.1.2 is the cyclic group $\mathbf{Z}/N\mathbf{Z}$ with shift $x \mapsto x+1$; this is the prototypical example of a finite dynamical system.

Example 2.1.5 (Bernoulli systems). Every non-empty set Ω induces a dynamical system $(\Omega^{\mathbf{Z}}, T)$, where T is the left shift $T(x_n)_{n \in \mathbf{Z}} := (x_{n+1})_{n \in \mathbf{Z}}$. This is the prototypical example of a very *pseudorandom* system, with plenty of mixing (e.g. the shift map tends to move a pair of two points randomly around the space).

Example 2.1.6 (Boolean Bernoulli system). This is isomorphic to a special case of Example 2.1.5, in which $X = 2^{\mathbf{Z}} := \{A : A \subset \mathbf{Z}\} \equiv \{0,1\}^{\mathbf{Z}}$ is the power set of the integers, and $TA := A - 1 := \{a - 1 : a \in A\}$ is the left shift. (Here we endow $\{0, 1\}$ with the discrete topology.)

Example 2.1.7 (Baker's map). Here, $X := [0, 1)^2$ and

$$T(x, y) := \left(\{2x\}, \frac{y + \lfloor 2x \rfloor}{2} \right),$$

where $\lfloor x \rfloor$ is the greatest integer function, and $\{x\} := x - \lfloor x \rfloor$ is the fractional part. This is isomorphic to Example 2.1.6, as can be seen by inspecting the effect of T on the binary expansions of x and y.

The map T^n can be interpreted as an isomorphism in several different categories:

  (1) as a set isomorphism (i.e., a bijection) $T^n : X \to X$ from points $x \in X$ to points $T^n x \in X$;

  (2) as a Boolean algebra isomorphism $T^n : 2^X \to 2^X$ from sets $E \subset X$ to sets $T^n E := \{T^n x : x \in E\}$; or

  (3) as an algebra isomorphism $T^n : \mathbf{R}^X \to \mathbf{R}^X$ from real-valued functions $f : X \to \mathbf{R}$ to real-valued functions $T^n f : X \to \mathbf{R}$, defined by

(2.1)
$$T^n f(x) := f(T^{-n} x);$$

  (4) as an algebra isomorphism $T^n : \mathbf{C}^X \to \mathbf{C}^X$ of complex-valued functions, defined again by (2.1).

We will abuse notation and use the same symbol T^n to refer to all of the above isomorphisms; the specific meaning of T^n should be clear from the context in all cases. Our sign conventions here are chosen so that we have the pleasant identities

(2.2)
$$T^n \{x\} = \{T^n x\}, \qquad T^n 1_E = 1_{T^n E}$$

for all points x and sets E, where of course 1_E is the *indicator function* of E.

One of the main topics of study in dynamical systems is the asymptotic behaviour of T^n as $n \to \infty$. We can pose this question in any of the above categories, thus

  (1) For a given point $x \in X$, what is the behaviour of $T^n x$ as $n \to \infty$?

  (2) For a given set $E \subset X$, what is the behaviour of $T^n E$ as $n \to \infty$?

  (3) For a given real- or complex-valued function $f : X \to \mathbf{R}$ or $f : X \to \mathbf{C}$, what is the behaviour of $T^n f$ as $n \to \infty$?

These are of course very general and vague questions, but we will formalise them in many different ways later in the course.[4] The answer to these questions also depends very much on the dynamical system; thus a major focus of study in this subject is to seek classifications of dynamical systems which allow one to answer the above questions satisfactorily.[5]

---

[4] For instance, one can distinguish between worst-case, average-case, and best-case behaviour in x, E, f, or n.

[5] In particular, ergodic theory is a framework in which our understanding of the dichotomy between structure and randomness is most developed; see Section 2.1.2 of *Structure and Randomness*.

One can also ask for more *quantitative* versions of the above asymptotic questions, in which n ranges in a finite interval (e.g. $[N] := \{1, \ldots, N\}$ for some large integer N), as opposed to going off to infinity, and one wishes to estimate various numerical measurements of $T^n x$, $T^n E$, or $T^n f$ in this range.

In this very general setting, in which X is an unstructured set, and T is an arbitrary bijection, there is not much of interest one can say with regards to these questions. However, one obtains a surprisingly rich and powerful theory when one adds a little more structure to X and T (thus changing categories once more). In particular, we will study the following two structured versions of a dynamical system:

(I) *Topological dynamical systems* $(X, T) = (X, \mathcal{F}, T)$, in which $X = (X, \mathcal{F})$ is a compact metrisable (and thus *Hausdorff*) topological space, and T is a topological isomorphism (i.e., a homeomorphism); and

(II) *Measure-preserving systems* $(X, T) = (X, \mathcal{X}, \mu, T)$, in which $X = (X, \mathcal{X}, \mu)$ is a probability space,[6] and T is a probability space isomorphism, i.e., T and T^{-1} are both measurable, and $\mu(TE) = \mu(E)$ for all measurable $E \in \mathcal{X}$. For technical reasons we also require the measurable space $(X, \mathcal{X})$ to be *separable* (i.e., $\mathcal{X}$ is countably generated).

Remark 2.1.8. By *Urysohn's metrisation theorem*, a compact space is metrisable if and only if it is Hausdorff and *second countable*, thus providing a purely topological characterisation of a topological dynamical system.

Remark 2.1.9. It is common to add a bit more structure to each of these systems, for instance endowing a topological dynamical system with a metric, or endowing a measure-preserving system with the structure of a standard Borel space; we will see examples of this in later lectures.

The study of topological dynamical systems and measure-preserving systems is known as *topological dynamics* and *ergodic theory*, respectively. The two subjects are closely analogous at a heuristic level, and also have some more rigorous connections between them, so we shall pursue them in a somewhat parallel fashion in this course.

Remark 2.1.10. Observe that we assume compactness in (I) and finite measure in (II); these "boundedness" assumptions ensure that the dynamics somewhat resembles the (overly simple) case of a finite dynamical system.

---

[6]In this course we shall tilt towards a measure-theoretic perspective rather than a probabilistic one, thus it might be better to think of μ as a normalised finite measure rather than as a probability measure. On the other hand, we will rely crucially on the probabilistic notions of *conditional expectation* and *conditional independence* later in this course.

Dynamics on non-compact topological spaces or infinite measure spaces is a more complicated topic; see for instance [**Aa1997**]. (Thanks to Tamar Ziegler for this reference.)

Note that the action of the isomorphism T^n on sets E and functions f will be compatible with the topological or measure-theoretic structure:

(1) If $(X,T) = (X,\mathcal{F},T)$ is a topological dynamical system, then $T^n : \mathcal{F} \to \mathcal{F}$ is a topological isomorphism on open sets, and $T^n : C(X) \to C(X)$ is also a C^*-algebra isomorphism on the space $C(X)$ of real-valued (or complex-valued) continuous functions on X.

(2) If $(X,T) = (X,\mathcal{X},\mu,T)$ is a measure-preserving system, then $T^n : \mathcal{X} \to \mathcal{X}$ is a σ-algebra isomorphism on measurable sets, and $T^n : L^p(\mathcal{X},\mu) \to L^p(\mathcal{X},\mu)$ is a Banach space isomorphism on p^{th}-power integrable functions for $1 \le p \le \infty$. (For $p = \infty$, T^n is a von Neumann algebra isomorphism, whilst for $p = 2$, T^n is a Hilbert space isomorphism (i.e., a unitary transformation).)

We can thus see that tools from the analysis of Banach spaces, von Neumann algebras, and Hilbert spaces may have some relevance to ergodic theory; for instance, the spectral theorem for unitary operators is quite useful.

In the first half of this course, we will study topological dynamical systems and measure-preserving systems in great generality (with few assumptions on the structure of such systems), and then specialise to specific systems as appropriate. This somewhat abstract approach is broadly analogous to the combinatorial (as opposed to algebraic or arithmetic) approach to additive number theory. For instance, we will shortly be able to establish the following general result in topological dynamics (see Theorem 2.3.4):

Theorem 2.1.11 (Birkhoff recurrence theorem). *Let (X,T) be a topological dynamical system. Then there exists a point $x \in X$ which is* recurrent *in the sense that there exists a sequence $n_j \to \infty$ such that $T^{n_j}x \to x$ as $j \to \infty$.*

As a corollary, we will be able to obtain the more concrete result (see Section 2.4):

Theorem 2.1.12 (Weyl recurrence theorem). *Let $P : \mathbf{Z} \to \mathbf{R}/\mathbf{Z}$ be a polynomial (modulo 1). Then there exists a sequence $n_j \to \infty$ such that $P(n_j) \to P(0)$.*

This is already a somewhat non-trivial theorem; consider for instance the case $P(n) := \sqrt{2}n^2 \bmod 1$.

In a similar spirit, in Section 2.4 we will be able to prove the general topological dynamical result (see Theorem 2.4.1).

Theorem 2.1.13 (Topological van der Waerden theorem). *Let $(U_\alpha)_{\alpha \in A}$ be an open cover of a topological dynamical system (X, T), and let $k \geq 1$ be an integer. Then there exists an open set U in this cover and a shift $n \geq 1$ such that $U \cap T^n U \cap \ldots \cap T^{(k-1)n} U \neq \emptyset$. (Equivalently, there exist U, n, and a point x such that $x, T^n x, \ldots, T^{(k-1)n} x \in U$.)*

From this we will derive an (equivalent) combinatorial result:

Theorem 2.1.14 (Van der Waerden theorem). *Let $\mathbf{N} = U_1 \cup \ldots \cup U_m$ be a finite colouring of the natural numbers. Then one of the colour classes U_j contains arbitrarily long arithmetic progressions.*

More generally, topological dynamics is an excellent tool for establishing colouring theorems of Ramsey type.

Analogously, in Sections 2.10–2.15 we will be able to prove the following general ergodic theory result (see Theorem 2.10.3)):

Theorem 2.1.15 (Furstenberg multiple recurrence theorem). *Let (X, T) be a measure-preserving system, let $E \in \mathcal{X}$ be a set of positive measure, and let $k \geq 1$. Then there exists $n \geq 1$ such that $E \cap T^n E \cap \ldots \cap T^{(k-1)n} E \neq \emptyset$ (or equivalently, there exist $x \in X$ and $n \geq 1$ such that $x, T^n x, \ldots, T^{(k-1)n} \in E$).*

Similarly, if $f : X \to \mathbf{R}^+$ is a bounded measurable non-negative function which is not almost everywhere zero, and $k \geq 1$, then

$$(2.3) \qquad \liminf_{N \to \infty} \frac{1}{N} \sum_{n=1}^{N} \int_X f T^n f \ldots T^{(k-1)n} f > 0.$$

We will also deduce an equivalent (and highly non-trivial) combinatorial analogue (see Theorem 2.10.1):

Theorem 2.1.16 (Szemerédi's theorem). *Let $E \subset \mathbf{Z}$ be a set of positive upper density, thus $\limsup_{N \to \infty} \frac{|E \cap [-N,N]|}{2N+1} > 0$. Then E contains arbitrarily long arithmetic progressions.*

More generally, ergodic theory methods are extremely powerful in deriving *density Ramsey theorems*. Indeed, there are several theorems of this type which currently have no known non-ergodic theory proof.[7]

The first half of this course will be devoted to results of the above type, which apply to general topological dynamical systems or general measure-preserving systems. One important insight that will emerge from analysis of the latter is that in many cases, a large portion of the measure-preserving

---

[7]From general techniques in proof theory, one could, in principle, take an ergodic theory proof and mechanically convert it into what would technically be a non-ergodic proof, for instance avoiding the use of infinitary objects, but this is not really in the spirit of what most mathematicians would call a genuinely new proof.

system is irrelevant for the purposes of understanding long-time average behaviour; instead, there will be a smaller system, known as a *characteristic factor* for the system, which completely controls these asymptotic averages. A deep and powerful fact is that in many situations, this characteristic factor is extremely structured algebraically, even if the original system has no obvious algebraic structure whatsoever. Because of this, it becomes important to study algebraic dynamical systems, such as the group actions on homogeneous spaces described earlier, as it allows one to obtain more precise results.[8] This study will be the focus of the second half of the course, particularly in the important case of *nilsystems*—group actions arising from a nilpotent Lie group with discrete stabiliser. One of the key results here is *Ratner's theorem*, which describes the distribution of orbits $\{T^n x : n \in \mathbf{Z}\}$ in nilsystems, and also in a more general class of group actions on homogeneous spaces. While we will not prove Ratner's theorem in full generality, we will cover a few special cases of this theorem in Sections 2.16, 2.17.

In closing, I should mention that the topics I intend to cover in this course are only a small fraction of the vast area of ergodic theory and dynamical systems; for instance, there are parts of this field connected with complex analysis and fractals, ODE, probability and information theory, harmonic analysis, group theory, operator algebras, or mathematical physics about which I will say absolutely nothing here.

Notes. This lecture first appeared at

terrytao.wordpress.com/2008/01/08.

2.2. Three categories of dynamical systems

Before we begin our study of dynamical systems, topological dynamical systems, and measure-preserving systems (as defined in Section 2.1), it is convenient to give these three classes the structure of a *category*. One of the basic insights of category theory is that mathematical objects in a given class (such as dynamical systems) are best studied not in isolation, but in relation to each other, via *morphisms*. Furthermore, many other basic concepts pertaining to these objects (e.g. subobjects, factors, direct sums, irreducibility, etc.) can be defined in terms of these morphisms. One advantage of taking this perspective here is that it provides a unified way of defining these concepts for the three different categories of dynamical systems, topological dynamical systems, and measure-preserving systems that we will study in

---

[8]For instance, this algebraic structure was used to show that the limit in (2.3) actually converges, a result which does not seem accessible purely through the techniques used to prove the Furstenberg recurrence theorem.

this course, thus sparing us the need to give any of our definitions (except for our first one below) in triplicate.

Informally, a *morphism* between two objects in a class is any map which respects all the structures of that class. For the three categories we are interested in, the formal definition is as follows.

Definition 2.2.1 (Morphisms).

(1) A *morphism* $\phi : (X, T) \to (Y, S)$ between two dynamical systems is a map $\phi : X \to Y$ which intertwines T and S in the sense that $S \circ \phi = \phi \circ T$.

(2) A *morphism* $\phi : (X, \mathcal{F}, T) \to (Y, \mathcal{G}, S)$ between two topological dynamical systems is a morphism $\phi : (X, T) \to (Y, S)$ of dynamical systems which is also continuous, thus $\phi^{-1}(U) \in \mathcal{F}$ for all $U \in \mathcal{G}$.

(3) A *morphism* $\phi : (X, \mathcal{X}, \mu, T) \to (Y, \mathcal{Y}, \nu, S)$ between two measure-preserving systems is a morphism $\phi : (X, T) \to (Y, S)$ of dynamical systems which is also measurable (thus $\phi^{-1}(E) \in \mathcal{X}$ for all $E \in \mathcal{Y}$) and measure-preserving (thus $\mu(\phi^{-1}(E)) = \nu(E)$ for all $E \in \mathcal{Y}$). Equivalently, $\nu = \phi_*(\mu)$ is the *push-forward* of μ by ϕ.

When it is clear what category we are working in, and what the shifts are, we shall often refer to a system by its underlying space; thus for instance a morphism $\phi : (X, \mathcal{X}, \mu, T) \to (Y, \mathcal{Y}, \nu, S)$ might be abbreviated as $\phi : X \to Y$.

If a morphism $\phi : X \to Y$ has an inverse $\phi^{-1} : Y \to X$ which is also a morphism, we say that ϕ is an *isomorphism*, and that X and Y are *isomorphic* or *conjugate*.

It is easy to see that morphisms obey the axioms of a (concrete) category, or in other words that the identity map $\mathrm{id}_X : X \to X$ on a system is always a morphism, and the composition $\psi \circ \phi : X \to Z$ of two morphisms $\phi : X \to Y$ and $\psi : Y \to Z$ is again a morphism.

Let us give some simple examples of morphisms.

Example 2.2.2 (Shift). If (X, T) is a dynamical system, a topological dynamical system, or a measure-preserving dynamical system, then $T^n : X \to X$ is an isomorphism for any integer n. (Indeed, one can view the map $X \mapsto T^n$ as a *natural transformation* from the identity functor on the category of dynamical systems (or topological dynamical systems, etc.) to itself, although we will not take this perspective here.)

Example 2.2.3 (Subsystems). Let (X, T) be a dynamical system, and let E be a subset of X which is T-invariant in the sense that $T^n E = E$ for all n. Then the restriction $(E, T \restriction_E)$ of (X, T) to E is itself a dynamical system, and the inclusion map $\iota : E \to X$ is a morphism. In the category of topological dynamical systems $(X, \mathcal{F}, T)$, we have the same assertion so long as E is *closed* (hence compact, since X is compact). In the category of measure-preserving systems $(X, \mathcal{X}, \mu, T)$, we have the same assertion so long as E has full measure (thus $E \in \mathcal{X}$ and $\mu(E) = 1$). We thus see that subsystems are not very common in measure-preserving systems and will in fact play very little role there; however, subsystems (and specifically, *minimal* subsystems) will play a fundamental role in topological dynamics.

Example 2.2.4 (Skew shift). Let $\alpha \in \mathbf{R}$ be a fixed real number. Let (X, T) be the dynamical system $X := (\mathbf{R}/\mathbf{Z})^2$, $T : (x_1, x_2) \mapsto (x_1 + \alpha, x_2 + x_1)$, let (Y, S) be the dynamical system $Y := \mathbf{R}/\mathbf{Z}$, $S : y \mapsto y + \alpha$, and let $\pi : X \to Y$ be the projection map $\pi : (x_1, x_2) \to x_1$. Then π is a morphism. If one converts X and Y into either a topological dynamical system or a measure-preserving system in the obvious manner, then π remains a morphism. Observe that π foliates the big space X "upstairs" into "vertical" fibres $\pi^{-1}(\{y\})$, $y \in Y$, indexed by the small "horizontal" space "downstairs"; the shift S on the factor space Y downstairs determines how the fibres move (the shift T upstairs sends each vertical fibre $\pi^{-1}(\{y\})$ to another vertical fibre $\pi^{-1}(\{Sy\})$, but does not govern the dynamics *within* each fibre. More generally, any *factor map* (i.e., a surjective morphism) exhibits this type of behaviour.[9]

Example 2.2.5 (Universal pointed dynamical system). Let $\mathbf{Z} = (\mathbf{Z}, +1)$ be the dynamical system given by the integers with the standard shift $n \mapsto n+1$. Then given any other dynamical system (X, T) with a distinguished point $x \in X$, the orbit map $\phi : n \mapsto T^n x$ is a morphism from $\mathbf{Z}$ to X. This allows us to lift most questions about dynamical systems (with a distinguished point x) to those for a single "universal" dynamical system, namely the integers (with distinguished point 0). One cannot pull off the same trick directly with topological dynamical systems or measure-preserving systems, because $\mathbf{Z}$ is non-compact and does not admit a shift-invariant probability measure. As we shall see later, the former difficulty can be resolved by passing to a universal compactification of the integers, namely the *Stone-Čech compactification* $\beta\mathbf{Z}$ (or equivalently, the space of *ultrafilters* on the integers), though with the important caveat that this compactification is not metrisable. To resolve the second difficulty (with the assistance of a distinguished set rather than a distinguished point), see the next example.

---

[9]Another example of a factor map is the map $\pi : \mathbf{Z}/N\mathbf{Z} \to \mathbf{Z}/M\mathbf{Z}$ between two cyclic groups (with the standard shift $x \mapsto x + 1$) given by $\pi : x \mapsto x \bmod M$. This is a well-defined factor map when M is a factor of N, which may help explain the terminology. If we wanted to adhere strictly to the category-theoretic philosophy, we should use *epimorphisms* rather than surjections, but we will not require this subtle distinction here.

Example 2.2.6 (Universal dynamical system with distinguished set). Recall the boolean Bernoulli system $(2^{\mathbf{Z}}, U)$ (Example 2.1.6). Given any other dynamical system (X, T) with a distinguished set $A \subset X$, the *recurrence map* $\phi : X \to 2^{\mathbf{Z}}$ defined by $\phi(x) := \{n \in \mathbf{Z} : T^n x \in A\}$ is a morphism. Observe that $A = \phi^{-1}(B)$, where B is the *cylinder set* $B := \{E \in 2^{\mathbf{Z}} : 0 \in E\}$. Thus we can push forward an arbitrary dynamical system (X, T, A) with distinguished set to a universal dynamical system $(2^{\mathbf{Z}}, U, B)$. Actually one can restrict $(2^{\mathbf{Z}}, U, B)$ to the subsystem $(\phi(X), U \!\restriction_{\phi(X)}, B \cap \phi(X))$, which is easily seen to be shift-invariant. In the category of topological dynamical systems, the above assertions still hold (giving $2^{\mathbf{Z}}$ the *product topology*), so long as A is clopen. In the category of measure-preserving systems $(X, \mathcal{X}, \mu, T)$, the above assertions hold as long as A is measurable, $2^{\mathbf{Z}}$ is given the product σ-algebra, and the *push-forward measure* $\phi_*(\mu)$.

Now we begin our analysis of dynamical systems. When studying other mathematical objects (e.g. groups or representations), often one of the first steps in the theory is to decompose general objects into "irreducible" ones, and then hope to classify the latter. Let us see how this works for dynamical systems (X, T) and topological dynamical systems $(X, \mathcal{F}, T)$. (For measure-preserving systems, the analogous decomposition will be the *ergodic decomposition*, which we will discuss in Section 2.9.5.)

Definition 2.2.7. A *minimal dynamical system* is a system (X, T) which has no proper subsystems (Y, S). A *minimal topological dynamical system*[10] is a system $(X, \mathcal{F}, T)$ with no proper subsystems $(Y, \mathcal{G}, S)$.

For a dynamical system, it is not hard to see that for any $x \in X$, the orbit $Y = T^{\mathbf{Z}} x = \{T^n x : n \in \mathbf{Z}\}$ is a minimal system, and conversely that all minimal systems arise in this manner; in particular, every point is contained in a minimal orbit. It is also easy to see that any two minimal systems (i.e., orbits) are either disjoint or coincident. Thus every dynamical system can be uniquely decomposed into the disjoint union of minimal systems. Also, every orbit $T^{\mathbf{Z}} x$ is isomorphic to $\mathbf{Z}/\text{Stab}(x)$, where $\text{Stab}(x) := \{n \in \mathbf{Z} : T^n x = x\}$ is the *stabiliser group* of x. Since we know all the subgroups of $\mathbf{Z}$, we conclude that every minimal system is either equivalent to a cyclic group shift $(\mathbf{Z}/N\mathbf{Z}, x \mapsto x+1)$ for some $N \geq 1$, or to the integer shift $(\mathbf{Z}, x \mapsto x+1)$.

---

[10]One could make the same definition for measure-preserving systems, but it tends to be a bit vacuous—given any measure-preserving system that contains points of measure zero, one can make it trivially smaller by removing the orbit $T^{\mathbf{Z}} x := \{T^n x : n \in \mathbf{Z}\}$ of any point x of measure zero. One could place a topology on the space X and demand that it be compact, in which case minimality just means that the probability measure μ has full support.

Thus we have completely classified all dynamical systems up to isomorphism as the arbitrary union of these minimal examples.[11]

For topological dynamical systems, it is still true that any two minimal systems are either disjoint or coincident (why?), but the situation nevertheless is more complicated. First of all, orbits need not be closed (take for instance the circle shift $(\mathbf{R}/\mathbf{Z}, x \mapsto x + \alpha)$ with α irrational). If one considers the *orbit closure* $\overline{T^{\mathbf{Z}}x}$ of a point x, then this is now a subsystem (why?), and every minimal system is the orbit closure of any of its elements (why?), but in the converse direction, not all orbit closures are minimal. Consider for instance the boolean Bernoulli system $(2^{\mathbf{Z}}, A \mapsto A - 1)$ with $x = \mathbf{N} := \{0, 1, 2, \ldots\} \in 2^{\mathbf{Z}}$ being the natural numbers. Then the orbit $T^{\mathbf{Z}}x$ of x consists of all the half-lines $\{a, a + 1, \ldots, \} \in 2^{\mathbf{Z}}$ for $a \in \mathbf{Z}$, but it is not closed; it has the point $\mathbf{Z} \in 2^{\mathbf{Z}}$ and the point $\emptyset \in 2^{\mathbf{Z}}$ as limit points (recall that $2^{\mathbf{Z}}$ is given the product (i.e., pointwise) topology). Each of these points is an invariant point of T and thus forms its own orbit closure, which is obviously minimal.[12]

Thus we see that finite dynamical systems do not quite form a perfect model for topological dynamical systems. A slightly better (but still imperfect) model would be that of *non-invertible* finite dynamical systems (X, T), in which $T : X \to X$ is now just a function rather than a permutation. Then we can still verify that all minimal orbits are given by disjoint cycles, but they no longer necessarily occupy all of X; it is quite possible for the orbit $T^{\mathbf{N}}x = \{T^n x : n \in \mathbf{N}\}$ of a point x to start outside of any of the minimal cycles, although it will eventually be absorbed in one of them.

In the above examples, the limit points of an orbit formed their own minimal orbits. In some cases, one has to pass to limits multiple times before one reaches a minimal orbit. For instance, consider the boolean Bernoulli system again, but now consider the point

$$y := \bigcup_{n=0}^{\infty} [4^n, 2 \times 4^n] = [1, 2] \cup [4, 8] \cup [16, 32] \cup \ldots \in 2^{\mathbf{Z}},$$

where we use the notation $[N, M] := \{n \in \mathbf{Z} : N \leq n \leq M\}$. Observe that the point x defined earlier is not in the orbit $T^{\mathbf{N}}y$, but lies in the orbit closure, as it is the limit of $T^{4^n}y$. On the other hand, the orbit closure of x does not contain y. So the orbit closure of x is a subsystem of that of y, and then inside the former system one has the minimal systems $\{\mathbf{Z}\}$ and $\{\emptyset\}$. It is not hard to iterate this type of example and see that we can have quite intricate hierarchies of systems.

---

[11]In the case of finite dynamical systems, the integer shift does not appear, and we have recovered the classical fact that every permutation is uniquely decomposable into the product of disjoint cycles.

[12]This argument shows that x itself is not contained in any minimal system—why?

Exercise 2.2.1. Construct a topological dynamical system $(X, \mathcal{F}, T)$ and a sequence of orbit closures $\overline{T^{\mathbf{Z}} x_n}$ in X which form a proper nested sequence, thus

$$\overline{T^{\mathbf{Z}} x_1} \supsetneq \overline{T^{\mathbf{Z}} x_2} \supsetneq \overline{T^{\mathbf{Z}} x_3} \supsetneq \ldots .$$

Hint: Take a countable family of nested Bernoulli systems, and find a way to represent each one as an orbit closure.

Despite this apparent complexity, we can always terminate such hierarchies of subsystems at a minimal system:

Lemma 2.2.8. *Every topological dynamical system* $(X, \mathcal{F}, T)$ *contains a minimal dynamical system.*

Proof. Observe that the intersection of any chain of subsystems of X is again a subsystem (here we use the *finite intersection property* of compact sets to guarantee that the intersection is non-empty, and we also use the fact that the arbitrary intersection of closed or T-invariant sets is again closed or T-invariant). The claim then follows from *Zorn's lemma*.[13]   □

Exercise 2.2.2. Recall that every compact metrisable space is *second countable* and thus has a countable topological base. Suppose we are given an explicit enumeration $V_1, V_2, \ldots$ of such a base. Then find a proof of Lemma 2.2.8 which avoids the axiom of choice.

It would be nice if we could use Lemma 2.2.8 to decompose topological dynamical systems into the union of minimal subsystems, as we did in the case of non-topological dynamical systems. Unfortunately this does not work so well; the problem is that the complement of a minimal system is an open set rather than a closed set, and so we cannot cleanly separate a minimal system from its complement.[14]

We will study minimal dynamical systems in detail in the next few lectures. I'll close now with some examples of minimal systems.

Example 2.2.9 (Cyclic group shift). The cyclic group shift $(\mathbf{Z}/N\mathbf{Z}, x \mapsto x+1)$, where N is a positive integer, is a minimal system, and these are the only discrete minimal topological dynamical systems. More generally, if x is a periodic point of a topological dynamical system (thus $T^N x = x$ for some $N \geq 1$), then the closed orbit of x is isomorphic to a cyclic group shift and is thus minimal.

---

[13] We will always assume the axiom of choice throughout this course.

[14] In any case, the preceding examples already show that there can be some points in a system that are not contained in any minimal subsystem. Also, in contrast with non-invertible non-topological dynamical systems, our examples also show that a closed orbit can contain multiple minimal subsystems, so we cannot reduce to some sort of "nilpotent" system that has only one minimal system.

Example 2.2.10 (Torus shift). Consider a torus shift $((\mathbf{R}/\mathbf{Z})^d, x \mapsto x + \alpha)$, where $\alpha \in \mathbf{R}^d$ is a fixed vector. It turns out that this system is minimal if and only if[15] α is *totally irrational*, which means that $n \cdot \alpha$ is not an integer for any non-zero $n \in \mathbf{Z}^d$.

Example 2.2.11 (Morse sequence). Let $A = \{a, b\}$ be a two-letter alphabet, and consider the Bernoulli system $(A^{\mathbf{Z}}, T)$ formed from doubly infinite words

$$\ldots x_{-2} x_{-1} . x_0 x_1 x_2 \ldots$$

in A with the left-shift. Now define the sequence of finite words

$$w_1 := a.b;$$
$$w_2 := abba.baab;$$
$$w_3 := abbabaabbaababba.baababbaabbabaab;$$
$$\ldots$$

by the recursive formula

$$w_1 := a.b; \quad w_{i+1} := f(w_i),$$

where $f(w)$ denotes the word formed from w by replacing each occurrence of a and b by *abba* and *baab*, respectively. These words w_i converge pointwise to an infinite word

$$w = \ldots abbabaababbabaabbaababba.baababbaabbabaababbabaab \ldots .$$

Exercise 2.2.3. Show that w is not a periodic element of $A^{\mathbf{Z}}$, but that the orbit $\overline{T^{\mathbf{Z}} w}$ is both closed and minimal. *Hint*: Find large subwords of w which appear *syndetically*, which means that the gaps between each appearance are bounded. In fact, all subwords of w appear syndetically. One can also work with a more explicit description of w involving the number of non-zero digits in the binary expansion of the index. (This set is an example of a *substitution minimal set*.)

Exercise 2.2.4. Let $(X, \mathcal{F}, T)$ and $(Y, \mathcal{G}, S)$ be topological dynamical systems. Define the *product* of these systems to be $(X \times Y, \mathcal{F} \times \mathcal{G}, T \times S)$, where $X \times Y$ is the Cartesian product, $\mathcal{F} \times \mathcal{G}$ is the product topology, and $T \times S$ is the map $(x, y) \mapsto (Tx, Sy)$. Note that there are obvious projection morphisms from this product system to the two original systems. Show that this product system is indeed a product in the sense of category theory, thus any other system that maps to the two original systems factors uniquely through the product. Establish analogous claims in the categories of dynamical systems and measure-preserving systems.

---

[15]The "if" part is slightly non-trivial; see Corollary 1.3.2; but the "only if" part is easy, and is left as an exercise.

Exercise 2.2.5. Let $(X, \mathcal{F}, T)$ and $(Y, \mathcal{G}, S)$ be topological dynamical systems. Define the *disjoint union* of these systems to be $(X \uplus Y, \mathcal{F} \uplus \mathcal{G}, T \uplus S)$ where $(X \uplus Y, \mathcal{F} \uplus \mathcal{G})$ is the disjoint union of $(X, \mathcal{F})$ and $(Y, \mathcal{G})$, and $T \uplus S$ is the map which agrees with T on X and agrees with S on Y. Note that there are obvious embedding morphisms from the two original systems into the disjoint union. Show that the disjoint union is a coproduct in the sense of category theory, thus any system that is mapped to from the two origina systems factors uniquely through the disjoint union. Are analogous claims true for the categories of dynamical systems and measure-preserving systems?

Notes. This lecture first appeared at

terrytao.wordpress.com/2008/01/10.

Thanks to Andy P. and Sean Prendiville for corrections.

2.3. Minimal dynamical systems, recurrence, and the Stone-Čech compactification

We now begin the study of *recurrence* in topological dynamical systems $(X, \mathcal{F}, T)$, that is, how often a non-empty open set U in X returns to intersect itself, or how often a point x in X returns to be close to itself. Not every set or point needs to return to itself; consider for instance what happens to the shift $x \mapsto x + 1$ on the compactified integers $\{-\infty\} \cup \mathbf{Z} \cup \{+\infty\}$. Nevertheless, we can always show that at least one set (from any open cover) returns to itself:

Theorem 2.3.1 (Simple recurrence in open covers). *Let $(X, \mathcal{F}, T)$ be a topological dynamical system, and let $(U_\alpha)_{\alpha \in A}$ be an open cover of X. Then there exists an open set U_α in this cover such that $U_\alpha \cap T^n U_\alpha \neq \emptyset$ for infinitely many n.*

Proof. By compactness of X, we can refine the open cover to a finite subcover. Now consider an orbit $T^{\mathbf{Z}} x = \{T^n x : n \in \mathbf{Z}\}$ of some arbitrarily chosen point $x \in X$. By the infinite pigeonhole principle, one of the sets U_α must contain an infinite number of the points $T^n x$ counting multiplicity; in other words, the recurrence set $S := \{n : T^n x \in U_\alpha\}$ is infinite. Letting n_0 be an arbitrary element of S, we thus conclude that $U_\alpha \cap T^{n_0 - n} U_\alpha$ contains $T^{n_0} x$ for every $n \in S$, and the claim follows. $\square$

Exercise 2.3.1. Conversely, use Theorem 2.3.1 to deduce the infinite pigeonhole principle (i.e., that whenever $\mathbf{Z}$ is coloured into finitely many colours, one of the colour classes is infinite). *Hint*: Look at the orbit closure of c inside $A^{\mathbf{Z}}$, where A is the set of colours and $c : \mathbf{Z} \to A$ is the colouring function.

Now we turn from recurrence of sets to recurrence of individual points, which is a somewhat more difficult, and highlights the role of minimal dynamical systems (as introduced in Section 2.2) in the theory. We will study the subject using two (largely equivalent) approaches, the first being the more traditional "epsilon and delta" approach, and the second using the *Stone-Čech compactification* $\beta\mathbf{Z}$ of the integers (or equivalently, via *ultrafilters*).

Before we begin, it will be notationally convenient[16] to place a metric d on our compact metrisable space X. There are of course infinitely many metrics that one could place here, but they are all coarsely equivalent in the following sense: if d, d' are two metrics on X, then for every $\delta > 0$ there exists an $\varepsilon > 0$ such that $d'(x, y) < \delta$ whenever $d(x, y) < \varepsilon$, and similarly with the role of d and d' reversed. This claim follows from the standard fact that continuous functions between compact metric spaces are uniformly continuous. Because of this equivalence, it will not actually matter for any of our results what metric we place on our spaces. For instance, we could endow a Bernoulli system $A^{\mathbf{Z}}$, where A is itself a compact metrisable space (and thus $A^{\mathbf{Z}}$ is compact by *Tychonoff's theorem*), with the metric

$$(2.4) \qquad d((a_n)_{n\in\mathbf{Z}}, (b_n)_{n\in\mathbf{Z}}) := \sum_{n\in\mathbf{Z}} 2^{-|n|} d_A(a_n, b_n),$$

where d_A is some arbitrarily selected metric on A. Note that this metric is not shift-invariant.

Exercise 2.3.2. Show that if A contains at least two points, then the Bernoulli system $A^{\mathbf{Z}}$ (with the standard shift) cannot be endowed with a shift-invariant metric. *Hint*: Find two distinct points which converge to each other under the shift map.

Fix a metric d. For each n, the shift $T^n : X \to X$ is continuous, and hence uniformly continuous since X is compact, thus for every $\delta > 0$ there exists $\varepsilon > 0$ depending on δ and n such that $d(T^n x, T^n y) < \delta$ whenever $d(x, y) < \varepsilon$. However, we caution that the T^n need not be uniformly *equicontinuous*; the quantity ε appearing above can certainly depend on n. Indeed, they need not even be equicontinuous. For instance, this will be the case for the Bernoulli shift with the metric (2.4) (why?), and more generally for any system that exhibits "mixing" or other chaotic behaviour. At the other extreme, in the case of *isometric* systems—systems in which T preserves the metric d—the shifts T^n are all isometries, and thus are clearly uniformly equicontinuous. (We will study isometric systems further in Section 2.6.)

---

[16] As an exercise, the reader is encouraged to recast all the material here in a manner which does not explicitly mention a metric.

We can now classify points x in X based on the dynamics of the orbit $T^{\mathbf{Z}}x := \{T^n x : n \in \mathbf{Z}\}$:

Definition 2.3.2 (Points in a topological dynamical system).

(1) x is *invariant* if $Tx = x$.

(2) x is *periodic* if $T^n x = x$ for some non-zero n.

(3) x is *almost periodic* if for every $\varepsilon > 0$, the set $\{n \in \mathbf{Z} : d(T^n x, x) < \varepsilon\}$ is *syndetic* (i.e., it has bounded gaps);

(4) x is *recurrent* if for every $\varepsilon > 0$, the set $\{n \in \mathbf{Z} : d(T^n x, x) < \varepsilon\}$ is infinite. Equivalently, there exists a sequence n_j of integers with $|n_j| \to \infty$ such that $\lim_{j \to \infty} T^{n_j} x = x$.

It is clear that every invariant point is periodic, that every periodic point is almost periodic, and every almost periodic point is recurrent. These inclusions are all strict. For instance, in the circle shift system $(\mathbf{R}/\mathbf{Z}, x \mapsto x + \alpha)$ with $\alpha \in \mathbf{R}$ irrational, it turns out that every point is almost periodic, but no point is periodic.

Exercise 2.3.3. In the boolean Bernoulli system $(2^{\mathbf{Z}}, A \mapsto A - 1)$, show that the discrete Cantor set

$$(2.5) \qquad x := \bigcup_{N=1}^{\infty} \left\{ \sum_{n=0}^{N} \epsilon_n 10^n : \epsilon_n \in \{-1, 0, +1\} \right\}$$

is recurrent but not almost periodic.

In a general topological dynamical system, it is quite possible to have points which are non-recurrent (as the example of the compactified integer shift already shows). But if we restrict to a *minimal* dynamical system, things get much better:

Lemma 2.3.3. *If $(X, \mathcal{F}, T)$ is a minimal topological dynamical system, then every element of X is almost periodic (and hence recurrent).*

Proof. Suppose for contradiction that we can find a point x of X which is not almost periodic. This means that we can find $\varepsilon > 0$ such that the set $\{n : d(T^n x, x) < \varepsilon\}$ is not syndetic. Thus, for any $m > 0$, we can find an n_m such that $d(T^n x, x) \geq \varepsilon$ for all $n \in [n_m - m, n_m + m]$ (say).

Since X is compact, the sequence $T^{n_m} x$ must have at least one limit point y. But then one verifies (using the continuity of the shift operators) that

$$(2.6) \qquad d(T^h y, x) = \lim_{m \to \infty} d(T^{n_m + h} x, x) \geq \varepsilon$$

for all h. This means that the orbit closure $\overline{T^{\mathbf{Z}} y}$ of y does not contain x, contradicting the minimality of X. The claim follows. $\qquad \square$

Exercise 2.3.4. If x is a point in a topological dynamical system, show that x is almost periodic if and only if it lies in a minimal system. Because of this, almost periodic points are sometimes referred to as *minimal* points.

Combining Lemma 2.3.3 with Lemma 2.2.8, we immediately obtain

Theorem 2.3.4 (Birkhoff recurrence theorem). *Every topological dynamical system contains at least one point x which is almost periodic (and hence recurrent).*

Note that this is stronger than Theorem 2.3.1, as can be seen by considering the element U_α of the open cover which contains the almost periodic point. Indeed, we now have obtained a stronger conclusion, namely that the set of return times $\{n : T^n U_\alpha \cap U_\alpha \neq \emptyset\}$ is not only infinite, it is syndetic.

Exercise 2.3.5. State and prove a version of the Birkhoff recurrence theorem in which the map $T : X \to X$ is continuous but not assumed to be invertible. (Of course, all references to $\mathbf{Z}$ now need to be replaced with $\mathbf{N}$.)

The Birkhoff recurrence theorem does not seem particularly strong, as it only guarantees existence of a single recurrent (or almost periodic) point. For general systems, this is inevitable, because it can happen that the majority of the points are non-recurrent (look at the compactified integer shift system, for instance). However, suppose the system is a group quotient $(G/\Gamma, x \mapsto gx)$. To make this a topological dynamical system, we need G to be a topological group, and Γ to be a cocompact subgroup of G (such groups are also sometimes referred to as *uniform* subgroups). Then we see that the system is a *homogeneous space*: given any two points $x, y \in G/\Gamma$, there exists a group element $h \in G$ such that $hx = y$. Thus we expect any two points in G/Γ to behave similarly to each other. Unfortunately, this does not quite work in general, because the action of h need not preserve the shift $x \mapsto gx$, as there is no reason that h commutes with g. But suppose that g is a *central* element of G, i.e., it commutes with every element of G; this is for instance the case if G is abelian. Then the action of h is now an isomorphism on the dynamical system $(G/\Gamma, x \mapsto gx)$. In particular, if $hx = y$, we see that x is almost periodic (or recurrent) if and only if y is. We thus conclude:

Theorem 2.3.5 (Kronecker type approximation theorem). *Suppose $(G/\Gamma, x \mapsto gx)$ is a topological group quotient dynamical system such that g lies in the centre $Z(G)$ of G. Then every point in this system is almost periodic (and hence recurrent).*

Applying this theorem to the torus shift $((\mathbf{R}/\mathbf{Z})^d, x \mapsto x + \alpha)$, where $\alpha = (\alpha_1, \dots, \alpha_d) \in \mathbf{R}^d$ is a vector, we thus obtain that for any $\varepsilon > 0$, the set

$$(2.7) \qquad \{n \in \mathbf{Z} : \operatorname{dist}(n\alpha, \mathbf{Z}^d) < \varepsilon\}$$

is syndetic (and in particular, infinite). This should be compared with the classical Kronecker approximation theorem.

It is natural to ask what happens when g is not central. If G is a Lie group and the action of g on the Lie algebra $\mathfrak{g}$ is unipotent rather than trivial, then Theorem 2.3.5 still holds; this follows from *Ratner's theorem*, which we will discuss in Sections 2.16–2.17. But the claim is not true for all group quotients. Consider for instance the Bernoulli shift system $(X,T) = ((\mathbf{Z}/2\mathbf{Z})^{\mathbf{Z}}, T)$, which is isomorphic to the boolean Bernoulli shift system. As the previous examples have already shown, this system contains both recurrent and non-recurrent elements. On the other hand, it is intuitive that this system has a lot of symmetry, and indeed we can view it as a group quotient $(G/\Gamma, x \mapsto gx)$. Specifically, G is the *lamplighter group* $G = \mathbf{Z}/2\mathbf{Z} \wr \mathbf{Z}$. To describe this group, we observe that the group $(\mathbf{Z}/2\mathbf{Z})^{\mathbf{Z}}$ acts on X by addition, whilst the group $\mathbf{Z}$ acts on X via the shift map T. The lamplighter group $G := (\mathbf{Z}/2\mathbf{Z})^{\mathbf{Z}} \times \mathbf{Z}$ then acts by both addition and shift:

$$(2.8) \qquad (a,n) : x \mapsto T^n x + a \ \text{ for all } (a,n) \in G.$$

In order for this to be a group action, we endow G with the multiplication law

$$(2.9) \qquad (a,n)(b,m) := (a + T^n b, n + m).$$

One easily verifies that this really does make G into a group, and if we give G the product topology, it becomes a *topological group*. Clearly, G acts transitively on the compact space X, and so $X \equiv G/\Gamma$ for some cocompact subgroup Γ (which turns out to be isomorphic to $\mathbf{Z}$—why?). By construction, the shift map T can be expressed using the group element $(0,1) \in G$, and so we have turned the Bernoulli system into a group quotient. Since this system contains non-recurrent points (e.g. the indicator function of the natural numbers), we see that Theorem 2.3.5 does not hold for arbitrary group quotients.

2.3.1. The ultrafilter approach.

Now we turn to a different approach to topological recurrence, which relies on compactifying the underlying group $\mathbf{Z}$ that acts on topological dynamical systems. By doing so, all the epsilon management issues (cf. Section 1.5 of *Structure and Randomness*) go away, and the subject becomes very algebraic in nature. On the other hand, some subtleties arise also; for instance, the compactified object $\beta\mathbf{Z}$ is not a group, but merely a left-continuous semigroup.

This approach is based on *ultrafilters* or (equivalently) via the *Stone-Čech compactification*. Let us recall how this compactification works.

Theorem 2.3.6 (Stone-Čech compactification). *Every locally compact Hausdorff (LCH) space X can be embedded in a compact Hausdorff space βX in which X is an open dense set. (In particular, if X is already compact, then $\beta X = X$.) Furthermore, any continuous function $f : X \to Y$ between LCH spaces extends uniquely to a continuous function $\beta f : \beta X \to \beta Y$.*

Proof. (Sketch) This proof uses the intuition that βX should be the "finest" compactification of X. Recall that a compactification of an LCH space X is any compact Hausdorff space containing X as an open dense set. We say that one compactification Y of X is *finer* than another Z if there is a surjective[17] continuous map from Y to Z that is the identity on X. For instance, the two-point compactification $\{-\infty\} \cup \mathbf{Z} \cup \{+\infty\}$ of the integers is finer than the one-point compactification $\mathbf{Z} \cup \{\infty\}$. This is clearly a partial ordering; also, the *inverse limit* of any chain (totally ordered set) of compactifications can be verified (by *Tychonoff's theorem*) to still be a compactification. Hence, by *Zorn's lemma*,[18] there is a maximal compactification βX. To verify the extension property for continuous functions $f : X \to Y$, note (by replacing Y with βY if necessary) that we may take Y to be compact. Let Z be the closure of the graph $X' := \{(x, f(x)) : x \in X\}$ in $(\beta X) \times Y$. Clearly, X' is homeomorphic to X, and so Z is a compactification of X. Also, there is an obvious surjective continuous map from Z to βX; thus by maximality, this map must be a homeomorphism, so Z is the graph of a continuous function $\beta f : \beta X \to \beta Y$, and the claim follows (the uniqueness of βf is easily established). $\square$

Exercise 2.3.6. Let X be discrete (and thus clearly LCH), and let βX be the Stone-Čech compactification. For any $p \in \beta X$, let $[p] \in 2^{2^X}$ be the collection of all sets $A \subset X$ such that $\beta 1_A(p) = 1$. Show that $[p]$ is an *ultrafilter*, or in other words that it obeys the following four properties:

(1) $\emptyset \notin [p]$.

(2) If $U \in [p]$ and $V \in 2^X$ are such that $U \subset V$, then $V \in [p]$.

(3) If $U, V \in [p]$, then $U \cap V \in [p]$.

(4) If $U, V \in 2^X$ are such that $U \cup V = X$, then at least one of U and V lies in $[p]$.

Furthermore, show that the map $p \mapsto [p]$ is a homeomorphism between βX and the space of ultrafilters, which we endow with the topology induced from the product topology on $2^{2^X} \equiv \{0,1\}^{2^X}$, where we give $\{0,1\}$ the discrete topology (one can place some other topologies here also). Thus we see

---

[17]Note that as X is dense in Y, and Z is Hausdorff, this surjection is unique.

[18]There is a technical step one needs to verify to apply this lemma, namely the moduli space of compactifications of X is a set rather than a class. We leave this to the reader.

that in the discrete case, we can represent the Stone-Čech compactification explicitly via ultrafilters.

It is easy to see that $\beta(g \circ f) = (\beta g) \circ (\beta f)$ whenever $f : X \to Y$ and $g : Y \to Z$ are continuous maps between LCH spaces. In the language of category theory, we thus see that β is a *covariant functor* from the category of LCH spaces to the category of compact Hausdorff spaces.[19]

Exercise 2.3.7. Let X and Y be two LCH spaces. Show that the *disjoint union* $(\beta X) \uplus (\beta Y)$ of βX and βY is isomorphic to $\beta(X \uplus Y)$. (Indeed, this isomorphism is a *natural isomorphism*.) In the language of category theory, this means that β preserves coproducts.[20]

Note that if $f : X \to Y$ is continuous, then $\beta f : \beta X \to \beta Y$ is continuous also; since X is dense in βX, we conclude that[21]

$$(2.10) \qquad \beta f(p) = \lim_{x \to p} f(x)$$

for all $p \in \beta X$, where x is constrained to lie in X. In particular, the limit on the right exists for any continuous $f : X \to Y$, and thus if X is discrete, it exists for any (!) function $f : X \to Y$. Each p can then be viewed as a recipe for taking limits of arbitrary functions in a consistent fashion (although different p's can give different limits, of course). It is this ability to take limits without needing to check for convergence and without running into contradictions that makes the Stone-Čech compactification a useful tool here.[22]

The integers $\mathbf{Z}$ are discrete, and thus are clearly LCH. Thus we may form the compactification $\beta \mathbf{Z}$. The addition operation $+ : \mathbf{Z} \times \mathbf{Z} \to \mathbf{Z}$ can then be extended to $\beta \mathbf{Z}$ by the plausible-looking formula

$$(2.11) \qquad p + q := \lim_{n \to p} \lim_{m \to q} n + m$$

for all $p, q \in \beta \mathbf{Z}$, where n, m range in the integers $\mathbf{Z}$. Note that the double limit is guaranteed to exist by (2.10). Equivalently, we have

$$(2.12) \qquad \lim_{l \to p+q} f(l) = \lim_{n \to p} \lim_{m \to q} f(n + m)$$

---

[19]The above theorem does not explicitly define βX, but it is not hard to see that this compactification is unique up to homeomorphism, so the exact form of βX is somewhat moot. However, it is possible to create an ultrafilter-based description of βX for general LCH spaces X, though we will not do so here.

[20]Unfortunately, β does not preserve products, which leads to various subtleties, such as the non-commutativity of the compactification of commutative groups.

[21]Here and in the sequel, limits such as $\lim_{x \to p}$ are interpreted in the usual topological sense, thus (2.10) means that for every neighbourhood V of $\beta f(p)$, there exists a neighbourhood U of p such that $f(x) \in V$ for all $x \in U$.

[22]See also Section 1.5 of *Structure and Randomness* for further discussion.

for all functions $f : \mathbf{Z} \to X$ into an LCH space X; one can derive (2.12) from
(2.11) by applying $\beta f : \beta \mathbf{Z} \to \beta X$ to both sides of (2.11) and using (2.10)
and the continuity of βf repeatedly. This addition operation clearly extends
that of $\mathbf{Z}$ and is associative, thus we have turned $\beta \mathbf{Z}$ into a semigroup. We
caution however that this semigroup is not commutative, due to the usual
difficulty that double limits in (2.11) cannot be exchanged. (We will prove
non-commutativity shortly.) For similar reasons, $\beta \mathbf{Z}$ is not a group; the
obvious candidate for a negation operation $-p := \lim_{n \to p} -n$ is well defined,
but does not actually invert addition. The operation $(p, q) \mapsto p + q$ is
continuous in p for fixed q (why?), but is not necessarily continuous in q for
fixed p—again, due to the exchange of limits problem. Thus $\beta \mathbf{Z}$ is merely a
left-continuous semigroup. If however p is an integer, then the first limit in
(2.11) disappears, and one easily shows that $q \mapsto q + p$ is continuous in this
case (and for similar reasons one also recovers commutativity, $q + p = p + q$).

Exercise 2.3.8. Let us endow the two-point compactification $\{-\infty\} \cup \mathbf{Z} \cup$
$\{+\infty\}$ with the semigroup structure $+$ in which $x + (+\infty) = +\infty$ and
$x + (-\infty) = -\infty$ for all $x \in \{-\infty\} \cup \mathbf{Z} \cup \{+\infty\}$ (compare with (2.11)).
Show that there is a unique continuous map $\pi : \beta \mathbf{Z} \to \mathbf{Z} \cup \{-\infty\} \cup \{+\infty\}$
which is the identity on $\mathbf{Z}$, and that this map is a surjective semigroup
homomorphism. Using this homomorphism, conclude that

(1) $\beta \mathbf{Z}$ is not commutative. Furthermore, show that the centre $Z(\beta \mathbf{Z})$
$:= \{p \in \beta \mathbf{Z} : p + q = q + p \text{ for all } q \in \beta \mathbf{Z}\}$ is exactly equal to $\mathbf{Z}$.

(2) Show that if $p, q \in \beta \mathbf{Z}$ are such that $p + q \in \mathbf{Z}$, then $p, q \in \mathbf{Z}$.
("Once you go to infinity, you can never return.") Conclude in
particular that $\beta \mathbf{Z}$ is not a group.[23]

Remark 2.3.7. More generally, we can take any LCH left-continuous semigroup S and compactify it to obtain a compact Hausdorff left-continuous
semigroup βS. Observe that if $f : S \to S'$ is a homomorphism between two
LCH left-continuous semigroups, then $\beta f : \beta S \to \beta S'$ is also a homomorphism. Thus, from the viewpoint of category theory, β can be regarded as
a covariant functor from the category of LCH left-continuous semigroups to
the category of CH left-continuous semigroups.

The left-continuous non-commutative semigroup structure of $\beta \mathbf{Z}$ may
appear to be terribly weak when compared against the jointly continuous
commutative group structure of $\mathbf{Z}$, but $\beta \mathbf{Z}$ has a decisive trump card over
$\mathbf{Z}$: it is *compact*. We will see the power of compactness a little later in this
lecture.

---

[23]Note that this conclusion could already be obtained using the coarser one-point compactification $\mathbf{Z} \cup \{\infty\}$ of the integers.

A topological dynamical system $(X, \mathcal{F}, T)$ yields an action $n \mapsto T^n$ of the integers $\mathbf{Z}$. But we can automatically extend this action to an action $p \mapsto T^p$ of the compactified integers $\beta\mathbf{Z}$ by the formula

$$(2.13) \qquad\qquad T^p x := \lim_{n \to p} T^n x.$$

(Note that X is already compact, so that the limit in (2.13) stays in X.) One easily checks from (2.12) that this is indeed an action of $\beta\mathbf{Z}$ (thus $T^p T^q = T^{p+q}$ for all $p, q \in \mathbf{Z}$). The map $T^p x$ is continuous in p by construction; however we caution that it is no longer continuous in x (it is the exchange-of-limits problem once more!). Indeed, the map $T^p : X \to X$ can be quite nasty from an analytic viewpoint; for instance, it is possible for this map to not be Borel measurable.[24] But as we shall see, the *algebraic* properties of T^p are very good, and suffice for applications to recurrence, because once one has compactified the underlying semigroup $\beta\mathbf{Z}$, the need for point-set topology (and for all the epsilons that come with it) mostly disappears. For instance, we can now replace orbit closures by orbits:

Lemma 2.3.8. *Let $(X, \mathcal{F}, T)$ be a topological dynamical system, and let $x \in X$. Then*

$$\overline{T^{\mathbf{Z}}(x)} = T^{\beta\mathbf{Z}} x := \{T^p x : p \in \beta\mathbf{Z}\}.$$

Proof. Since $\beta\mathbf{Z}$ is compact, $T^{\beta\mathbf{Z}} x$ is compact also. Since $\mathbf{Z}$ is dense in $\beta\mathbf{Z}$, $T^{\mathbf{Z}} x$ is dense in $T^{\beta\mathbf{Z}} x$. The claim follows. □

From (2.13) we see that T^p is some sort of "limiting shift" operation. To get some intuition, let us consider the compactified integer shift $(\{-\infty\} \cup \mathbf{Z} \cup \{+\infty\}, x \mapsto x+1)$, and look at the orbit of the point 0. If one only shifts by integers $n \in \mathbf{Z}$, then $T^n 0$ can range across the region $\mathbf{Z}$ in the system but cannot reach $-\infty$ or $+\infty$. But now let $p \in \beta\mathbf{Z}\backslash\mathbf{Z}$ be any limit point of the positive integers $\mathbf{Z}^+$. (Note that at least one such limit point must exist, since $\mathbf{Z}^+$ is not compact. Indeed, in the language of Exercise 2.3.8, the set of all such limit points is $\pi^{-1}(+\infty)$.) Then from (2.13) we see that $T^p 0 = +\infty$. Similarly, if $q \in \beta\mathbf{Z}\backslash\mathbf{Z}$ is a limit point of the negative integers $\mathbf{Z}^-$ then $T^q 0 = -\infty$. Now, since $+\infty$ invariant, we have $T^q(+\infty) = +\infty$ by (2.13) again, and thus $T^q T^p 0 = +\infty$, while $T^p T^q 0 = -\infty$. In particular, we see that $p + q \neq q + p$, demonstrating non-commutativity in $\beta\mathbf{Z}$ (again, compare with Exercise 2.3.8). Informally, the problem here is that in (2.11), $n + m$ will go to $+\infty$ if we let m go to $+\infty$ first and then $n \to -\infty$ next, but if we take $n \to -\infty$ first and then $m \to +\infty$ next, $n + m$ instead goes to $-\infty$.

---

[24]This is the price one pays for introducing beasts generated by the axiom of choice into one's mathematical ecosystem.

Exercise 2.3.9. Let $A \subset \mathbf{Z}$ be a set of integers.

(1) Show that βA can be canonically identified with the closure of A in $\beta \mathbf{Z}$, in which case βA becomes a *clopen subset* of $\beta \mathbf{Z}$.

(2) Show that A is infinite if and only if $\beta A \not\subset \mathbf{Z}$.

(3) Show that A is syndetic if and only if $\beta A \cap (\beta \mathbf{Z} + p) \neq \emptyset$ for every $p \in \beta \mathbf{Z}$. (Since βA is clopen, this condition is also equivalent to requiring $\beta A \cap (\mathbf{Z} + p) \neq \emptyset$ for every $p \in \beta \mathbf{Z}$.)

(4) A set of integers A is said to be *thick* if it contains arbitrarily long intervals $[a_n, a_n + n]$; thus syndetic and thick sets always intersect each other. Show that A is thick if and only if there exists $p \in \beta \mathbf{Z}$ such that $\beta \mathbf{Z} + p \subset \beta A$. (Again, this condition is equivalent to requiring $\mathbf{Z} + p \subset \beta A$ for some p.)

Recall that a system is *minimal* if and only if it is the orbit closure of every point in that system. We thus have a purely algebraic description of minimality:

Corollary 2.3.9. *Let* $(X, \mathcal{F}, T)$ *be a topological dynamical system. Then* X *is minimal if and only if the action of* $\beta \mathbf{Z}$ *is transitive; thus for every* $x, y \in \mathbf{Z}$ *there exists* $p \in \beta \mathbf{Z}$ *such that* $T^p x = y$.

One also has purely algebraic descriptions of almost periodicity and recurrence:

Exercise 2.3.10. Let $(X, \mathcal{F}, T)$ be a topological dynamical system, and let x be a point in X.

(1) Show that x is almost periodic if and and only if for every $p \in \beta \mathbf{Z}$ there exists $q \in \beta \mathbf{Z}$ such that $T^q T^p x = x$. (In particular, Lemma 2.3.3 is now an immediate consequence of Corollary 2.3.9.)

(2) Show that x is recurrent if and only if there exists $p \in \beta \mathbf{Z} \backslash \mathbf{Z}$ such that $T^p x = x$.

Note that $\beta \mathbf{Z}$ acts on itself by addition, $p : q \mapsto p + q$, with the action being continuous when p is an integer. Thus one can view $\beta \mathbf{Z}$ itself as a topological dynamical system, with the caveat that $\beta \mathbf{Z}$ is not metrisable or even first countable (see Exercise 2.3.13). Nevertheless, it is still useful to think of $\beta \mathbf{Z}$ as behaving like a topological dynamical system. For instance:

Definition 2.3.10. An element $p \in \beta \mathbf{Z}$ is said to be *minimal* or *almost periodic* if for every $q \in \beta \mathbf{Z}$ there exists $r \in \beta \mathbf{Z}$ such that $r + q + p = p$.

Equivalently, p is minimal if $\beta \mathbf{Z} + p$ is a minimal left-ideal of $\beta \mathbf{Z}$, which explains the terminology.

Exercise 2.3.11. Show that for every $p \in \beta\mathbf{Z}$ there exists $q \in \beta\mathbf{Z}$ such that $q + p$ is minimal. *Hint*: Adapt the proof of Lemma 2.2.8. Also, show that if p is minimal, then $q + p$ and $p + q$ are also minimal for any $q \in \beta\mathbf{Z}$. This shows that minimal elements of $\beta\mathbf{Z}$ exist in abundance. However, observe from Exercise 2.3.6 that no integer can be minimal.

Exercise 2.3.12. Show that if $p \in \beta\mathbf{Z}$ is minimal, and x is a point in a topological dynamical system $(X, \mathcal{F}, T)$, then $T^p x$ is almost periodic. Conversely, show that x is almost periodic if and only if $x = T^p x$ for some minimal p. This gives an alternative (and more "algebraic") proof of the Birkhoff recurrence theorem.

Exercise 2.3.13. Show that no element of $\beta\mathbf{Z}\backslash\mathbf{Z}$ can be written as a limit of a sequence in $\mathbf{Z}$. *Hint*: If a sequence $n_j \in \mathbf{Z}$ converged to a limit $p \in \beta\mathbf{Z}$, one must have $\beta f(p) = \lim_{j \to \infty} f(n_j)$ for all functions $f : \mathbf{Z} \to K$ mapping into a compact Hausdorff space K. Conclude in particular that $\beta\mathbf{Z}$ is not metrisable, first countable, or sequentially compact.

Notes. This lecture first appeared at

terrytao.wordpress.com/2008/01/13.

Thanks to Richard Séguin, R.A., Eric, Liu Xiao Chuan, and Sean Prendiville for corrections.

2.4. Multiple recurrence

In Section 2.3, we established single recurrence properties for both open sets and for sequences inside a topological dynamical system $(X, \mathcal{F}, T)$. In this lecture, we generalise these results to multiple recurrence. More precisely, we shall prove

Theorem 2.4.1 (Multiple recurrence in open covers). *Let $(X, \mathcal{F}, T)$ be a topological dynamical system, and let $(U_\alpha)_{\alpha \in A}$ be an open cover of X. Then there exists U_α such that for every $k \geq 1$, we have $U_\alpha \cap T^{-r}U_\alpha \cap \ldots \cap T^{-(k-1)r}U_\alpha \neq \emptyset$ for infinitely many r.*

Note that this theorem includes Theorem 2.3.1 as the special case $k = 2$. This theorem is also equivalent to the following well-known combinatorial result [**vdW1927**]:

Theorem 2.4.2 (Van der Waerden's theorem). *Suppose the integers $\mathbf{Z}$ are finitely coloured. Then one of the colour classes contains arbitrarily long arithmetic progressions.*

Exercise 2.4.1. Show that Theorem 2.4.1 and Theorem 2.4.2 are equivalent.

Exercise 2.4.2. Show that Theorem 2.4.2 fails if "arbitrarily long" is replaced by "infinitely long". Deduce that a similar strengthening of Theorem 2.4.1 also fails.

Exercise 2.4.3. Use Theorem 2.4.2 to deduce a finitary version: Given any positive integers m and k, there exists an integer N such that whenever $\{1, \ldots, N\}$ is coloured into m colour classes, one of the colour classes contains an arithmetic progression of length k. *Hint*: Use a "compactness and contradiction" argument, as in Section 1.3 of *Structure and Randomness*.

We also have a stronger version of Theorem 2.4.1:

Theorem 2.4.3 (Multiple Birkhoff recurrence theorem). *Let $(X, \mathcal{F}, T)$ be a topological dynamical system. Then for any $k \geq 1$ there exists a point $x \in X$ and a sequence $r_j \to \infty$ of integers such that $T^{ir_j}x \to x$ as $j \to \infty$ for all $0 \leq i \leq k - 1$.*

These results already have some application to equidistribution of explicit sequences. Here is a simple example (which is also a consequence of *Weyl's polynomial equidistribution theorem*, Theorem 2.6.26):

Corollary 2.4.4. *Let α be a real number. Then there exists a sequence $r_j \to \infty$ of integers such that $\mathrm{dist}(r_j^2\alpha, \mathbf{Z}) \to 0$ as $j \to \infty$.*

Proof. Consider the skew shift system $X = (\mathbf{R}/\mathbf{Z})^2$ with $T(x, y) := (x + \alpha, y + x)$. By Theorem 2.4.3, there exists $(x, y) \in X$ and a sequence $n_j \to \infty$ such that $T^{r_j}(x, y)$ and $T^{2r_j}(x, y)$ both converge to (x, y). If we then use the easily verified identity

$$(2.14) \qquad (x, y) - 2T^{r_j}(x, y) + T^{2r_j}(x, y) = (0, r_j^2\alpha),$$

we obtain the claim. $\qquad\square$

Exercise 2.4.4. Use Theorem 2.4.1 or Theorem 2.4.2 in place of Theorem 2.4.3 to give an alternative derivation of Corollary 2.4.4.

Exercise 2.4.5. Prove Theorem 1.3.1.

As in Section 2.3, we will give both a traditional topological proof and an ultrafilter-based proof of Theorems 2.4.1 and 2.4.3; the reader is invited to see how the various proofs are ultimately equivalent to each other.

2.4.1. Topological proof of van der Waerden's theorem. We begin by giving a topological proof of Theorem 2.4.1, due to Furstenberg and Weiss [**FuWe1978**], which is secretly a translation of van der Waerden's original "colour focusing" combinatorial proof of Theorem 2.4.2 into the dynamical setting. To prove Theorem 2.4.1, it suffices to verify the following slightly weaker statement.

Theorem 2.4.5. *Let $(X, \mathcal{F}, T)$ be a topological dynamical system, and let $(U_\alpha)_{\alpha \in A}$ be an open cover of X. Then for every $k \geq 1$ there exists an open set U_α which contains an arithmetic progression $x, T^r x, T^{2r} x, \ldots, T^{(k-1)r} x$ for some $x \in X$ and $r > 0$.*

To see how Theorem 2.4.5 implies Theorem 2.4.1, first observe from compactness that we can take the open cover to be a finite cover. Then by the infinite pigeonhole principle, it suffices to establish Theorem 2.4.1 for each $k \geq 1$ separately. For each such k, Theorem 2.4.5 gives a single arithmetic progression $x, T^r x, \ldots, T^{(k-1)r} x$ inside one of the U_α. By replacing the system (X, T) with the product system $(X \times \mathbf{Z}/N\mathbf{Z}, (x, m) \mapsto (Tx, m + 1))$ for some large N and replacing the open cover $(U_\alpha)_{\alpha \in A}$ of X with the open cover $(U_\alpha \times \{m\})_{\alpha \in A, m \in \mathbf{Z}/N\mathbf{Z}}$ of $X \times \mathbf{Z}/N\mathbf{Z}$, one can make the spacing r in the arithmetic progression larger than any specified integer N. Thus by another application of the infinite pigeonhole principle, one of the U_α contains arithmetic progressions with arbitrarily large step r, and the claim follows.

Now we need to prove Theorem 2.4.5. By Lemma 2.2.8, to establish this theorem for minimal dynamical systems, we will need to note that for minimal systems, Theorem 2.4.5 automatically implies the following stronger-looking statement:

Theorem 2.4.6. *Let $(X, \mathcal{F}, T)$ be a minimal topological dynamical system, let U be a non-empty open set in X, and let $k \geq 1$. Then U contains an arithmetic progression $x, T^r x, \ldots, T^{(k-1)r} x$ for some $x \in X$ and $r \geq 1$.*

Indeed, the deduction of Theorem 2.4.6 from Theorem 2.4.5 is immediate from the following useful fact (cf. Lemma 2.3.3):

Lemma 2.4.7. *Let $(X, \mathcal{F}, T)$ be a minimal topological dynamical system, and let U be a non-empty open set in X. Then X can be covered by a finite number of translates $T^n U$ of U.*

Proof. The set $X \backslash \bigcup_{n \in \mathbf{Z}} T^n U$ is a proper closed invariant subset of X, which must therefore be empty since X is minimal. The claim then follows from the compactness of X. □

Remark 2.4.8. Of course, the claim is highly false for non-minimal systems; consider for instance the case when T is the identity. More generally, if X is non-minimal, consider an open set U which is the complement of a proper subsystem of X.

Now we need to prove Theorem 2.4.5. We do this by induction on k. The case $k = 1$ is trivial, so suppose $k \geq 2$ and the claim has already been proven for $k - 1$. By the above discussion, we see that Theorem 2.4.6 is also true for $k - 1$.

Now fix a minimal system $(X, \mathcal{F}, T)$ and an open cover $(U_\alpha)_{\alpha \in A}$, which we can take to be finite. We need to show that one of the U_α contains an arithmetic progression $x, T^r x, \ldots, T^{(k-1)r} x$ of length k. To do this, we first need an auxiliary construction.

Lemma 2.4.9 (Construction of colour focusing sequence). *Let the notation and assumptions be as above. Then for any $J \geq 0$ there exists a sequence $x_0, \ldots, x_J$ of points in X, a sequence $U_{\alpha_0}, \ldots, U_{\alpha_J}$ of sets in the open cover (not necessarily distinct), and a sequence $r_1, \ldots, r_J$ of positive integers such that $T^{i(r_{a+1} + \ldots + r_b)} x_b \in U_{\alpha_a}$ for all $0 \leq a \leq b \leq J$ and $1 \leq i \leq k-1$.*

Proof. We induct on J. The case $J = 0$ is trivial. Now suppose inductively that $J \geq 1$, and that we have already constructed $x_0, \ldots, x_{J-1}$, $U_{\alpha_0}, \ldots, U_{\alpha_{J-1}}$, and $r_1, \ldots, r_{J-1}$ with the required properties. Now let V be a suitably small neighbourhood of x_{J-1} (depending on all the above data) to be chosen later. By Theorem 2.4.6 for $k - 1$, V contains an arithmetic progression $y, T^{r_J} y, \ldots, T^{(k-2)r_J} y$ of length $k - 1$. If one sets $x_J := T^{-r_J} y$, and lets U_{α_J} be an arbitrary set in the open cover containing x_J, then we observe that
(2.15)
$$T^{i(r_{a+1} + \ldots + r_J)} x_J = T^{i(r_{a+1} + \ldots + r_{J-1})} (T^{(i-1)r_J} y) \in T^{i(r_{a+1} + \ldots + r_{J-1})}(V)$$

for all $0 \leq a < J$ and $1 \leq i \leq k-1$. If V is a sufficiently small neighbourhood of x_{J-1}, we thus see (from the continuity of the $T^{i(r_{a+1} + \ldots + r_{J-1})}$) that we verify all the required properties needed to close the induction. $\square$

We apply the above lemma with J equal to the number of sets in the open cover. By the pigeonhole principle, we can thus find $0 \leq a < b \leq J$ such that $U_{\alpha_a} = U_{\alpha_b}$. If we then set $x := x_b$ and $r := r_{a+1} + \ldots + r_b$, we obtain Theorem 2.4.5 as required.

Remark 2.4.10. It is instructive to compare the $k = 2$ case of the above arguments with the proof of Theorem 2.3.1. (For a comparison of this type of proof with the more classical combinatorial proof, see [**Ta2007**].)

2.4.2. Ultrafilter proof of van der Waerden's theorem. We now give a translation of the above proof into the language of *ultrafilters* (or more precisely, the language of *Stone-Čech compactifications*). This language may look a little strange, but it will be convenient when we study more general colouring theorems in the next lecture. As before, we will prove Theorem 2.4.5 instead of Theorem 2.4.1 (thus we only need to find one progression, rather than infinitely many). The key proposition is the following.

Proposition 2.4.11 (Ultrafilter version of van der Waerden's theorem). *Let p be a minimal element of $\beta\mathbf{Z}$. Then for any $k \geq 1$ there exists $q \in \beta(\mathbf{Z} \times \mathbf{N})$*

such that

(2.16) $$\lim_{(n,r)\to q} n + ir + p = p \quad \text{for all } 0 \leq i \leq k-1.$$

Suppose for the moment that this proposition is true. Applying it with some minimal element p of $\beta\mathbf{Z}$ (which must exist, thanks to Exercise 2.3.11), we obtain $q \in \beta(\mathbf{Z} \times \mathbf{N})$ obeying (2.16). If we let $x := T^p y$ for some arbitrary $y \in X$, we thus obtain

(2.17) $$\lim_{(n,r)\to q} T^{n+ir}x = x \text{ for all } 0 \leq i \leq k-1.$$

If we let U_α be an element of the open cover that contains x, we thus see that $T^{n+ir}x \in U_\alpha$ for all $0 \leq i \leq k-1$ and all $(n,r) \in \mathbf{Z} \times \mathbf{N}$ which lie in a sufficiently small neighbourhood of q. Since an LCH space is always dense in its Stone-Čech compactification, the space of all (n,r) with this property is non-empty, and Theorem 2.4.5 follows.

Proof of Proposition 2.4.11. We induct on k. The case $k = 1$ is trivial (one could take e.g. $q = (0,1)$), so suppose $k > 1$ and that the claim has already been proven for $k - 1$. Then we can find $q' \in \beta(\mathbf{Z} \times \mathbf{N})$ such that

(2.18) $$\lim_{(n,r)\to q'} n + ir + p = p$$

for all $0 \leq i \leq k - 2$.

Now consider the expression

(2.19) $$p_{i,a,b} := \lim_{(n_1,r_1)\to q'} \ldots \lim_{(n_b,r_b)\to q'} i(r_{a+1} + \ldots + r_b) + m_b + p$$

for any $1 \leq a \leq b$ and $1 \leq i \leq k - 1$, where

(2.20) $$m_b := \sum_{i=1}^{b} n_i - r_i.$$

Applying (2.18) to the (n_b, r_b) limit in (2.19), we obtain the recursion $p_{i,a,b} = p_{i,a,b-1}$ for all $b > a$. Iterating this, we conclude that

(2.21) $$p_{i,a,b} = p_{i,a,a} = p_{0,a,a}$$

for all $1 \leq i \leq k - 1$. For $i = 0$, (2.21) need not hold, but instead we have the easily verified identity

(2.22) $$p_{0,a,b} = p_{0,b,b}.$$

Now let $p_* \in \beta\mathbf{Z}\backslash\mathbf{Z}$ be arbitrary (one could pick $p_* := p$, for instance) and define $p' := \lim_{a\to p_*} p_{0,a,a} = \lim_{b\to p_*} p_{0,b,b}$. Observe from (2.19) that all the $p_{i,a,b}$ lie in the closed set $\beta\mathbf{Z} + p$, and so p' does also. Since p is minimal, there must exist $p'' \in \beta\mathbf{Z}$ such that $p = p'' + p'$. Expanding this out using (2.21) or (2.22), we conclude that

$$(2.23) \qquad \lim_{h \to p''} \lim_{a \to p_*} \lim_{b \to p_*} h + p_{i,a,b} = p$$

for all $0 \le i \le k - 1$. Applying (2.19), we conclude that

$$(2.24) \qquad \lim_{h \to p''} \lim_{a \to p_*} \lim_{b \to p_*} \lim_{(n_1,r_1) \to q'} \cdots \lim_{(n_b,r_b) \to q'} n + ir + p = p,$$

where $n := h + m_b$ and $r := r_{a+1} + \ldots + r_b$. Now, define $q \in \beta(\mathbf{Z} \times \mathbf{N})$ to be the limit

$$(2.25) \qquad q := \lim_{h \to p''} \lim_{a \to p_*} \lim_{b \to p_*} \lim_{(n_1,r_1) \to q'} \cdots \lim_{(n_b,r_b) \to q'} (n, r);$$

then we obtain Proposition 2.4.11 as desired. $\qquad \square$

Exercise 2.4.6. Strengthen Proposition 2.4.11 with the additional conclusion $\lim_{(n,r) \to q} r \notin \mathbf{N}$. Using this stronger version, deduce Theorem 2.4.1 directly without using the trick of multiplying X by a cyclic shift system that was used to deduce Theorem 2.4.1 from Theorem 2.4.5.

Theorem 2.4.1 can be generalised to multiple commuting shifts:

Theorem 2.4.12 (Multiple recurrence in open covers). *Let $(X, \mathcal{F})$ be a compact topological space, and let $T_1, \ldots, T_k : X \to X$ be commuting homeomorphisms. Let $(U_\alpha)_{\alpha \in A}$ be an open cover of X. Then there exists U_α such that $T_1^{-r} U_\alpha \cap \ldots \cap T_k^{-r} U_\alpha \ne \emptyset$ for infinitely many r.*

Exercise 2.4.7. By adapting one of the above arguments, prove Theorem 2.4.12.

Exercise 2.4.8. Use Theorem 2.4.12 to establish the following *multidimensional van der Waerden theorem* (due to Gallai): If a lattice $\mathbf{Z}^d$ is finitely coloured, and $v_1, \ldots, v_d \in \mathbf{Z}^d$, then one of the colour classes contains a pattern of the form $n + rv_1, \ldots, n + rv_d$ for some $n \in \mathbf{Z}^d$ and some non-zero r.

Exercise 2.4.9. Show that Theorem 2.4.12 can fail, even for $k = 3$ and $T_1 = \mathrm{id}$, if the shift maps T_j are not assumed to commute. *Hint*: First show that in the free group F_2 on two generators a, b, and for any word $w \in F_2$ and non-zero integer r, the three words $w, a^n w, b^n w$ cannot all begin with the same generator after reduction. This can be used to disprove a non-commutative multidimensional van der Waerden theorem, which can in turn be used to disprove a non-commutative version of Theorem 2.4.12.

2.4.3. Proof of multiple Birkhoff's theorem. We now use van der Waerden's theorem and an additional Baire category argument to deduce Theorem 2.4.3 from Theorem 2.4.1. The key new ingredient is

Lemma 2.4.13 (Semicontinuous functions are usually continuous). *Let* (X, d) *be a metric space, and let* $F : X \to \mathbf{R}$ *be semicontinuous. Then the set of points* x *where* F *is discontinuous is a set of the first category (i.e., a countable union of nowhere dense sets). In particular, by the* Baire category theorem, *if* X *is complete and non-empty, then* F *is continuous at at least one point.*

Proof. Without loss of generality we can take F to be upper semicontinuous. Suppose F is discontinuous at some point x. Then, by upper continuity, there exists a rational number q such that

$$(2.26) \qquad \liminf_{y \to x} F(y) < q \leq F(x).$$

In other words, x lies in the boundary of the closed set $\{x : F(x) \geq q\}$. As we know, boundaries of closed sets are always nowhere dense, and the claim follows. $\qquad\square$

Now we prove Theorem 2.4.3. Without loss of generality we can take X to be minimal. Let us place a metric d on the space X. Define the function $F : X \to \mathbf{R}^+$ by the formula

$$(2.27) \qquad F(x) := \inf_{n \geq 1} \sup_{1 \leq i \leq k-1} d(T^{in}x, x).$$

It will suffice to show that $F(x) = 0$ for at least one x (notice that if the infimum is actually attained at zero for some n, then x is a periodic point and the claim is obvious). Suppose for contradiction that F is always positive. Observe that F is upper semicontinuous, and so by Lemma 2.4.13 there exists a point of continuity of F. In particular there exists a non-empty open set U such that F is bounded away from zero.

By uniform continuity of T^n, we see that if F is bounded away from zero on U, it is also bounded away from zero on $T^n V$ for any n (though the bound from below depends on n). Applying Lemma 2.4.7, we conclude that F is bounded away from zero on all of X, thus there exists $\varepsilon > 0$ such that $F(x) > \varepsilon$ for all $x \in X$. But this contradicts Theorem 2.4.1 (or Theorem 2.4.5), using the balls of radius $\varepsilon/2$ as the open cover. This contradiction completes the proof of Theorem 2.4.3.

Exercise 2.4.10. Generalise Theorem 2.4.3 to the case in which T is merely assumed to be continuous, rather than be a homeomorphism. *Hint*: Let $\tilde{X} \subset X^{\mathbf{Z}}$ denote the space of all sequences $(x_n)_{n \in \mathbf{Z}}$ with $x_{n+1} = Tx_n$ for all n, with the topology induced from the product space $X^{\mathbf{Z}}$. Use a limiting argument to show that $\tilde{X}$ is non-empty. Then turn $\tilde{X}$ into a topological dynamical system and apply Theorem 2.4.3.

Exercise 2.4.11. Generalise Theorem 2.4.3 to multiple commuting shifts (analogously to how Theorem 2.4.12 generalises Theorem 2.4.1).

Exercise 2.4.12. Combine Exercises 2.4.10 and 2.4.11 by obtaining a generalisation of Theorem 2.4.3 to multiple non-invertible commuting shifts.

Exercise 2.4.13. Let $(X, \mathcal{F}, T)$ be a minimal topological dynamical system, and let $k \geq 1$. Call a point x in X k-*fold recurrent* if there exists a sequence $n_j \to \infty$ such that $T^{in_j}x \to x$ for all $0 \leq i \leq k-1$. Show that the set of k-fold recurrent points in X is *residual* (i.e., the complement is of the first category). In particular, the set of k-fold recurrent points is dense.

Exercise 2.4.14. In the boolean Bernoulli system $(2^{\mathbf{Z}}, A \mapsto A+1)$, show that the set A consisting of all non-zero integers which are divisible by 2 an even number of times is almost periodic. Conclude that there exists a minimal topological dynamical system $(X, \mathcal{F}, T)$ such that not every point in X is 3-fold recurrent (in the sense of the previous exercise). (Compare this with the arguments in the previous lecture, which imply that every point in X is 2-fold recurrent.)

Exercise 2.4.15. Suppose that a sequence of continuous functions $f_n : X \to \mathbf{R}$ on a metric space converges pointwise everywhere to another function $f : X \to \mathbf{R}$. Show that f is continuous on a residual set.

Exercise 2.4.16. Let $(X, \mathcal{F}, T)$ be a minimal topological dynamical system, and let $f : X \to \mathbf{R}$ be a function which is T-invariant, thus $Tf = f$. Show that if f is continuous at even one point x_0, then it has to be constant. *Hint*: x_0 is in the orbit closure of every point in X.

Notes. This lecture first appeared at

terrytao.wordpress.com/2008/01/15.

Thanks to Nilay and an anonymous commenter for corrections. Ed Dean (answering a question of Richard Borcherds) pointed out the recent paper [**Ge2008**] (building on the earlier paper [**Gi1987**]) that uses proof mining techniques to convert the topological dynamics proof of van der Waerden's theorem into a quantitative argument that gives essentially the same bounds as the classical combinatorial proof of that theorem.

2.5. Other topological recurrence results

In this lecture, we use topological dynamics methods to prove some other Ramsey-type theorems, and more specifically, the polynomial van der Waerden theorem, the hypergraph Ramsey theorem, Hindman's theorem, and the

Hales-Jewett theorem. In proving these statements, I have decided to focus on the ultrafilter-based proofs, rather than the combinatorial or topological proofs, though of course these styles of proof are also available for each of the above theorems.

2.5.1. The polynomial van der Waerden theorem. We first prove a significant generalisation of van der Waerden's theorem (Theorem 2.4.2):

Theorem 2.5.1 (Polynomial van der Waerden theorem). *Let $(P_1, \ldots, P_k)$ be a tuple of integer-valued polynomials $P_1, \ldots, P_k : \mathbf{Z} \to \mathbf{Z}$ (or tuple for short) with $P_1(0) = \ldots = P_k(0)$. Then whenever the integers are finitely coloured, one of the colour classes will contain a pattern of the form $n + P_1(r), \ldots, n + P_k(r)$ for some $n \in \mathbf{Z}$ and $r \in \mathbf{N}$.*

This result is due to Bergelson and Leibman [**BeLe1996**], who proved it using "epsilon and delta" topological dynamical methods. A combinatorial proof was obtained more recently in [**Wa2000**]. In these notes, I will translate the Bergelson-Leibman argument to the ultrafilter setting.

Note that the case $P_j(r) := (j-1)r$ recovers the ordinary van der Waerden theorem. But the result is significantly stronger; it implies for instance that one of the colour classes contains arbitrarily many shifted geometric progressions $n+r, n+r^2, \ldots, n+r^k$, which does not obviously follow from the van der Waerden theorem. The result here only claims a single monochromatic pattern $n + P_1(r), \ldots, n + P_k(r)$, but it is not hard to amplify this theorem to show that at least one colour class contains infinitely many such patterns.

Remark 2.5.2. The theorem can fail if the hypothesis $P_1(0) = \ldots = P_k(0)$ is dropped; consider for instance the case $k = 2$, $P_1(r) = 0$, $P_2(r) = 2r + 1$, and with the integers partitioned (or coloured) into the odd and even integers. More generally, the theorem fails whenever there exists a modulus N such that the polynomials $P_1, \ldots, P_k$ are never simultaneously equal modulo N. This turns out to be the only obstruction; this is a somewhat difficult recent result of Bergelson, Leibman, and Lesigne [**BeLeLe2007**].

Exercise 2.5.1. Show that the polynomial

$$P(r) := (r^2 - 2)(r^2 - 3)(r^2 - 6)(r^2 - 7)(r^3 - 3)$$

has a root modulo N for every positive integer N, but has no root in the integers. Thus we see that the Bergelson-Leibman-Lesigne result is stronger than the polynomial van der Waerden theorem; it does not seem possible to directly use the latter to conclude that in every finite colouring of the integers, one of the classes contains the pattern $n, n + P(r)$.

Here are the topological dynamics and ultrafilter versions of the above theorem.

Theorem 2.5.3 (Polynomial van der Waerden theorem, topological dynamics version). *Let $(P_1, \ldots, P_k)$ be a tuple with $P_1(0) = \ldots = P_k(0)$. Let $(U_\alpha)_{\alpha \in A}$ be an open cover of a topological dynamical system $(X, \mathcal{F}, T)$. Then there exists a set U_α in this cover such that $T^{P_1(r)}U \cap \ldots \cap T^{P_k(r)}U \neq \emptyset$ for at least one $r > 0$.*

Theorem 2.5.4 (Polynomial van der Waerden theorem, ultrafilter version). *Let $(P_1, \ldots, P_k)$ be a tuple with $P_1(0) = \ldots = P_k(0)$, and let $p \in \beta\mathbf{Z}$ be a minimal ultrafilter. Then there exists $q \in \beta(\mathbf{Z} \times \mathbf{N})$ such that*

$$(2.28) \qquad \lim_{(n,r) \to q} n + P_i(r) + p = p \quad \text{for all } 1 \leq i \leq k.$$

Exercise 2.5.2. Show that Theorem 2.5.1 and Theorem 2.5.3 are equivalent, and that Theorem 2.5.4 implies Theorem 2.5.3 (or Theorem 2.5.1). (For the converse implication, see Exercise 2.5.21.)

As in Section 2.4, we shall prove Theorem 2.5.4 by induction. However, the induction will be more complicated than just inducting on the number k of polynomials involved, or on the degree of these polynomials, but will instead involve a more complicated measure of the "complexity" of the polynomials being measured. Let us say that a tuple $(P_1, \ldots, P_k)$ *obeys the vdW property* if the conclusion of Theorem 2.5.4 is true for this tuple. Thus, for instance, from Proposition 2.4.11 we know that any tuple of *linear* polynomials which vanish at the origin will obey the vdW property.

Our goal is to show that every tuple of polynomials which simultaneously vanish at the origin has the vdW property. The strategy will be to reduce from any given tuple to a collection of "simpler" tuples. We first begin with an easy observation that one can always shift one of the polynomials to be zero:

Lemma 2.5.5 (Translation invariance). *Let Q be any integer-valued polynomial. Then a tuple $(P_1, \ldots, P_k)$ obeys the vdW property if and only if $(P_1 - Q, \ldots, P_k - Q)$ has the vdW property.*

Proof. Let $p \in \beta\mathbf{Z}$ be minimal. If $(P_1 - Q, \ldots, P_k - Q)$ has the vdW property, then we can find $q \in \beta(\mathbf{Z} \times \mathbf{N})$ such that

$$(2.29) \qquad \lim_{(n,r) \to q} n + P_i(r) - Q(r) + p = p \quad \text{for all } 1 \leq i \leq k.$$

If we then define $q' := \lim_{(n,r) \to q}(n - Q(r), r) \in \beta(\mathbf{Z} \times \mathbf{N})$, one easily verifies that (2.28) holds (with q replaced by q'), and the claim holds. The converse implication is similar. $\square$

Now we come to the key inductive step.

Lemma 2.5.6 (Inductive step). *Let $(P_0, P_1, \ldots, P_k)$ be a tuple with $P_0 = 0$, and let Q be another integer-valued polynomial. Suppose that for every finite set of integers $h_1, \ldots, h_m$, the tuple $(P_i(\cdot + h_j) - P_i(h_j) - Q(\cdot))_{1 \leq i \leq k; 1 \leq j \leq m}$ has the vdW property. Then $(0, P_1, \ldots, P_k)$ also has the vdW property.*

Proof. This will be a reprise of the proof of Proposition 2.4.11. Given any finite number of pairs $(n_1, r_1), \ldots, (n_{b-1}, r_{b-1}) \in \mathbf{Z} \times \mathbf{N}$ with $b \geq 1$, we see from the hypothesis that there exists $q_b \in \beta(\mathbf{Z} \times \mathbf{N})$ (depending on these pairs) such that

$$(2.30) \quad \lim_{(n_b, r_b) \to q_b} n_b + P_i(r_{a+1} + \ldots + r_b) - P_i(r_{a+1} + \ldots + r_{b-1}) - Q(r_b) + p = p$$

for all $0 \leq a < b$.

Now, for every $0 \leq a \leq b$ and $0 \leq i \leq k$, consider the expression $p_{a,b,i} \in \beta \mathbf{Z} + p$ defined by

$$(2.31) \quad p_{a,b,i} := \lim_{(n_1, r_1) \to q_1} \cdots \lim_{(n_b, r_b) \to q_b} P_i(r_{a+1} + \ldots + r_b) + m_b + p,$$

where $q_1, \ldots, q_b$ are defined recursively as above and

$$(2.32) \quad m_b := \sum_{i=1}^{b} n_i - Q(r_i).$$

From (2.30) we see that

$$(2.33) \quad p_{a,b,i} = p_{a,b-1,i}$$

for all $0 \leq a < b$ and $1 \leq i \leq k$, and thus

$$(2.34) \quad p_{a,b,i} = p_{a,a,i} = p_{a,a,0}$$

in this case. For $i = 0$, we have the slightly different identity

$$(2.35) \quad p_{a,b,0} = p_{b,b,0}.$$

We let $p_* \in \beta \mathbf{Z}/\mathbf{Z}$ be arbitrary, and set $p' := \lim_{a \to p_*} p_{a,a,0} = \lim_{b \to p_*} p_{b,b,0} \in \beta \mathbf{Z} + p$. By the minimality of p, we can find $p'' \in \beta \mathbf{Z}$ such that $p'' + p' = p$. We thus have

$$(2.36) \quad \lim_{h \to p''} \lim_{a \to p_*} \lim_{b \to p_*} h + p_{a,b,i} = p$$

for all $0 \leq i \leq k$. If one then sets

$$(2.37) \quad q := \lim_{h \to p''} \lim_{a \to p_*} \lim_{b \to p_*} \lim_{(n_1, r_1) \to q_1} \cdots \lim_{(n_b, r_b) \to q_b} (n, r),$$

where $n := h + m_b$ and $r := r_{a+1} + \ldots + r_b$, one easily verifies (2.28) as required. $\qquad \square$

Let us see how this lemma is used in practice. Suppose we wish to show that the tuple $(0, r^2)$ has the vdW property (where we use r to denote the independent variable). Applying Lemma 2.5.6 with $Q(r) := r^2$, we reduce to showing that the tuples $((r + h_1)^2 - h_1^2 - r^2, \ldots, (r + h_m)^2 - h_m^2 - r_m^2)$ have the vdW property for all finite collections $h_1, \ldots, h_m$ of integers. But observe that all the polynomials in these tuples are linear polynomials that vanish at the origin. By the ordinary van der Waerden theorem, these tuples all have the vdW property, and so $(0, r^2)$ has the vdW property also.

A similar argument shows that the tuple $(0, r^2+P_1(r), \ldots, r^2+P_k(r))$ has the vdW property whenever $P_1, \ldots, P_k$ are linear polynomials that vanish at the origin. Applying Lemma 2.5.5, we see that $(Q_1(r), r^2 + P_1(r), \ldots, r^2 + P_k(r))$ obeys the vdW property when Q_1 is also linear and vanishing at the origin.

Now, let us consider a tuple $(Q_1(r), Q_2(r), r^2 + P_1(r), \ldots, r^2 + P_k(r))$, where Q_2 is also a linear polynomial that vanishes at the origin. The vdW property for this tuple follows from the previously established vdW properties by first applying Lemma 2.5.5 to reduce to the case $Q_1 = 0$, and then applying Lemma 2.5.6 with $Q = Q_2$. Continuing in this fashion, we see that a tuple $(Q_1(r), \ldots, Q_l(r), r^2 + P_1(r), \ldots, r^2 + P_k(r))$ will also obey the vdW property for any linear $Q_1, \ldots, Q_l, P_1, \ldots, P_k$ that vanish at the origin, for any k and l.

Now the vdW property for the tuple $(0, r^2, 2r^2)$ follows from the previously established cases and Lemma 2.5.6 with $Q(r) = r^2$.

Remark 2.5.7. It is possible to continue this inductive procedure (known as *PET induction*; the PET stands, variously, for "polynomial ergodic theorem" or "polynomial exhaustion theorem"); this is carried out in Exercise 2.5.3 below.

Exercise 2.5.3. Define the *top order monomial* of a non-zero polynomial $P(r) = a_d r^d + \ldots + a_0$ with $a_d \neq 0$ to be $a_d r^d$. Define the top order monomials of a tuple $(0, P_1, \ldots, P_k)$ to be the set of top order monomials of the $P_1, \ldots, P_k$, not counting multiplicity; for instance, the top order monomials of $(0, r^2, r^2 + r, 2r^2, 2r^2 + r)$ are $\{r^2, 2r^2\}$. Define the *weight vector* of a tuple $(P_1, \ldots, P_k)$ relative to one of its members P_i to be the infinite vector $(w_1, w_2, \ldots) \in \mathbf{Z}_{\geq 0}^{\mathbf{N}}$, where each w_d denotes the number of monomials of degree d in the top order monomials of $(P_1 - P_i, \ldots, P_k - P_i)$. Thus, for instance, the tuple $(0, r^2, r^2 + r, 2r^2, 2r^2 + r)$ has weight vector $(0, 2, 0, \ldots)$ with respect to 0, but weight vector $(1, 2, 0, \ldots)$ with respect to (say) r^2. Let us say that one weight vector $(w_1, w_2, \ldots)$ is larger than another $(w_1', w_2', \ldots)$ if there exists $d \geq 1$ such that $w_d > w_d'$ and $w_i = w_i'$ for all $i > d$.

(1) Show that the space of all weight vectors is a well-ordered set.

(2) Show that if $(0, P_1, \ldots, P_k)$ is a tuple with $k \geq 1$ and P_1 non-linear, and $h_1, \ldots, h_m$ are integers with $m \geq 1$, then the weight vector of $(P_i(\cdot + h_j) - P_i(h_j))_{1 \leq i \leq k; 1 \leq j \leq m}$ with respect to $P_1(\cdot + h_1)$ is strictly smaller than the weight vector of $(0, P_1, \ldots, P_k)$ with respect to P_1.

(3) Using the previous two claims, Lemma 2.5.5, and Lemma 2.5.6, deduce Theorem 2.5.4.

Exercise 2.5.4. Find a direct proof of Theorem 2.5.3 analogous to the "epsilon and delta" proof of Theorem 2.5.8 from the previous lecture. (You can look up [**BeLe1996**] if you are stuck.)

Exercise 2.5.5. Let $P_1, \ldots, P_k : \mathbf{Z} \to \mathbf{Z}^d$ be vector-valued polynomials (thus each of the d components of each of the P_i is a polynomial) which all vanish at the origin. Show that if $\mathbf{Z}^d$ is finitely coloured, then one of the colour classes contains a pattern of the form $n + P_1(r), \ldots, n + P_k(r)$ for some $n \in \mathbf{Z}^d$ and $r \in \mathbf{N}$.

Exercise 2.5.6. Show that for any polynomial sequence $P : \mathbf{Z} \to (\mathbf{R}/\mathbf{Z})^d$ taking values in a torus, there exist integers $n_j \to \infty$ such that $P(n_j)$ converges to $P(0)$. (One can also tweak the argument to make the n_j converge to positive infinity, by the "doubling up" trick of replacing $P(n)$ with $(P(n), P(-n))$.) On the other hand, show that this claim can fail with exponential sequences such as $P(n) := 10^n \alpha \bmod 1 \in \mathbf{R}/\mathbf{Z}$ for certain values of α. Thus we see that polynomials have better recurrence properties than exponentials.

2.5.2. Ramsey's theorem. Given any finite palette K of colours, a vertex set V, and an integer $k \geq 1$, define a K-*coloured hypergraph* $G = (V, E)$ of order k on V to be a function $E : \binom{V}{k} \to K$, where $\binom{V}{k} := \{e \subset V : |e| = k\}$ denotes the k-element subsets of V. Thus for instance a hypergraph of order 1 is a vertex colouring, a hypergraph of order 2 is an edge-coloured complete graph, and so forth. We say that a hypergraph G is *monochromatic* if the edge colouring function E is constant. If W is a subset of V, we refer to the hypergraph $G \restriction_W := (W, E \restriction_{\binom{W}{k}})$ as a *subhypergraph* of G.

We will now prove the following result:

Theorem 2.5.8 (Hypergraph Ramsey theorem). *Let K be a finite set, let $k \geq 1$, and let $G = (V, E)$ be a K-coloured hypergraph of order k on a countably infinite vertex set V. Then G contains arbitrarily large finite monochromatic subhypergraphs.*

Remark 2.5.9. There is a stronger statement known, namely that G contains an infinitely large monochromatic subhypergraph, but we will not prove this statement, known as the *infinite hypergraph Ramsey theorem*.

In the case $k = 1$, these statements are the pigeonhole principle and infinite pigeonhole principle, respectively, and are compared in Section 1.3 of *Structure and Randomness*.

Exercise 2.5.7. Show that Theorem 2.5.8 implies a finitary analogue: given any finite K and positive integers k, m, there exists N such that every K-coloured hyeprgraph of order k on $\{1, \ldots, N\}$ contains a monochromatic subhypergraph on m vertices. *Hint*: As in Exercise 2.5.4, one should use a compactness and contradiction argument (as in Section 1.3 of *Structure and Randomness*).

It is not immediately obvious, but Theorem 2.5.8 is a statement about a topological dynamical system, albeit one in which the underlying group is not the integers $\mathbf{Z}$, but rather the symmetric group $\mathrm{Sym}_0(V)$, defined as the group of bijections from V to itself that are the identity outside of a finite set. More precisely, we have

Theorem 2.5.10 (Hypergraph Ramsey theorem, topological dynamics version). *Let V be a countably infinite set, and let W be a finite subset of V, thus $\mathrm{Sym}_0(W) \times \mathrm{Sym}_0(V \backslash W)$ is a subgroup of $\mathrm{Sym}_0(V)$. Let $(X, \mathcal{F}, T)$ be a $\mathrm{Sym}_0(V)$-topological dynamical system; thus $(X, \mathcal{F})$ is compact metrisable and $T : \sigma \mapsto T^\sigma$ is an action of $\mathrm{Sym}_0(V)$ on X via homeomorphisms. Let $(U_\alpha)_{\alpha \in A}$ be an open cover of X such that each U_α is $\mathrm{Sym}_0(W) \times \mathrm{Sym}_0(V \backslash W)$-invariant. Then there exists an element U_α of this cover such that for every finite set $\Gamma \subset \mathrm{Sym}_0(V)$ there exists a group element $\sigma \in \mathrm{Sym}_0(V)$ such that $\bigcap_{\gamma \in \Gamma} (T^{\gamma \sigma})^{-1}(U_\alpha) \neq \emptyset$ (i.e., there exists $x \in X$ such that $T^{\gamma \sigma} x \in U_\alpha$ for all $\gamma \in \Gamma$).*

This claim should be compared with Theorem 2.5.3 or Theorem 2.4.1.

Exercise 2.5.8. Show that Theorem 2.5.8 and Theorem 2.5.10 are equivalent. *Hint*: At some point, you will need to use the fact that the quotient space $\mathrm{Sym}_0(V)/(\mathrm{Sym}_0(W) \times \mathrm{Sym}_0(V \backslash W))$ is isomorphic to $\binom{V}{|W|}$.

As before, though, we shall only illustrate the ultrafilter approach to Ramsey's theorem, leaving the other approaches to exercises. Here, we will not work on the compactified integers $\beta \mathbf{Z}$, but rather on the compactified[25] permutations $\beta \mathrm{Sym}_0(V)$. This is a semigroup with the usual multiplication law

$$(2.38) \qquad pq := \lim_{\sigma \to p} \lim_{\rho \to q} \sigma \rho.$$

---

[25] We will view $\mathrm{Sym}_0(V)$ here as a discrete group; one could also give this group the topology inherited from the product topology on V^V, leading to a slightly coarser (and thus less powerful) compactification, which is still sufficient for the arguments here.

We say that $p \in \beta\mathrm{Sym}_0(V)$ is *minimal* if $\beta\mathrm{Sym}_0(V)p$ is a minimal left-ideal of $\beta\mathrm{Sym}_0(V)$. One can show (by repeating Exercise 2.3.11) that every left ideal $\beta\mathrm{Sym}_0(V)p$ contains at least one minimal element; in particular, minimal elements exist.

Note that if W is a k-element subset of V, then there is an image map $\pi_W : \mathrm{Sym}_0(V) \to \binom{V}{k}$ which maps a permutation σ to its inverse image $\sigma^{-1}(W)$ of W. We can compactify this to a map[26] $\beta\pi_W : \beta\mathrm{Sym}_0(V) \to \beta\binom{V}{k}$. We can now formulate the ultrafilter version of Ramsey's theorem.

Theorem 2.5.11 (Hypergraph Ramsey theorem, ultrafilter version). *Let V be countably infinite, and let $p \in \beta\mathrm{Sym}_0(V)$ be minimal. Then for every finite set W, $\beta\pi_W$ is constant on $\beta\mathrm{Sym}_0(V)p$, thus $\beta\pi_W(qp) = \beta\pi_W(p)$ for all $q \in \beta\mathrm{Sym}_0(V)$.*

This result should be compared with Proposition 2.4.11 (or Theorem 2.5.4).

Exercise 2.5.9. Show that Theorem 2.5.11 implies both Theorem 2.5.8 and Theorem 2.5.10.

Proof of Theorem 2.5.11. By relabeling we may assume that $V = \{1, 2, 3, \ldots\}$ and $W = \{1, \ldots, k\}$ for some k.

Given any integers $1 \le a < i_1 < i_2 < \ldots < i_a$, let $\sigma_{i_1,\ldots,i_a} \in \mathrm{Sym}_0(V)$ denote the permutation that swaps j with i_j for all $1 \le j \le a$, but leaves all other integers unchanged. We select some non-principal ultrafilter $p_* := \beta V \backslash V$ and define the sequence $p_1, p_2, \ldots \in \beta\mathrm{Sym}_0(V)$ by the formula

$$(2.39) \qquad p_a := \lim_{i_1 \to p_*} \ldots \lim_{i_a \to p_*} \sigma_{i_1,\ldots,i_a} p.$$

(Note that the condition $a < i_1 < \ldots < i_a$ will be asymptotically true thanks to the choice of limits here.)

Let $a \ge k$, and let $\alpha \in \mathrm{Sym}_0(V)$ be a permutation which is the identity outside of $\{1, \ldots, a\}$. Then for every $\rho \in \mathrm{Sym}_0(V)$, we have the identity

$$(2.40) \qquad \pi_W(\alpha\sigma_{i_1,\ldots,i_a}\rho) = \pi_W(\sigma_{i_{j_1},\ldots,i_{j_k}}\rho),$$

where $j_1 < \ldots < j_k$ are the elements of $\alpha^{-1}(\{1, \ldots, k\})$ in order. Taking limits as $\rho \to p$, and then inserting the resulting formula into (2.39), we conclude (after discarding the trivial limits and relabeling the rest) that

$$(2.41) \qquad \beta\pi_W(\alpha p_a) = \lim_{i_1 \to p_*} \ldots \lim_{i_k \to p_*} \beta\pi_W(\sigma_{i_1,\ldots,i_k}p),$$

and in particular that $\beta\pi_W(\alpha p_a)$ is independent of α (if α is the identity outside of $\{1, \ldots, j\}$. Now let $p' := \lim_{a \to p_*} p_a$; then we have $\beta\pi_W(\alpha p')$ independent of p' for all $\alpha \in \mathrm{Sym}_0(V)$. Taking limits we conclude that $\beta\pi_W$

---

[26]Caution: $\beta\binom{V}{k}$ is *not* the same thing as $\binom{\beta V}{k}$; for instance the latter is not even compact.

is constant on $(\beta\mathrm{Sym}_0(V))p'$. But from construction we see that p' lies in the closed minimal ideal $(\beta\mathrm{Sym}_0(V))p$, thus $(\beta\mathrm{Sym}_0(V))p' = (\beta\mathrm{Sym}_0(V))p$. The claim follows. □

Exercise 2.5.10. Establish Theorem 2.5.8 directly by a combinatorial argument without recourse to topological dynamics or ultrafilters. (If you are stuck, I recommend reading the classic text [**GrRoSp1980**].)

Exercise 2.5.11. Establish Theorem 2.5.10 directly by a topological dynamics argument, using combinatorial arguments for the $k = 1$ case but then proceeding by induction afterwards (as in the proof of Theorem 2.4.5).

Remark 2.5.12. More generally, one can interpret the theory of graphs and hypergraphs on a vertex set V through the lens of dynamics of $\mathrm{Sym}_0(V)$ actions; I learned this perspective from Balazs Szegedy.

2.5.3. Idempotent ultrafilters and Hindman's theorem. Thus far, we have been using ultrafilter technology rather lightly, and indeed all of the arguments so far can be converted relatively easily to the topological dynamics formalism, or even a purely combinatorial formalism, with only a moderate amount of effort. But now we will exploit some deeper properties of ultrafilters, which are more difficult to replicate in other settings. In particular, we introduce the notion of an *idempotent* ultrafilter.

Definition 2.5.13 (Idempotent). Let $(S, \cdot)$ be a discrete semigroup, and let βS be given the usual semigroup operation $\cdot$. An element $p \in \beta S$ is *idempotent* if $p \cdot p = p$. (We of course define idempotence analogously if the group operation on S is denoted by $+$ instead of $\cdot$.)

Clearly, 0 is idempotent, but the remarkable fact is that many other idempotents exist as well. The key tool for creating them is

Lemma 2.5.14 (Ellis-Nakamura lemma [**El1958**]). *Let S be a discrete semigroup, and let K be a compact non-empty subsemigroup of βS. Then K contains at least one idempotent.*

Proof. A simple application of Zorn's lemma shows that K contains a compact non-empty subsemigroup K' which is minimal with respect to set inclusion. We claim that every element of K' is idempotent. To see this, let p be an arbitrary element of K'. Then observe that $K'p$ is a compact non-empty subsemigroup of K' and must therefore be equal to K'; in particular, $p \in K'p$. (Note that semigroups need not contain an identity.) In particular, the stabiliser $K'' := \{q \in K' : qp = p\}$ is non-empty. But one easily observes that this stabiliser is also a compact subsemigroup of K', and so $K'' = K'$. In particular, p must stabilise itself, i.e., it is idempotent. □

Remark 2.5.15. *A posteriori*, this result shows that the minimal non-empty subsemigroups K' are in fact just the singleton sets consisting of idempotents. But one cannot really see this without first deriving all of Lemma 2.5.14.

Idempotence turns out to be particularly powerful when combined with minimality, and to this end we observe the following corollary of the above lemma.

Corollary 2.5.16. *Let S be a discrete semigroup. For every $p \in \beta S$, there exists $q \in (\beta S)p$ which is both minimal and idempotent.*

Proof. By Exercise 2.3.11, there exists $r \in (\beta S)p$ which is minimal. It is then easy to see that every element of $(\beta S)r$ is minimal. Since $(\beta S)r \subset (\beta S)p$ is a compact non-empty subsemigroup of βS, the claim now follows from Lemma 2.5.14. $\square$

Remark 2.5.17. Somewhat amusingly, minimal idempotent ultrafilters require *three* distinct applications of Zorn's lemma to construct: one to define the compactified space βS, one to locate a minimal left ideal, and one to locate an idempotent inside that ideal! It seems particularly challenging therefore to define civilised substitutes for this tool which do not explicitly use the axiom of choice.

What can we do with minimal idempotent ultrafilters? One particularly striking example is *Hindman's theorem* [**Hi1974**]. Given any set A of positive integers, define $FS(A)$ to be the set of all finite sums $\sum_{n \in B} n$ from A, where B ranges over all finite non-empty subsets of A. (For instance, if $A = \{1, 2, 4, \ldots\}$ are the powers of 2, then $FS(A) = \mathbf{N}$.)

Theorem 2.5.18 (Hindman's theorem). *Suppose that the natural numbers $\mathbf{N}$ are finitely coloured. Then one of the colour classes contains a set of the form $FS(A)$ for some infinite set A.*

Remark 2.5.19. Theorem 2.5.18 implies *Folkman's theorem* [**Fo1970**], which has the same hypothesis but concludes that one of the colour classes contains sets of the form $FS(A)$ for arbitrarily large but finite sets A. In the converse direction, it does not seem possible to easily deduce Hindman's theorem from Folkman's theorem.

Exercise 2.5.12. Folkman's theorem in turn implies *Schur's theorem* [**Sc1916**], which asserts that if the natural numbers are finitely coloured, one of the colour classes contains a set of the form $FS(\{x, y\}) = \{x, y, x+y\}$ for some x, y (compare with the $k = 3$ case of van der Waerden's theorem). Using the Cayley graph construction, deduce Schur's theorem from Ramsey's theorem (the $k = 2$ case of Theorem 2.5.8). Thus we see that there are some connections between the various Ramsey-type theorems discussed here.

Proof of Theorem 2.5.18. By Corollary 2.5.16, we can find a minimal idempotent element p in $\beta\mathbf{N}$; note that as no element of $\mathbf{N}$ is minimal (cf. Exercise 2.3.8), we know that $p \notin \mathbf{N}$. Let $c : \mathbf{N} \to \{1,\ldots,m\}$ denote the given colouring function; then $\beta c(p)$ is a colour in $\{1,\ldots,m\}$. Since

$$(2.42) \qquad \lim_{n\to p} \beta c(n) = \beta c(p)$$

and

$$(2.43) \qquad \lim_{n\to p} \beta c(n+p) = \beta c(p+p) = \beta c(p),$$

we may find a positive integer n_1 such that $\beta c(n_1) = \beta c(n_1 + p) = \beta c(p)$. Now from (2.42), (2.43) and the similar calculations

$$(2.44) \qquad \lim_{n\to p} \beta c(n_1 + n) = \beta c(n_1 + p) = \beta c(p)$$

and

$$(2.45) \qquad \lim_{n\to p} \beta c(n_1 + n + p) = \beta c(n_1 + p + p) = \beta c(n_1 + p) = \beta c(p)$$

we can find an integer $n_2 > n_1$ such that $\beta c(n_2) = \beta c(n_2+p) = \beta c(n_2+n_1) = \beta c(n_2 + n_1 + p) = \beta c(p)$, thus $\beta c(m) = \beta c(m + p) = \beta c(p)$ for all $m \in FS(\{n_1, n_2\})$. Continuing inductively in this fashion, one can find $n_1 < n_2 < n_3 < \ldots$ such that $\beta c(m) = \beta c(m+p) = \beta c(p)$ for all $m \in FS(\{n_1,\ldots,n_k\})$ and all k. If we set $A := \{n_1, n_2, \ldots\}$, the claim follows. □

Remark 2.5.20. Purely combinatorial (and quite succinct) proofs of Hindman's theorem exist—see for instance the one in [**GrRoSp1980**]—but they generally rely on some *ad hoc* trickery. Here, the trickery has been encapsulated into the existence of minimal idempotent ultrafilters, which can be reused in other contexts (for instance, we will use it to prove the Hales-Jewett theorem below).

Exercise 2.5.13. Define an *IP-set* to be a set of positive integers which contains a subset of the form $FS(A)$ for some infinite A. Show that if an IP-set S is finitely coloured, then one of its colour classes is also an IP-set. *Hint*: S contains $FS(A)$ for some infinite $A = \{a_1, a_2, a_3, \ldots\}$. Show that the set $\bigcap_{n=1}^{\infty} \beta FS(\{a_n, a_{n+1}, \ldots\})$ is a compact non-empty semigroup and thus contains a minimal idempotent ultrafilter p. Use this p to repeat the proof of Theorem 2.5.18.

2.5.4. The Hales-Jewett theorem. Given a finite alphabet A, let $A^{<\omega}$ be the free semigroup generated by A, i.e., the set of all finite non-empty words using the alphabet A, with concatenation as the group operation. (E.g., if $A = \{a, b, c\}$, then $A^{<\omega}$ contains words such as abc and cbb, with $abc \cdot cbb = abccbb$.) If we add another letter $*$ to A (the "wildcard" letter), we create a larger semigroup $(A \cup \{*\})^{<\omega}$ (e.g., containing words such as $ab**c*$). We of course assume that $*$ was not already present in A. Given any letter $x \in A$, we have a semigroup homomorphism $\pi_x : (A \cup \{*\})^{<\omega} \to A^{<\omega}$ which substitutes every occurrence of the wildcard $*$ with x and leaves all other letters unchanged. (For instance, $\pi_a(ab**c*) = abaaca$.) Define a *combinatorial line* in $A^{<\omega}$ to be any set of the form $\{\pi_x(v) : x \in A\}$ for some $v \in (A\cup\{*\})^{<\omega}\backslash A^{<\omega}$. For instance, if $A = \{a, b, c\}$, then $\{abaaca, abbbcb, abcccc\}$ is a combinatorial line, generated by the word $v = ab**c*$.

We shall prove the following fundamental theorem [**HaJe1963**].

Theorem 2.5.21 (Hales-Jewett theorem). *Let A be a finite alphabet. If $A^{<\omega}$ is finitely coloured, then one of the colour classes contains a combinatorial line.*

Exercise 2.5.14. Show that the Hales-Jewett theorem has the following equivalent formulation: For every finite alphabet A and any $m \geq 1$ there exists N such that if A^N is partitioned into m classes, then one of the classes contains a combinatorial line.

Exercise 2.5.15. Assume the Hales-Jewett theorem. In this exercise we compare the strength of this theorem against other Ramsey-type theorems.

(1) Deduce van der Waerden's theorem (Theorem 2.4.2). *Hint*: The base k representation of the non-negative natural numbers provides a map from $\{0, \ldots, k-1\}^{<\omega}$ to $\mathbf{Z}_{\geq 0}$.

(2) Deduce the multidimensional van der Waerden's theorem of Gallai (Exercise 2.4.8).

(3) Deduce the *syndetic van der Waerden theorem* of Furstenberg [**Fu1977**]: If the integers are finitely coloured and k is a positive integer, then there are infinitely many monochromatic arithmetic progressions $n, n + r, \ldots, n + (k-1)r$ of length k, and furthermore the set of all the step sizes r which appear in such progressions is syndetic (i.e., it has bounded caps). *Hint*: Argue by contradiction, assuming that the set of all step sizes has arbitrarily long gaps, and use the Hales-Jewett theorem in a manner adapted to these gaps. (For an additional challenge, show that there exists a *single* colour class whose progressions of length k have spacings in a syndetic set for every k.)

(4) Deduce the *IP-van der Waerden theorem*: If the integers are finitely coloured, k is a positive integer, and S is an IP-set (see Exercise 2.5.13), show that there are infinitely many monochromatic arithmetic progressions whose step size lies in S. (For an additional challenge, show that one of the classes has the property that for every k, the spacings of the k-term progressions in that class forms an IP^*-set, i.e., it has non-empty intersection with every IP-set. There is an even stronger topological dynamics version of this statement, due to Furstenberg and Weiss [**FuWe1978**], which I will not describe here.)

(5) For any $d \geq 1$, define a *d-dimensional combinatorial subspace* of $A^{<\omega}$ to be a set of the form $\{\pi_{x_1,\ldots,x_d}(v) : x_1,\ldots,x_d \in A\}$, where $v \in (A \cup \{*_1,\ldots,*_d\})^{<\omega}$ is a word containing at least one copy of each of the d wildcards $*_1,\ldots,*_d$, and $\pi_{x_1,\ldots,x_d} : (A\cup\{*_1,\ldots,*_d\})^{<\omega} \to A^{<\omega}$ is the homomorphism that substitutes each wildcard $*_j$ with x_j. Show that if $A^{<\omega}$ is finitely coloured, then one of the colour classes contains arbitrarily high-dimensional combinatorial subspaces.

(6) Let F be a finite field. If the vector space $\lim_{n\to\infty} F^n$ (the inverse limit of the finite vector spaces F^n) is finitely coloured, show that one of the colour classes contains arbitrarily high-dimensional affine subspaces over F. (This *geometric Ramsey theorem* is due to [**GrLeRo1972**].)

We now give an ultrafilter-based proof of the Hales-Jewett theorem due to Blass [**Bl1993**]. As usual, the first step is to obtain a statement involving ultrafilters rather than colourings:

Proposition 2.5.22 (Hales-Jewett theorem, ultrafilter version). *Let A be a finite alphabet, and let p be a minimal idempotent element of the semigroup $\beta(A^{<\omega})$. Then there exists $q \in \beta(A \cup \{*\})^{<\omega})\backslash\beta(A^{<\omega})$ such that $\beta\pi_x(q) = p$ for all $x \in A$.*

Exercise 2.5.16. Deduce Theorem 2.5.21 from Proposition 2.5.22.

To prove Proposition 2.5.22, we need a variant of Corollary 2.5.16. Let $(S, \cdot)$ be a discrete semigroup, and let p and q be two idempotents in βS. We write $p \prec q$ if $pq = qp = p$.

Exercise 2.5.17. Show that $\prec$ is a partial ordering on the idempotents of βS, and that an idempotent is minimal in βS if and only if it is minimal with respect to $\prec$.

Lemma 2.5.23. *Let S be a discrete semigroup, and let p be an idempotent in βS. Then there exists a minimal idempotent q in βS such that $q \prec p$.*

Proof. By Exercise 2.3.11 (generalised to arbitrary discrete semigroups S), $(\beta S)p$ contains a minimal left ideal $(\beta S)r$. By Lemma 2.5.14, $(\beta S)r$ contains an idempotent s. Since $s \in (\beta S)p$ and p is idempotent, we conclude that $sp = s$. If we then set $q := ps$, we easily check that q is idempotent, that $q \prec p$, and (since q lies in the minimal left ideal $(\beta S)r$) it is minimal. The claim follows.   □

Proof of Proposition 2.5.22. Since p is an idempotent element of $\beta(A^{<\omega})$, it is also an idempotent element of $\beta(A \cup \{*\})^{<\omega}$. It need not be minimal in that semigroup, though. However, by Lemma 2.5.23, we can find a minimal idempotent q in $\beta(A \cup \{*\})^{<\omega}$ such that $q \prec p$.

Now let $x \in A$. Since $\pi_x : (A \cup \{*\})^{<\omega} \to A^{<\omega}$ is a homomorphism, $\beta\pi_x : \beta(A \cup \{*\})^{<\omega} \to \beta A^{<\omega}$ is also a homomorphism (why?). Since q is idempotent and $q \prec p$ (note that these are both purely *algebraic* statements), we conclude that $\beta\pi_x(q)$ is idempotent and $\beta\pi_x(q) \prec \beta\pi_x(p)$. But $\beta\pi_x(p) = p$ is minimal in $\beta A^{<\omega}$, hence by Exercise 2.5.17, we have $\beta\pi_x(q) = p$. The claim follows.   □

Exercise 2.5.18. Adapt the above proof to give an alternative proof of the ultrafilter version of van der Waerden's theorem (Proposition 2.5.22) which relies on idempotence rather than on induction on k. (If you are stuck, read the proof of [**Gl2003**, Proposition 1.55].)

Remark 2.5.24. Several of the above Ramsey-type theorems can be unified. For instance, the polynomial van der Waerden theorem and the Hales-Jewett theorem have been unified into the polynomial Hales-Jewett theorem of Bergelson and Leibman [**BeLe1999**] (see also [**Wa2000**]). This type of Ramsey theory is still an active subject, and we do not yet have a comprehensive and systematic theory (or a "universal" Ramsey theorem) that encompasses all known examples.

Exercise 2.5.19. Let X be an at most countable set (with the discrete topology), and let $\mathcal{F}$ be a family of subsets of X. Show that the following two statements are equivalent:

(1) Whenever X is finitely coloured, one of the colour classes contains a subset in $\mathcal{F}$.

(2) There exists $p \in \beta X$ such that every neighbourhood of p contains a subset in $\mathcal{F}$.

Exercise 2.5.20. Let X, Y be at most countable sets with the discrete topology, and let $f_1, \ldots, f_k : Y \to X$ be a finite collection of functions. Show that the following two statements are equivalent:

(1) Whenever X is finitely coloured, one of the colour classes contains a set $\{f_1(y), \ldots, f_k(y)\}$ for some $y \in Y$.

(2) There exists $q \in \beta Y$ such that $\beta f_1(q) = \ldots = \beta f_k(q)$.

Hint: Look at the closure of $\{(f_1(y), \ldots, f_k(y)) : y \in Y\}$ in $(\beta X)^k$.

Exercise 2.5.21. Using the previous exercise, deduce Theorem 2.5.4 from Theorem 2.5.1, and deduce Theorem 2.5.11 from Theorem 2.5.8.

Notes. This lecture first appeared at

terrytao.wordpress.com/2008/01/21.

Thanks to Yury, Liu Xiao Chuan, Nilay, and an anonymous commenter for corrections.

2.6. Isometric systems and isometric extensions

In this lecture, we move away from recurrence, and instead focus on the *structure* of topological dynamical systems. One remarkable feature of this subject is that starting from fairly "soft" notions of structure, such as topological structure, one can extract much more "hard" or "rigid" notions of structure, such as *geometric* or *algebraic* structure. The key concept needed to capture this structure is that of an *isometric system*, or more generally an *isometric extension*, which we shall discuss in this lecture. As an application of this theory we characterise the distribution of polynomial sequences in tori (a baby case of a variant of Ratner's theorem due to [**Gr1961**], which we will cover in Section 2.16).

2.6.1. Isometric systems. We begin with a key definition.

Definition 2.6.1 (Equicontinuous and isometric systems). Let $(X, \mathcal{F}, T)$ be a topological dynamical system.

(1) We say that the system is *isometric* if there exists a metric d on X such that the shift maps $T^n : X \to X$ are all isometries, thus $d(T^n x, T^n y) = d(x, y)$ for all n and all x, y. (Of course, once T is an isometry, all powers T^n are automatically isometries also, so it suffices to check the $n = 1$ case.)

(2) We say that the system is *equicontinuous* if there exists a metric d on X such that the shift maps $T^n : X \to X$ form a uniformly equicontinuous family, thus for every $\varepsilon > 0$ there exists $\delta > 0$ such that $d(T^n x, T^n y) \le \varepsilon$ whenever n, x, y are such that $d(x, y) \le \delta$. (As X is compact, equicontinuity and uniform equicontinuity are equivalent concepts.)

Example 2.6.2. The circle shift $x \mapsto x + \alpha$ on $\mathbf{R}/\mathbf{Z}$ is both isometric and equicontinuous. On the other hand, the Bernoulli shift on $\{0, 1\}^{\mathbf{Z}}$ is neither isometric nor equicontinuous (why?).

Example 2.6.3. Any finite dynamical system is both isometric and equicontinuous (as one can see by using the discrete metric).

Since all metrics are essentially equivalent on compact spaces, we see that the choice of metric is not actually important when checking equicontinuity, but it seems to be more important when checking for isometry. Nevertheless, there is actually no distinction between the two properties:

Exercise 2.6.1. Show that a topological dynamical system is isometric if and only if it is equicontinuous. *Hint*: One direction is obvious. For the other, if T^n is a uniformly equicontinuous family with respect to a metric d, consider the modified metric $\tilde{d}(x,y) := \sup_n d(T^n x, T^n y)$.

Remark 2.6.4. From this exercise we see that we can upgrade *topological* structure (equicontinuity) to *geometric* structure (isometry). The motif of studying topology through geometry pervades modern topology; witness for instance Perelman's proof of the Poincaré conjecture (Chapter 2 of Part II).

Exercise 2.6.2 (Ultrafilter characterisation of equicontinuity). Consider a topological dynamical system $(X, \mathcal{F}, T)$. Show that X is equicontinuous if and only if the maps $T^p : X \to X$ are homeomorphisms for every $p \in \beta\mathbf{Z}$.

Now we upgrade the geometric structure of isometry to the *algebraic* structure of being a compact abelian group action.

Definition 2.6.5 (Kronecker system). A topological dynamical system $(X, \mathcal{F}, T)$ is said to be a *Kronecker system* if it is isomorphic to a system of the form $(K, \mathcal{K}, S)$, where $(K, +, \mathcal{K})$ is a compact abelian metrisable topological group,[27] and $S : x \mapsto x + \alpha$ is a group rotation for some $\alpha \in K$.

Example 2.6.6. The circle rotation system is a Kronecker system, as is the standard shift $x \mapsto x+1$ on a cyclic group $\mathbf{Z}/N\mathbf{Z}$. Any product of Kronecker systems is again a Kronecker system.

Let us first observe that a Kronecker system is equicontinuous (and hence isometric). Indeed, the compactness of the topological group K (and the joint continuity of the addition law $+ : K \times K \to K$) easily ensures that the group rotations $g : x \mapsto x + g$ are uniformly equicontinuous as $g \in K$ varies. Since the shifts $T^n : x \mapsto x + n\alpha$ are all group rotations, the claim follows.

On the other hand, not every equicontinuous or isometric system is Kronecker. Consider for instance a finite dynamical system which is the disjoint union of two cyclic shifts of distinct order; it is not hard to see that this is not a Kronecker system. Nevertheless, it clearly contains Kronecker systems within it. Indeed, we have

---

[27] A *topological group* is a group with a topology such that the group operations $x \mapsto x^{-1}$ and $(x, y) \mapsto xy$ are continuous.

Proposition 2.6.7. *Every* minimal *equicontinuous (or isometric) system* $(X, \mathcal{F}, T)$ *is a Kronecker system, i.e., isomorphic to an abelian group rotation* $(K, \mathcal{K}, x \mapsto x + \alpha)$. *Furthermore, the orbit* $\{n\alpha : n \in \mathbf{Z}\}$ *is dense in* K.

Proof. By Exercise 2.6.1, we may assume that the system is isometric; thus we can find a metric d such that all the shift maps T^n are isometries. We view the T^n as lying inside the space $C(X \to X)$ of continuous maps from X to itself, endowed with the uniform topology. Let $G \subset C(X \to X)$ be the closure of the maps $\{T^n : n \in \mathbf{Z}\}$. One easily verifies that G is a closed metrisable topological group of isometries in $C(X \to X)$; from the Arzelà-Ascoli theorem we see that G is compact. Also, since T^n and T^m commute for every n and m, we see upon taking limits that G is abelian.

Now let $x \in X$ be an arbitrary point. Then we see that the image $\{f(x) : f \in G\}$ of G under the evaluation map $f \mapsto f(x)$ is a compact non-empty invariant subset of X, and thus equal to all of X by minimality. If we then define the stabiliser $\Gamma := \{f \in G : f(x) = x\}$, we see that Γ is a closed (hence compact) subgroup of the abelian group G. Since $X = \{f(x) : f \in G\}$, we thus see that there is a continuous bijection $f\Gamma \mapsto f(x)$ from the quotient group $K := G/\Gamma$ (with the quotient topology) to X. Since both spaces here are compact Hausdorff, this map is a homeomorphism. This map is thus an isomorphism of topological dynamical systems between the Kronecker system K (with the group rotation given by $\alpha := T \mod \Gamma \in G/\Gamma$) and X. Since K is a compact metrisable (thanks to Hausdorff distance) topological group, the claim follows (relabeling the group operation as $+$). Note that the density of $\{n\alpha : n \in \mathbf{Z}\}$ in K is clear from construction. $\square$

Remark 2.6.8. Once one knows that X is homeomorphic to a Kronecker system with $\{n\alpha : n \in \mathbf{Z}\}$ dense, one can *a posteriori* return to the proof and conclude that the stabiliser Γ is trivial. But I do not see a way to establish that fact directly. In any case, when we move to isometric extensions below, the analogue of the stabiliser Γ can certainly be non-trivial.

To get from minimal isometric systems to non-minimal isometric systems, we can use

Proposition 2.6.9. *Any isometric system* $(X, \mathcal{F}, T)$ *can be partitioned into the union of disjoint minimal isometric systems.*

Proof. Since minimal systems are automatically disjoint, it suffices to show that every point $x \in X$ is contained in a minimal dynamical system, or equivalently that the orbit closure $\overline{T^{\mathbf{Z}} x}$ is minimal. If this is not the case, then there exists $y \in \overline{T^{\mathbf{Z}} x}$ such that x does not lie in the orbit closure of y. But by definition of orbit closure, we can find a sequence n_j such that $T^{n_j} x$

converges to y. By the isometry property, this implies that $T^{-n_j}y$ converges to x, and so x is indeed in the orbit closure of y, a contradiction. $\square$

Thus every equicontinuous or isometric system can be expressed as a union of disjoint Kronecker systems.

We can use the algebraic structure of isometric systems to obtain much quicker (and slightly stronger) proofs of various recurrence theorems. For instance, we can give a short proof of (a slight strengthening of) the multiple Birkhoff recurrence theorem (Theorem 2.4.3) as follows.

Proposition 2.6.10 (Multiple Birkhoff's theorem for isometric systems). *Let $(X, \mathcal{F}, T)$ be an isometric system. Then for every $x \in X$ there exists a sequence $n_j \to \infty$ such that $T^{kn_j}x \to x$ for every integer k.*

Proof. By Proposition 2.6.9 followed by Proposition 2.6.7, it suffices to check this for Kronecker systems $(K, \mathcal{K}, x \mapsto x + \alpha)$ in which $\{n\alpha : n \in \mathbf{Z}\}$ is dense in K. But then we can find a sequence n_j such that $n_j\alpha \to 0$ in K, and thus (since K is a topological group) $kn_j\alpha \to 0$ in K for all k. The claim follows. $\square$

The above argument illustrates one of the reasons why it is desirable to have an algebraic structural theory of various types of dynamical systems; it makes it much easier to answer many interesting questions regarding systems of this kind, such as those involving recurrence.

2.6.2. The Kronecker factor. We have seen that isometric systems are basically Kronecker systems (or unions thereof). Of course, not all systems are isometric. However, it turns out that every system contains a maximal isometric *factor*. Recall that a factor of a topological dynamical system $(X, \mathcal{F}, T)$ is a surjective morphism $\pi : X \to Y$ from X to another topological dynamical system $(Y, \mathcal{G}, S)$. (We shall sometimes abuse notation and refer to $\pi : X \to Y$ as the factor, when it is really the quadruplet $(\pi, Y, \mathcal{G}, S)$.) We say that one factor $\pi : X \to Y$ *refines* (or is *finer than*) another factor $\pi' : X \to Y'$ if we can factorise $\pi' = f \circ \pi$ for some continuous map $f : Y \to Y'$. (Note from surjectivity that this map, if it exists, is unique.) We say that two factors are *equivalent* if they refine each other. Observe that modulo equivalence, refinement is a partial ordering on factors.

Example 2.6.11. The identity factor id : $X \to X$ is finer than any other factor of X, which in turn is finer than the trivial factor pt : $X \to$ pt that maps to a point.

Exercise 2.6.3. Show that any factor of a minimal topological dynamical system is again minimal.

We note two useful operations on factors. Firstly, given two factors $\pi : X \to Y = Y$ and $\pi' : X \to Y'$, one can define their *join* $\pi \vee \pi' : X \to Y \vee Y'$, where $Y \vee Y' := \{(\pi(x), \pi'(x)) : x \in X\} \subset Y \times Y'$ is the compact subspace of the product system $Y \times Y'$, and $\pi \vee \pi' : X \to Y \vee Y'$ is the surjective morphism $\pi \vee \pi' : x \mapsto (\pi(x), \pi'(x))$. One can verify that $\pi \vee \pi'$ is the least common refinement of π and π', hence the name.

Secondly, given a chain $(\pi_\alpha)_{\alpha \in A}$ of factors $\pi_\alpha : X \to Y_\alpha$ (thus π_α refines π_β for all $\alpha > \beta$), one can form their *inverse limit* $\pi = \lim_{\leftarrow}(\pi_\alpha)_{\alpha \in A} : X \to Y = \lim_{\leftarrow}(Y_\alpha)_{\alpha \in A}$ by first letting $f_{\alpha\beta} : Y_\alpha \to Y_\beta$ be the factoring maps for all $\alpha > \beta$, observing that $f_{\beta\gamma} \circ f_{\alpha\beta} = f_{\alpha\gamma}$ for all $\alpha > \beta > \gamma$, and then defining $Y \subset \prod_\alpha Y_\alpha$ to be the compact subspace of the product system $\prod_\alpha Y_\alpha$ defined as

$$(2.46) \qquad Y := \{(y_\alpha)_{\alpha \in A} : f_{\alpha\beta}(y_\alpha) = y_\beta \text{ whenever } \alpha > \beta\},$$

and then setting $\pi : x \mapsto (\pi_\alpha(x))_{\alpha \in A}$. One easily verifies that π is indeed a factor of X, and it is the least upper bound of the π_α.

Next, we observe that these operations interact well with the isometry property:

Exercise 2.6.4. Let $\pi : X \to Y$ and $\pi' : X \to Y'$ be two factors such that Y and Y' are both isometric. Then $\pi \vee \pi' : X \to Y \vee Y'$ is also isometric.

Lemma 2.6.12. *Let $(\pi_\alpha)_{\alpha \in A}$ be a totally ordered set of factors $\pi_\alpha : X \to Y_\alpha$ with $Y_\alpha = (Y_\alpha, \mathcal{G}_\alpha, S_\alpha)$ isometric. Then the inverse limit $\pi : X \to Y$ of the π_α is such that Y is also isometric.*

Proof. Observe that we have factor maps $f_\alpha : Y \to Y_\alpha$ which are surjective morphisms, which themselves factor as $f_\beta = f_{\alpha\beta} \circ f_\alpha$ for $\alpha > \beta$ and some surjective morphisms $f_{\alpha\beta} : Y_\alpha \to Y_\beta$. Let us fix a metric d on Y. For each $\alpha \in A$, consider the compact subset $\Delta_\alpha := \{(y, y') \in Y \times Y : f_\alpha(y) = f_\alpha(y')\}$ of $Y \times Y$. These sets decrease as α increases, and their intersection is the diagonal $\{(y, y) : y \in Y\}$ (why?). Applying the *finite intersection property* in the compact sets $\{(y, y') \cap \Delta_\alpha : d(y, y') \geq \varepsilon\}$, we conclude that for every $\varepsilon > 0$ there exists α such that $d(y, y') < \varepsilon$ whenever $f_\alpha(y) = f_\alpha(y')$.

Now suppose for contradiction that Y is not isometric, and hence not uniformly equicontinuous. Then there are sequences $y_j, y'_j \in Y$ with $d(y_j, y'_j) \to 0$, an $\varepsilon > 0$, and a sequence n_j of integers such that $d(S^{n_j} y_j, S^{n_j} y'_j) > \varepsilon$. By compactness we may assume that y_j, y'_j both converge to the same point. But by the preceding discussion, we can find $\alpha \in A$ such that $d(y, y') < \varepsilon/4$ whenever $f_\alpha(y) = f_\alpha(y')$. In other words, for any z in Y_α, the fibre $f_\alpha^{-1}(\{z\})$ has diameter at most $\varepsilon/4$.

Now let $z_j := f_\alpha(y_j)$ and $z'_j := f_\alpha(y'_j)$. Then z_j and z'_j converge to the same point z in Y_α, and so by equicontinuity of Y_α, $d(S_\alpha^{n_j} z_j, S_\alpha^{n_j} z'_j)$ goes to zero. By compactness and passing to a subsequence we can assume that $S_\alpha^{n_j} z_j$ and $S_\alpha^{n_j} z_j'$ both converge to some point z_* in Y_α. On the other hand, from the preceding discussion and the triangle inequality, we see that the fibres $f_\alpha^{-1}(\{S_\alpha^{n_j} z_j\})$ and $f_\alpha^{-1}(\{S_\alpha^{n_j} z_j'\})$ are separated by a distance at least $\varepsilon/2$ in Y. On the other hand, the distance between $f_\alpha^{-1}(\{S_\alpha^{n_j} z_j\})$ and $f_\alpha^{-1}(\{z_*\})$ must go to zero as $j \to \infty$ (as a simple sequential compactness argument shows), and similarly the distance between $f_\alpha^{-1}(\{S_\alpha^{n_j} z_j'\})$ and $f_\alpha^{-1}(\{z_*\})$ goes to zero. Since $f_\alpha^{-1}(\{z_*\})$ has diameter at most $\varepsilon/4$, we obtain a contradiction. The claim follows. $\square$

Combining Exercise 2.6.2 and Lemma 2.6.12 with Zorn's lemma (and noting that with the trivial factor pt $: X \to$ pt, the image pt is clearly isometric) we obtain

Corollary 2.6.13 (Existence of maximal isometric factor). *For every topological dynamical system $(X, \mathcal{F}, T)$ there is a factor $\pi : X \to Y$ with Y isometric, which is maximal with respect to refinement among all such factors with this property. This factor is unique up to equivalence.*

By Proposition 2.6.7 and Exercise 2.6.3, the maximal isometric factor of a minimal system is a Kronecker system, and we refer to it as the *Kronecker factor* of that minimal system X.

Exercise 2.6.5 (Explicit description of Kronecker factor). Let $(X, \mathcal{F}, T)$ be a minimal topological dynamical system, and let $Q \subset X \times X$ be the set

$$(2.47) \qquad Q := \bigcap_V \overline{(T \times T)^{\mathbf{Z}}(V)},$$

where V ranges over all open neighbourhoods of the diagonal $\{(x, x) : x \in X\}$ of $X \times X$, and $T \times T : (x, y) \mapsto (Tx, Ty)$ is the product shift. Let $\sim$ be the finest equivalence relation on X such that the set $R_\sim := \{(x, y) \in X \times X : x \sim y\}$ is closed and contains Q. (The existence and uniqueness of $\sim$ can be established by intersecting $R_\sim$ over all candidates $\sim$ together.) Show that the projection map $\pi : X \to X/\sim$ to the equivalence classes of $\sim$ (with the quotient topology) is (up to isomorphism) the Kronecker factor of X.

The Kronecker factor is also closely related to the concept of an eigenfunction. We say that a continuous function $f : X \to \mathbf{C}$ is an *eigenfunction* of a topological dynamical system $(X, \mathcal{F}, T)$ if it is not identically zero and we have $Tf = \lambda f$ for some $\lambda \in \mathbf{C}$, which we refer to as an *eigenvalue* for T.

Exercise 2.6.6. Let $(X, \mathcal{F}, T)$ be a minimal topological dynamical system.

(1) Show that if λ is an eigenvalue for T, then λ lies in the unit circle $S^1 := \{z \in \mathbf{C} : |z| = 1\}$, and furthermore there exists a unimodular

eigenfunction $g : X \to S^1$ with this eigenvalue. *Hint*: The zero set of an eigenfunction is a closed shift-invariant subset of X.

(2) Show that for every eigenvalue λ, the eigenspace $\{f \in C(X) : Tf = \lambda f\}$ is one-dimensional, i.e., all eigenvalues have geometric multiplicity 1. *Hint*: First establish this in the case $\lambda = 1$.

(3) If $g : X \to S^1$ is a unimodular eigenfunction with non-trivial eigenvalue $\lambda \neq 1$, show that $g : X \to g(X)$ is an isometric factor of X, where $g(X) \subset S^1$ is given the shift $z \mapsto \overline{\lambda}z$. Conclude in particular that $g = c\chi \circ \pi$, where $\pi : X \to K$ is the Kronecker factor, $\chi : K \to S^1$ is a character of K, and c is a constant. Conversely, show that all functions of the form $c\chi \circ \pi$ are eigenfunctions.[28]

We will see eigenfunctions (and various generalisations of the eigenfunction concept) playing a decisive role in the structure theory of measure-preserving systems, which we will get to in a few lectures.

2.6.3. Isometric extensions. To cover more general systems than just the isometric systems, we need the more flexible concept of an *isometric extension*.

Definition 2.6.14 (Extensions). If $\pi : X \to Y = (Y, \mathcal{G}, S)$ is a factor of $(X, \mathcal{F}, T)$, we say that $(X, \mathcal{F}, T)$ is an *extension* of $(Y, \mathcal{G}, S)$, and refer to $\pi : X \to Y$ as the *projection map* or *factor map*. We refer to the (compact) spaces $\pi^{-1}(\{y\})$ for $y \in Y$ as the *fibres* of this extension.

Example 2.6.15. The skew shift (Example 2.2.4) is an extension of the circle shift, with the fibres being the "vertical" circles. All systems are extensions of a point, and (somewhat trivially) are also extensions of themselves.

Definition 2.6.16 (Isometric extensions). Let $(X, \mathcal{F}, T)$ be an extension of a topological dynamical system $(Y, \mathcal{G}, S)$ with projection map $\pi : X \to Y$. We say that this extension is *isometric* if there exists a metric $d_y : \pi^{-1}(\{y\}) \times \pi^{-1}(\{y\}) \to \mathbf{R}^+$ on each fiber $\pi^{-1}(\{y\})$ with the following properties:

(a) (Isometry) For every $y \in Y$ and $x, x' \in \pi^{-1}(\{y\})$, we have

$$d_{Sy}(Tx, Tx') = d_y(x, x').$$

(b) (Continuity) The function $d : \bigcup_{y \in Y} \pi^{-1}(\{y\}) \times \pi^{-1}(\{y\}) \to \mathbf{R}^+$ formed by gluing together all the d_y is continuous (where we view the domain as a compact subspace $\{(x, x') \in X \times X : \pi(x) = \pi(x')\}$ of $X \times X$).

---

[28]From this, it is possible to reconstruct the Kronecker factor canonically from the eigenfunctions of X; we leave the details to the reader.

(c) (Isometry, again) The metric spaces $(\pi^{-1}(\{y\}), d_y)$ and $(\pi^{-1}(\{y'\}),$ $d_{y'})$ are isometric for any $y, y' \in Y$.

Example 2.6.17. The skew shift is an isometric extension of the circle shift, where we give each fibre the standard metric.

Example 2.6.18. A topological dynamical system is an isometric extension of a point if and only if it is isometric.

Exercise 2.6.7. If X is minimal, show that properties (a), (b) in Definition 2.6.16 automatically imply property (c). Furthermore, in this case show that the isometry group $\text{Isom}(\pi^{-1}(\{y\}))$ of any fibre acts transitively on that fibre. Show however that property (c) can fail even when properties (a) and (b) hold if X is not assumed to be minimal.

Exercise 2.6.8 (Topological characterisation of isometric extensions). Let $(X, \mathcal{F}, T)$ be a extension of a minimal topological dynamical system $(Y, \mathcal{G}, S)$ with factor map $\pi : X \to Y$, and let d be a metric on X. Show that X is an isometric extension if and only if the shift maps T^n are uniformly equicontinuous relative to π in the sense that for every $\varepsilon > 0$ there exists $\delta > 0$ such that for every $x, y \in X$ with $\pi(x) = \pi(y)$ and $d(x, y) < \delta$, we have $d(T^n x, T^n y) < \varepsilon$ for all n.

An important subclass of isometric extensions is the *group extensions*. Recall that an *automorphism* of a topological dynamical system is an isomorphism of that system to itself, i.e., a homeomorphism that commutes with the shift.

Definition 2.6.19 (Group extensions). Let $(X, \mathcal{F}, T)$ be a topological dynamical system. Suppose that we have a compact group G of automorphisms of X (where we endow G with the uniform topology). Then the quotient space $Y := G \backslash X = \{Gx : x \in X\}$ is also a compact metrisable space, and one easily sees that the projection map $\pi : X \mapsto Y$ is a factor map. We refer to X as a *group extension* of Y (or of any other system isomorphic to Y). We refer to G as the *structure group* of the extension. We say that the group extension is an *abelian group extension* if G is abelian.

Example 2.6.20 (Cocycle extensions). If G is a compact topological metrisable group, $(Y, \mathcal{G}, S)$ is a topological dynamical system, and $\sigma : Y \to G$ is a continuous map, then we define the *cocycle extension* $X = Y \times_\sigma G$ to be the product space $Y \times G$ with the shift $T : (y, \zeta) \mapsto (Sy, \sigma(y)\zeta)$, and with the factor map $\pi : (y, \zeta) \mapsto y$. One easily verifies that X is a group extension of Y with structure group G. The converse is not quite true for topological reasons; not every G-bundle can be globally trivialised, although one can still describe general group extensions by patching together cocycle extensions on local trivialisations.

Example 2.6.21. The skew shift is a cocycle extension (and hence group extension) $Y \times_\sigma (\mathbf{R}/\mathbf{Z})$ of the circle shift Y, with $\sigma(y) := y$ being the identity map. Any Kronecker system is an abelian group extension of a point.

Exercise 2.6.9. Show that every group extension is an isometric extension. *Hint*: The group G acts equicontinuously on itself, and thus isometrically on itself by choosing the right metric, as in Exercise 2.6.1.

Exercise 2.6.10. Let $(Y, \mathcal{G}, S)$ be a topological dynamical system, and G a compact topological metrisable group. We say that two cocycles σ, σ' : $Y \to G$ are *cohomologous* if we have $\sigma'(y) = \phi(Sy)\sigma(y)\phi(y)^{-1}$ for some continuous map $\phi : Y \to G$. Show that if σ, σ' are cohomologous, then the cocycle extensions $Y \times_\sigma G$ and $Y \times_{\sigma'} G$ are isomorphic. Understanding exactly which cocycles are cohomologous to each other is a major topic of study in dynamical systems (though not one which we will pursue here).

In view of Proposition 2.6.7 and Exercise 2.6.15, it is reasonable to ask whether every minimal isometric extension is a group extension. The answer is no (though actually constructing a counterexample is a little tricky). The reason is that we can form intermediate systems between a system $Y = G \backslash X$ and a group extension X of that system by quotienting out a subgroup. Indeed, if H is a closed subgroup of the structure group G, then $H \backslash X$ is a factor of X and an isometric extension of $G \backslash X$, but need not be a group extension of $G \backslash X$ (basically because G/H need not be a group). But this is the only obstruction to obtaining an analogue of Proposition 2.6.7:

Lemma 2.6.22. *Suppose that X is an isometric extension of another topological dynamical system Y with projection map $\pi : X \to Y$. Suppose also that X is minimal. Then there exists a group extension Z of Y with structure group G (thus $Y \equiv G \backslash Z$) and a closed subgroup H of G such that X is isomorphic to $H \backslash Z$, and π is (after applying the isomorphisms) the projection map from $H \backslash Z$ to $G \backslash Z$; thus we have the commutative diagram*

$$
\begin{array}{ccc}
Z & \to & X = H \backslash Z \\
& \searrow & \downarrow \\
& & Y = G \backslash Z
\end{array}
$$

(2.48)

Proof. For each $y \in Y$, let V_y be the metric space $\pi^{-1}(\{y\})$ with the metric d_y given by Definition 2.6.19. Thus for any integer n and any $y \in Y$, T^n is an isometry from V_y to $V_{S^n y}$; taking limits, we see that for any $p \in \beta\mathbf{Z}$, T^p is an isometry from V_y to $V_{S^p y}$. Also, the T^p clearly commute with the shift T. Fix a point $y_0 \in Y$ and set $G := \mathrm{Isom}(V_{y_0})$.

Let W be the space of all pairs (y, f) where $y \in Y$ and f is an isometry from V_{y_0} to V_y. This is a compact metrisable space with a shift $U : (y, f) \mapsto (Sy, T \circ f)$ and an action $g : (y, f) \mapsto (y, f \circ g^{-1})$ of G that commutes with U.

We let Z be the orbit closure in W of the G-orbit $\{y_0\} \times G$ under the shift U. If we fix a point $x_0 \in V_{y_0}$, then Z projects onto X by the map $f \mapsto f(y_0)$, and onto Y by the map $(y, f) \mapsto y$; these maps of course commute with the projection $\pi : x \mapsto \pi(x)$ from X to Y. Because X is minimal (and thus equal to all of its orbit closures), one sees that all of these projections are surjective morphisms, thus Z extends both Y and X. Also, one verifies that Z is a group extension over Y with structure group G, and a group extension over X with structure group given by the stabiliser $H := \{g \in G : gx_0 = x_0\}$. The claim follows. $\square$

Exercise 2.6.11. Show that if a minimal extension $\pi : X \to Y$ is finite, then it is automatically an abelian group extension. *Hint*: Recall from Section 2.2 that minimal finite systems are equivalent to shifts on a cyclic group.

An important feature of isometric or group extensions is that they tend to preserve recurrence properties of the system. We will see this phenomenon prominently when we turn to the ergodic theory analogue of isometric extensions, but for now let us give a simple illustrative result in this direction:

Proposition 2.6.23. *Let $(X, \mathcal{F}, T)$ be an isometric extension of $(Y, \mathcal{G}, S)$ with factor map $\pi : X \to Y$, and let y be a recurrent point of Y (see Definition 2.3.2 for a definition). Then every point x in the fibre $\pi^{-1}(\{y\})$ is a recurrent point in X.*

Proof. It will be convenient to use ultrafilters. In view of Lemma 2.6.22, it suffices to prove the claim for group extensions (note that recurrence is preserved under morphisms). Since y is recurrent, there exists $p \in \beta\mathbf{Z}\backslash\mathbf{Z}$ such that $S^p y = y$ (see Exercise 2.3.10). Thus $\pi(T^p x) = \pi(x)$. Since $Y = G\backslash X$, this implies that $T^p x = gx$ for some $g \in G$. We can iterate this (recalling that G commutes with T) to conclude that $T^{np} x = g^n x$ for all positive integers n. But by considering the action of g on G, we know (from Theorem 2.3.4) that we have $g^{n_j} h \to h$ for some $h \in G$ and $n_j \to +\infty$; canceling the h, and then applying to x, we conclude that $g^{n_j} x \to x$, and thus $T^{n_j p} x \to x$. If we write $q := \lim_{j \to r} n_j p$ for some $r \in \beta\mathbf{N}\backslash\mathbf{N}$, we conclude that $T^q x = x$ and so x is recurrent as desired. $\square$

2.6.4. Application: Distribution of polynomial sequences in tori. Now we apply the above theory to the following specific problem:

Problem 2.6.24. Let $P : \mathbf{Z} \to (\mathbf{R}/\mathbf{Z})^d$ be a polynomial sequence in a d-dimensional torus, thus $P(n) = \sum_{j=0}^{k} c_j n^j$ for some $c_0, \ldots, c_k \in (\mathbf{R}/\mathbf{Z})^d$. Compute the orbit closure $\overline{P(\mathbf{Z})} = \overline{\{P(n) : n \in \mathbf{Z}\}}$.

(We will be vague here about what "compute" means.)

Example 2.6.25. Is the orbit $\{(\sqrt{2}n \bmod 1, \sqrt{3}n^2 \bmod 1) : n \in \mathbf{Z}\}$ dense in the two-dimensional torus $(\mathbf{R}/\mathbf{Z})^2$?

The answer should of course depend on the polynomial P; for instance, if P is constant, then the orbit closure is clearly a point. Similarly, if the polynomial P has a constraint of the form $m \cdot P = c$ for some non-zero $m \in \mathbf{Z}^d$ and $c \in \mathbf{R}/\mathbf{Z}$, then the orbit closure is clearly going to be contained inside the proper subset $\{x \in (\mathbf{R}/\mathbf{Z})^d : m \cdot x = c\}$ of the torus. For instance, $\{(\sqrt{2}n^2 \bmod 1, 2\sqrt{2}n^2 \bmod 1) : n \in \mathbf{Z}\}$ is clearly not dense in the two-dimensional torus, as it is contained in the closed one-dimensional subtorus $\{(x, 2x) : x \in \mathbf{R}/\mathbf{Z}\}$.

In the above example, it is clear that the problem of computing the orbit closure of $(\sqrt{2}n^2 \bmod 1, 2\sqrt{2}n^2 \bmod 1)$ reduces to computing the orbit closure of $(\sqrt{2}n^2 \bmod 1)$. More generally, if a polynomial $P : \mathbf{Z} \to (\mathbf{R}/\mathbf{Z})^d$ obeys a constraint $m \cdot P = c$ for some non-zero *irreducible* $m \in \mathbf{Z}^d$ (i.e., m does not factor as $m = qm'$ for some $q > 1$ and $m' \in \mathbf{Z}^d$, or equivalently, the greatest common divisor of the coefficients of m is 1), then some elementary number theory shows that the set $\{x \in (\mathbf{R}/\mathbf{Z})^d : m \cdot x = c\}$ is isomorphic (after an invertible affine transformation with integer coefficients on the torus) to the standard subtorus $(\mathbf{R}/\mathbf{Z})^{d-1}$.

Exercise 2.6.12. Prove the above claim. *Hint*: The Euclidean algorithm may come in handy.

Because of this, we see that whenever we have a constraint of the form $m \cdot P = c$ with m irreducible, we can reduce Problem 2.6.24 to an instance of Problem 2.6.24 with one lower dimension. What if m is not irreducible? A typical example of this would be when[29] $P(n) := (\sqrt{2}n^2, 2\sqrt{2}n^2 + \frac{1}{2}n)$. Here, we have the constraint $(-4, 2) \cdot P(n) = 0$, which constrains P to the union of two one-dimensional tori, rather than a single one-dimensional torus. But we can eliminate this multiplicity by the trick of working with the odd and even components $\{P(2n + 1) : n \in \mathbf{Z}\}$ and $\{P(2n) : n \in \mathbf{Z}\}$, respectively. One observes that each component obeys an irreducible constraint, namely $(-2, 1) \cdot P(2n) = 0$ and $(-2, 1) \cdot P(2n + 1) = \frac{1}{2}$, respectively, and so by the preceding discussion, the problem of computing the orbit closures for each of these components reduces to that of computing an orbit closure in a torus of one lower dimension.

Exercise 2.6.13. More generally, show that if P obeys a constraint $m \cdot P(n) = c$ with m not necessarily irreducible, then there exists an integer $q \geq 1$ such that the orbits $\{P(qn+r) : n \in \mathbf{Z}\}$ obey a constraint $m' \cdot P(qn+r) = c_r$ with m' irreducible.

---

[29]I am going to drop the "mod 1" terms to remove clutter.

From Exercises 2.6.12 and 2.6.13, we see that every time we have a constraint of the form $m \cdot P(n) = c$ for some non-zero m, we can reduce Problem 2.6.24 to one or more copies of this problem in one lower dimension. So, without loss of generality (and by inducting on dimension) we may assume that no such constraint exists. (We will see this "induction on dimension" type of argument also in Section 2.16, when we study Ratner-type theorems in more detail.)

Now that all the "obvious" restrictions on the orbit have been removed, one might expect $P(n)$ to be uniformly distributed throughout the torus. Happily, this is indeed the case (at least at the topological level):

Theorem 2.6.26 (Equidistribution theorem). *Let $P : \mathbf{Z} \to (\mathbf{R}/\mathbf{Z})^d$ be a polynomial sequence which does not obey any constraint of the form $m \cdot P(n) = c$ with $m \in \mathbf{Z}^d$ non-zero. Then the orbit $P(\mathbf{Z})$ is dense in $(\mathbf{R}/\mathbf{Z})^d$ (i.e., the orbit closure is the whole torus).*

Remark 2.6.27. The recurrence theorems we have already encountered (e.g. Corollary 2.4.4 or Theorem 2.5.1) do not seem to directly establish this result, instead giving the weaker result that every element in $P(\mathbf{Z})$ is a limit point.

Exercise 2.6.14. Assuming Theorem 2.6.26, show that the answer to Problem 2.6.24 is always "a finite union of subtori", regardless of what the coefficients of P are.

Theorem 2.6.26 can be proven using Weyl's theory of equidistribution (Theorem 1.3.1), which is based on bounds of exponential sums; but we shall instead present a topological dynamics argument based on some ideas of Furstenberg [**Fu1981**]. Amusingly, this argument will use some *global* topology (specifically, *winding numbers*) and not just *local* (point-set) topology.

To prove this theorem, we begin with the linear one-dimensional case, in which one considers the orbit closure of $\{n\alpha + \beta : n \in \mathbf{Z}\}$ for some $\alpha, \beta \in \mathbf{R}/\mathbf{Z}$. The constant term β only affects this closure by a translation, and we can ignore it. One then easily checks that the orbit closure $\overline{\{n\alpha : n \in \mathbf{Z}\}}$ is a closed subgroup of $\mathbf{R}/\mathbf{Z}$. Fortunately, we have a classification of these objects:

Lemma 2.6.28. *Let H be a closed subgroup of $\mathbf{R}/\mathbf{Z}$. Then either $H = \mathbf{R}/\mathbf{Z}$, or H is a cyclic group of the form $H = \{x \in \mathbf{R}/\mathbf{Z} : Nx = 0\}$ for some $N \geq 1$.*

Proof. If H is not all of $\mathbf{R}/\mathbf{Z}$, then its complement, being a non-empty open set, is the union of disjoint open intervals. Let x be the boundary of one of these intervals; then x lies in the closed set H. Translating the group H by

x, we conclude that 0 is also the boundary of one of these intervals. Since $H = -H$, we thus see that 0 is an isolated point in H. If we then let y be the non-zero element of H closest to the origin (the case when $H = \{0\}$ can of course be checked separately), we check (using the Euclidean algorithm) that y generates H, and the claim easily follows. $\square$

Exercise 2.6.15. Using the above lemma, prove Theorem 2.6.26 in the case when $d = 1$ and P is linear.

Exercise 2.6.16. Obtain another proof of Lemma 2.6.28 using Fourier analysis and the fact that the only non-trivial subgroups of $\mathbf{Z}$ (the Pontryagin dual of $\mathbf{R}/\mathbf{Z}$) are the groups $N \cdot \mathbf{Z}$ for $N \geq 1$.

Now we consider the linear case in higher dimensions. The key lemma is the following.

Lemma 2.6.29. *Let H be a closed subgroup of $(\mathbf{R}/\mathbf{Z})^d$ for some $d \geq 1$ such that $\pi(H) = (\mathbf{R}/\mathbf{Z})^{d-1}$, where $\pi : (\mathbf{R}/\mathbf{Z})^d \to (\mathbf{R}/\mathbf{Z})^{d-1}$ is the canonical projection. Then either $H = (\mathbf{R}/\mathbf{Z})^d$ or $H = \{x \in (\mathbf{R}/\mathbf{Z})^d : m \cdot x = 0\}$ for some $m \in \mathbf{Z}^d$ with final coefficient non-zero.*

Proof. The fibre $H \cap \pi^{-1}(\{0\})$ is isomorphic to a closed subgroup of $\mathbf{R}/\mathbf{Z}$, so we can apply Lemma 2.6.28. If this subgroup is full, then it is not hard to see that $H = (\mathbf{R}/\mathbf{Z})^d$, so suppose instead that $H \cap \pi^{-1}(\{0\})$ is isomorphic to the cyclic group of order N. We then apply the homomorphism $f_N : (x_1, \ldots, x_d) \to (x_1, \ldots, x_{d-1}, Nx_d)$, and observe that $H_N := f_N(H)$ is a closed subgroup of $(\mathbf{R}/\mathbf{Z})^d$ whose fibres are a point, i.e., H_N is a graph $\{(x, \phi(x)) : x \in (\mathbf{R}/\mathbf{Z})^{d-1}\}$ for some $\phi : (\mathbf{R}/\mathbf{Z})^{d-1} \to \mathbf{R}/\mathbf{Z}$. Observe that the projection map $(x, \phi(x)) \mapsto x$ is a continuous bijection from the compact Hausdorff space H_N to the compact Hausdorff space $(\mathbf{R}/\mathbf{Z})^{d-1}$, and is thus a homeomorphism; in particular, ϕ is continuous. Since H_N is a group, ϕ must be a homomorphism. It is then a standard exercise to conclude that ϕ is linear, and hence takes the form $(x_1, \ldots, x_{d-1}) \mapsto m_1 x_1 + \ldots + m_{d-1} x_{d-1}$ for some integers $m_1, \ldots, m_{d-1}$. The claim follows by some routine algebra. $\square$

Exercise 2.6.17. Using the above lemma, prove Theorem 2.6.26 in the case when d is arbitrary and P is linear.

We now turn to the polynomial case. The basic idea is to re-express $P(n)$ in terms of the orbit $T^n x$ of some topological dynamical system on a torus. We have already seen this happen with the skew shift $((\mathbf{R}/\mathbf{Z})^2, (x, y) \mapsto (x + \alpha, y + x))$, where the orbits $T^n x$ exhibit quadratic behaviour in n. More generally, an iterated skew shift such as

$$(2.49) \qquad ((\mathbf{R}/\mathbf{Z})^d, (x_1, \ldots, x_d) \mapsto (x_1 + \alpha, x_2 + x_1, \ldots, x_d + x_{d-1}))$$

generates orbits $T^n x$ whose final coefficient contains degree d terms such as $\frac{n(n-1)\ldots(n-d+1)}{d!} \alpha$. What we would like to do is find criteria under which we could demonstrate that systems such as (2.49) are *minimal*; this would mean that every orbit closure in that system is dense, which would clearly be relevant for proving results such as Theorem 2.6.26.

To do this, we will exploit the fact that systems such as (2.49) can be built as towers of isometric extensions; for instance, the system (2.49) is an isometric extension over the same system (2.49) associated to $d-1$ (which, in the case $d = 1$, is simply a point). Now, isometric extensions do not always preserve minimality; for instance, if one takes a trivial cocycle extension $Y \times_0 G$, then the system is certainly non-minimal, as every horizontal slice $Y \times \{g\}$ of that system is a subsystem. More generally, any cocycle extension which is cohomologous to the trivial cocycle (see Exercise 2.6.10) will not be minimal. However, it turns out that if one has a topological obstruction to triviality, then minimality is preserved. We will formulate this fact using the machinery of *winding numbers*. Recall that every continuous map $f : \mathbf{R}/\mathbf{Z} \to \mathbf{R}/\mathbf{Z}$ has a *winding number* $[f] \in \mathbf{Z}$, which can be defined as the unique integer such that f is homotopic to the linear map $x \mapsto [f]x$. Note that the map $f \mapsto [f]$ is linear, and also that $[f]$ is unchanged if one continuously deforms f.

We now give a variant of a lemma of Furstenberg [**Fu1981**].

Lemma 2.6.30. *Let $(Y, \mathcal{G}, S)$ be a minimal topological dynamical system. Let $\sigma : Y \to (\mathbf{R}/\mathbf{Z})^d$ be a cocycle such that for every non-zero $m \in \mathbf{Z}^d$ there exists a loop $\gamma : \mathbf{R}/\mathbf{Z} \to Y$ such that $S \circ \gamma$ is homotopic to γ and $[m \cdot \sigma \circ \gamma] \neq 0$. Then $Y \times_\sigma \mathbf{R}/\mathbf{Z}$ is also minimal.*

Proof. We induct on d. The case $d = 0$ is trivial, so suppose $d \geq 1$ and the claim has already been proven for $d - 1$. Suppose for contradiction that $Y \times_\sigma (\mathbf{R}/\mathbf{Z})^d$ contains a proper minimal subsystem Z. Then $\pi(Z)$ is a subsystem of Y, and must therefore equal all of Y, by minimality of Y. Now we use the action of $(\mathbf{R}/\mathbf{Z})^d$ on $Y \times_\sigma (\mathbf{R}/\mathbf{Z})^d$, which commutes with the shift $T : (y, \zeta) \mapsto (Sy, \sigma(y) + \zeta)$. For every $\theta \in (\mathbf{R}/\mathbf{Z})^d$, we see that $\theta + Z$ is also a minimal subsystem, and so is either equal to Z or disjoint from Z. If we let $H := \{\theta \in (\mathbf{R}/\mathbf{Z})^d : \theta + Z = Z\}$, we conclude that H is a closed subgroup of $(\mathbf{R}/\mathbf{Z})^d$.

We now claim that the projection of H to $(\mathbf{R}/\mathbf{Z})^{d-1}$ must be all of $(\mathbf{R}/\mathbf{Z})^{d-1}$. For if this were not the case, we could project Z down to $Y \times_{\sigma'} (\mathbf{R}/\mathbf{Z})^{d-1}$, where $\sigma' : Y \to (\mathbf{R}/\mathbf{Z})^{d-1}$ is the projection of σ, and obtain a proper subsystem of that extension. But by induction hypothesis we see that $Y \times_{\sigma'} (\mathbf{R}/\mathbf{Z})^{d-1}$ is minimal, a contradiction, thus proving the claim.

We can now apply Lemma 2.6.29. If H is all of $(\mathbf{R}/\mathbf{Z})^d$ then Z is all of $Y \times_\sigma (\mathbf{R}/\mathbf{Z})^d$, a contradiction. Thus we have $H = \{\zeta \in (\mathbf{R}/\mathbf{Z})^d : m \cdot \zeta = 0\}$ for some non-zero $m \in \mathbf{Z}^d$, and thus Z must take the form

(2.50) $$Z = \{(y, \zeta) \in Y \times_\sigma (\mathbf{R}/\mathbf{Z})^d : m \cdot \zeta = \phi(y)\}$$

for some $\phi : Y \to \mathbf{R}/\mathbf{Z}$. Arguing as in the proof of Lemma 2.6.29 we can show that Y is homeomorphic to the image of Z under the map $(y, \zeta) \mapsto (y, m \cdot \zeta)$ and so ϕ must be continuous. Since Z is shift-invariant, we must have the equation

(2.51) $$\phi(Sy) = \phi(y) + m \cdot \sigma(y).$$

We apply this for y in the loop γ associated to m by hypothesis, and take degrees to conclude that

(2.52) $$[\phi \circ S \circ \gamma] = [\phi \circ \gamma] + [m \cdot \sigma \circ \gamma].$$

But as $S \circ \gamma$ is homotopic to γ, we have $[\phi \circ S \circ \gamma] = [\phi \circ \gamma]$ and thus $[m \cdot \sigma \circ \gamma] = 0$, contradicting the hypothesis. $\qquad\square$

Exercise 2.6.18. Using the above lemma and an induction on d, show that the system (2.49) is minimal whenever α is irrational. (The key, of course, is to make a good choice for the loop γ that makes all computations easy.)

Exercise 2.6.19. More generally, show that the product of any finite number of systems of the form (2.49) remains minimal, as long as the numbers α that generate each factor system are linearly independent with respect to each other and to 1 over the rationals $\mathbf{Q}$.

It is now possible to deduce Theorem 2.6.26 from Exercise 2.6.19 and a little bit of linear algebra. We sketch the ideas as follows. Firstly we take all the non-constant coefficients that appear in P and look at the space they span, together with 1, over the rationals $\mathbf{Q}$. This is a finite-dimensional space, and so has a basis containing 1 which is linearly independent over $\mathbf{Q}$. The non-constant coefficients of P are rational linear combinations of elements of this basis; by dividing the basis elements by some suitable integer (and using the trick of passing from $P(n)$ to $P(qn + r)$ if necessary) we can ensure that the coefficients of P are in fact integer linear combinations of basis elements. This allows us to write P as an affine-linear combination (with integer coefficients) of the coefficients of an orbit in the type of product system considered in Exercise 2.6.19. If this affine transformation has full rank, then we are done; otherwise, the affine transformation maps to some subspace of the torus of the form $\{x : m \cdot x = c\}$, contradicting the hypothesis on P. Theorem 2.6.26 follows.

Notes. This lecture first appeared at

terrytao.wordpress.com/2008/01/24.

Thanks to Nilay, mmailliw/william, Zaher Hani, Sugata, and Liu Xiao Chuan for corrections.

2.7. Structural theory of topological dynamical systems

In our final lecture on topological dynamics, we discuss a remarkable theorem of Furstenberg [**Fu1963**] that classifies a major type of topological dynamical system—*distal* systems—in terms of highly structured (from an algebraic point of view) systems, namely towers of isometric extensions. This theorem is also a model for an important analogous result in ergodic theory, the *Furstenberg-Zimmer structure theorem*, to which we will turn in a few lectures. We will not be able to prove Furstenberg's structure theorem for distal systems here in full, but we hope to illustrate some of the key points and ideas.

2.7.1. Distal systems. Furstenberg's theorem concerns a significant generalisation of the equicontinuous (or isometric) systems, namely the *distal* systems.

Definition 2.7.1 (Distal systems). Let $(X, \mathcal{F}, T)$ be a topological dynamical system, and let d be an arbitrary metric on X (it is not important which metric one picks here). We say that two points x, y in X are *proximal* if we have $\liminf_{n \to \infty} d(T^n x, T^n y) = 0$. We say that X is distal if no two distinct points $x \neq y$ in X are proximal, or equivalently if for every distinct x, y there exists $\varepsilon > 0$ such that $d(T^n x, T^n y) \geq \varepsilon$ for all n.

It is obvious that every isometric or equicontinuous system is distal, but the converse is not true, as the following example shows:

Example 2.7.2. If $\alpha \in \mathbf{R}$, then the skew shift

$$((\mathbf{R}/\mathbf{Z})^2, (x, y) \mapsto (x + \alpha, y + x))$$

turns out to be not equicontinuous. Indeed, if we start with a pair of nearby points $(0, 0), (0, 1/2n)$ for some large n and apply T^n, we end up with $(n\alpha, \frac{n(n-1)}{2}\alpha)$ and $(\alpha, \frac{n(n-1)}{2}\alpha + \frac{1}{2})$, thus demonstrating failure of equicontinuity. On the other hand, the system is still distal: given any pair of distinct points $(x, y), (x', y')$, either $x \neq x'$ (in which case the horizontal separation between $T^n(x, y)$ and $T^n(x', y')$ is bounded from below) or $x = x'$ (in which case the vertical separation is bounded from below).

Exercise 2.7.1. Show that any non-trivial Bernoulli system $\Omega^{\mathbf{Z}}$ is not distal.

Distal systems interact nicely with the action $p \mapsto T^p$ of the compactified integers $\beta \mathbf{Z}$:

Exercise 2.7.2. Let $(X, \mathcal{F}, T)$ be a topological dynamical system.

(1) Show that two points x, y in X are proximal if and only if $T^p x = T^p y$ for some $p \in \beta \mathbf{Z}$.

(2) Show that X is distal if and only if all the maps T^p for $p \in \beta \mathbf{Z}$ are injective.

(3) If X is distal, show that $T^p = \mathrm{id}$ whenever $p \in \beta \mathbf{Z}$ is idempotent. *Hint*: Use part (2).

(4) If X is distal, show that the set of transformations $G := \{T^p : p \in \beta \mathbf{Z}\}$ on X forms a group G, known as the *Ellis group* of X. *Hint*: Use part (3), together with Lemma 2.5.14.

(5) Show that G is a compact subset of X^X (with the product topology), and that G acts transitively on X if and only if X is minimal.

Exercise 2.7.3. Show that an inverse limit of a totally ordered set $(Y_\alpha)_{\alpha \in A}$ of distal factors is still distal. (This turns out to be slightly easier than Lemma 2.6.12.)

Exercise 2.7.4. Show that every topological dynamical system has a maximal distal factor. *Hint*: Repeat the proof of Corollary 2.6.13.

Exercise 2.7.5. Show that any distal system can be partitioned into disjoint minimal distal systems. *Hint*: One can of course adapt the proof of Proposition 2.6.9; but there is a slicker way to do it by exploiting the Ellis group.

Note that the skew shift system, while not isometric, does have a nontrivial isometric factor, namely the circle shift $(\mathbf{R}/\mathbf{Z}, x \mapsto x + \alpha)$ with the projection map $\pi : (x, y) \mapsto x$. It turns out that this phenomenon is general:

Theorem 2.7.3 (Baby Furstenberg structure theorem). *Let $(X, \mathcal{F}, T)$ be minimal, distal, and non-trivial (i.e., not a point). Then X has a non-trivial isometric factor $\pi : X \to Y$.*

This result—a toy case of Furstenberg's full structure theorem—is already rather difficult to establish. We will not give Furstenberg's original proof here (though see Exercise 2.7.13 below), but will at least sketch how the factor $\pi : X \to Y$ is constructed. A key object in the construction is the symmetric function $F : X \times X \to \mathbf{R}^+$ defined by the formula

(2.53) $$F(x, y) := \inf_{n \in \mathbf{Z}} d(T^n x, T^n y).$$

Example 2.7.4. We again consider the skew shift

$$((\mathbf{R}/\mathbf{Z})^2, (x,y) \mapsto (x+\alpha, y+x))$$

with α irrational. For the sake of concreteness let us choose the taxicab metric $d((x,y),(x',y')) := \|x-x'\|_{\mathbf{R}/\mathbf{Z}} + \|y-y'\|_{\mathbf{R}/\mathbf{Z}}$, where $\|x\|_{\mathbf{R}/\mathbf{Z}}$ is the distance from x to the integers. Then one can check that $F((x,y),(x',y'))$ is equal to $\|x-x'\|_{\mathbf{R}/\mathbf{Z}}$ when $x-x'$ is irrational, and equal to $\|x-x'\|_{\mathbf{R}/\mathbf{Z}} + \frac{1}{q}\|q(y-y')\|_{\mathbf{R}/\mathbf{Z}}$ when $x-x'$ is rational, where q is the least positive integer such that $q(x-x')$ is an integer. Thus F is highly discontinuous, but it is at least upper semicontinuous in each of its two variables.[30]

Exercise 2.7.6. Let G be the Ellis group of a minimal distal system X.

(1) For any $x,y \in X$, show that $F(x,y) = \inf_{g \in G} d(gx, gy)$. In particular, $F(gx, gy) = F(x,y)$ for all $g \in G$.

(2) For any $x,y \in X$, show that the set $\{(gx, gy) : g \in G\}$ is a minimal subsystem of $X \times X$ (with the product shift $(x,y) \mapsto (Tx, Ty)$). Conclude in particular that if $F(x,y) < a$, then the set $\{n \in \mathbf{Z} : d(T^n x, T^n y) < a)\}$ is syndetic.

(3) If $x,y \in X$ and $a > 0$ is such that $F(x,y) < a$, show that there exists ε such that $F(x,z) < a$ whenever $F(y,z) < \varepsilon$.

(4) Let $X_F = (X, \mathcal{F}_F)$ be the space X whose topology is generated by the basic open sets $U_{a,x} := \{y \in X : F(x,y) < a\}$. (That this is a base follows from (3).) Equivalently, X_F is equipped with the weakest topology in which F is upper semicontinuous in each variable. Show that X_F is a weaker topological space than X (i.e., the identity map from X to X_F is continuous); in particular, X_F is compact. Also show that all the maps in G are homeomorphisms on X_F.

If the space X_F defined in Exercise 2.7.6 were Hausdorff, then the system $(X_F, \mathcal{F}_F, T)$ would be equicontinuous, by Exercise 2.6.2. Unfortunately, X_F is not Hausdorff in general. However, it turns out that we can "quotient out" the non-Hausdorff nature of X_F. Define the equivalence relation $\sim$ on X_F by declaring $x \sim y$ if we have $F(x,z) = F(y,z)$ for all z outside of a set of the first category in X. This is clearly an equivalence relation, and so we can create the quotient space $Y := X_F/\sim$; since X embeds into X_F, we thus have a factor map $\pi : X \to Y$. It is a deep fact (which we will not prove here) that this quotient space is non-trivial and Hausdorff, and that $\sim$ is preserved by the shift T and even by the Ellis group G (thus if $x \sim y$ and $g \in G$ then $gx \sim gy$). Because of this, G continues to act on

---

[30] Actually, the upper semicontinuity of F holds for arbitrary topological dynamical systems, since F is the infimum of continuous functions.

Y homeomorphically, and so by Exercise 2.6.2, $\pi : X \to Y$ is a non-trivial isometric factor of X as desired.

Exercise 2.7.7. Show that in the case of the skew shift (Example 2.7.4), this construction recovers the factor that was discussed just before Theorem 2.7.3. (The trickiness of this exercise should already give you some idea of the difficulty level of Theorem 2.7.3.)

2.7.2. The Furstenberg structure theorem for distal systems. We have already noted that isometric systems are distal systems. More generally, we have

Exercise 2.7.8. Show that an isometric extension of a distal system is still distal. *Hint*: Example 2.7.2 is a good model case.

Thus, for instance, the iterated skew shifts that appear in (2.49) are distal. Also, recall from Exercise 2.7.7 that the inverse limit of distal systems is again distal. It turns out that these are the *only* ways to generate distal systems, in the following sense [**Fu1963**]:

Theorem 2.7.5 (Furstenberg's structure theorem for distal systems). *Let $(X, \mathcal{F}, T)$ be a distal system. Then there exists an ordinal α and a factor Y_β for every $\beta \leq \alpha$ with the following properties:*

(1) *$Y_\emptyset$ is a point.*

(2) *For every successor ordinal $\beta + 1 \leq \alpha$, $Y_{\beta+1}$ is an isometric extension of Y_β.*

(3) *For every limit ordinal $\beta \leq \alpha$, Y_β is an inverse limit of the Y_γ for $\gamma < \beta$.*

(4) *Y_α is equal to X.*

The collection of factors $(Y_\beta)_{\beta \leq \alpha}$ is sometimes known as a "Furstenberg tower".

Theorem 2.7.5 follows by applying Zorn's lemma with the following key proposition:

Proposition 2.7.6 (Key inductive step). *Let $(X, \mathcal{F}, T)$ be a distal system, and let Y be a proper factor of X (i.e., the factor map is not an isomorphism). Then there exists another factor Z of X which is a proper isometric extension of Y.*

Note that Theorem 2.7.3 is the special case of Proposition 2.7.6 when Y is a point. Indeed, Proposition 2.7.6 is proven in the same way as Theorem 2.7.3, but with several additional technicalities which I will not discuss here; see [**Fu1963**] for details.

Exercise 2.7.9. Deduce Theorem 2.7.5 from Proposition 2.7.6 and Zorn's lemma.

Remark 2.7.7. It is known that in Theorem 2.7.5, one can take the ordinal α to be countable, and conversely that for every countable ordinal α, there exists a system whose smallest Furstenberg tower has height α; see [**BeFo1996**].

Remark 2.7.8. Several generalisations and extensions of Furstenberg's structure theorem are known, but they are somewhat technical to state and will not be detailed here; see [**Gl2000**] for a discussion.

2.7.3. Weak mixing and isometric factors. We have seen that distal systems always contain non-trivial isometric factors. What about more general systems? It turns out that there is in fact a nice dichotomy between systems with non-trivial isometric factors, and those without.

Definition 2.7.9 (Topological transitivity). A topological dynamical system $(X, \mathcal{F}, T)$ is *topologically transitive* if, for every pair U, V of non-empty open sets, there exists an integer n such that $T^n U \cap V \neq \emptyset$.

Exercise 2.7.10. Show that a topological dynamical system is topologically transitive if and only if it is equal to the orbit closure of one of its points.[31]

Exercise 2.7.11. Show that any factor of a topologically transitive system is again topologically transitive.

Definition 2.7.10 (Topological weak mixing). A topological dynamical system $(X, \mathcal{F}, T)$ is *topologically weakly mixing* if the product system $X \times X$ is topologically transitive.

Exercise 2.7.12. A system is said to be *topologically mixing* if for every pair U, V of non-empty open sets, one has $T^n U \cap V \neq \emptyset$ for all sufficiently large n. Show that topological mixing implies topological weak mixing. (The converse is false, but actually constructing a counterexample is somewhat tricky.)

Example 2.7.11. No circle shift $(\mathbf{R}/\mathbf{Z}, x \mapsto x + \alpha)$ is topologically weak mixing (or topologically mixing), even though such shifts are minimal (and hence transitive) when α is irrational. On the other hand, any Bernoulli shift is easily seen to be topologically mixing (and hence topologically weak mixing).

We have the following dichotomy, first proven in [**KeRo1969**] (using ideas from [**Fu1963**]).

---

[31]Compare this with minimal systems, which is the orbit closure of *any* of its points. Thus minimality is stronger than topological transitivity; for instance, the compactified integers $\{-\infty\} \cup \mathbf{Z} \cup \{+\infty\}$ with the usual shift is topologically transitive but not minimal.

Theorem 2.7.12 (Dichotomy between structure and randomness). *Suppose* $(X, \mathcal{F}, T)$ *is a minimal topological dynamical system. Then exactly one of the following statements is true:*

(1) *(Structure) X has a non-trivial isometric factor.*

(2) *(Randomness) X is topologically weakly mixing.*

Remark 2.7.13. Combining this with Exercise 2.6.6, we obtain an equivalent formulation of this theorem: A minimal system is topologically weakly mixing if and only if it has no non-trivial eigenfunctions.

Proof. We first prove the easy direction: that if X has a non-trivial isometric factor, then it is not topologically weakly mixing. In view of Exercise 2.7.11, it suffices to prove this when X itself is isometric. Let x, x' be two distinct points of Y, let r denote the distance between x and x' with respect to the metric that makes X isometric, and let B and B' be the open balls of radius $r/10$ centred at x and x' respectively. As X is isometric, we see for any integer n that $T^n B$ cannot intersect both B and B', or equivalently that $(T \times T)^n (B \times B)$ cannot intersect $B \times B'$. Thus X is not topologically transitive as desired.

Now we prove the difficult direction: if X is not topologically weakly mixing, then it has a non-trivial isometric factor. For this we use an argument from [**BlHoMa2000**], based on the earlier work [**McM1978**]. By Definition 2.7.10, there exist open non-empty sets U, V in $X \times X$ such that $(T \times T)^n U \cap V = \emptyset$ for all n. If we thus set $K := \overline{\bigcup_n (T \times T)^n U}$, we see that K is a compact proper $T \times T$-invariant subset of $X \times X$ with non-empty interior. On the other hand, the projection of K to either factor of $X \times X$ is a non-empty compact invariant subset of X and thus must be all of X.

We need to somehow use K to build an isometric factor of X. For this, we shall move from the topological dynamics setting to that of the ergodic theory setting. By Corollary 2.7.17 in the appendix, X admits an invariant Borel measure μ. The support of μ is a non-empty closed invariant subset of X, and is thus equal to all of X by minimality.

The space $L^1(X, \mu)$ is a metric space, with an isometric shift map $Tf := f \circ T^{-1}$. We define the map $\pi : X \to L^1(X, \mu)$ by the formula

$$(2.54) \qquad\qquad \pi(x) : y \mapsto 1_K(x, y)$$

for all $x \in X$, where 1_K is the indicator function of K. Because K has non-empty interior and non-empty exterior, and because μ has full support, it is not hard to show that π is non-constant. By the T-invariance of W, it also preserves the shift T. So if we can show that π is continuous, we see that $\pi(X)$ will be a non-trivial isometric factor of X and we will be done.

Let us first consider the scalar function $f(x) := \int_X 1_K(x,y)\, d\mu(y)$. From the dominated convergence theorem and the fact that K is closed, we see that f is upper semicontinuous, and continuous at at least one point, thanks to Lemma 2.4.13. On the other hand, since K is $T \times T$-invariant and μ is T-invariant, we see that f is T-invariant. Applying Exercise 2.4.16 we see that f is constant. On the other hand, as K is closed we have $\limsup_{x \to x_0} 1_K(x,y) \leq 1_K(x_0,y)$ for any $x_0 \in X$, and so by dominated convergence again we see that $1_K(x,\cdot)$ converges in L^1 to zero outside of the support of $1_K(x_0,\cdot)$. Combining this with the constancy of f we conclude that $1_K(x,\cdot)$ converges to $1_K(x_0,\cdot)$ in L^1 on all of X, and thus π is continuous as required. $\qquad\square$

Remark 2.7.14. Note how the measure-theoretic structure was used to obtain metric structure, by passing from the measure space (X,μ) to the metric space $L^1(X,\mu)$. This again shows that one can sometimes upgrade weak notions of structure (such as topological or measure-theoretic structure) to strong notions (such as geometric or algebraic structure).

Exercise 2.7.13. Use Theorem 2.7.12 to prove Theorem 2.7.3. *Hint*: Use Exercise 2.7.10.

Remark 2.7.15. It would be very convenient if one had a relative version of Theorem 2.7.12, namely that if X is an extension of Y, then X is either relatively topologically weakly mixing with respect to Y (which means that the relative product $X \times_Y X := \{(x,x') \in X \times X : \pi(x) = \pi(x')\}$ is topologically transitive), or else X has a factor Z which is a non-trivial isometric extension of Y; among other things, this would have given a new proof of Theorem 2.7.5, and in fact establish a somewhat stronger structural theorem. Unfortunately, this relative version fails; a counterexample (based on the Morse sequence, Example 2.2.11) can be found in [**Gl2003**, Exercise 1.19.3]. Nevertheless, the analogue of this claim does hold true in the measure-theoretic setting, as we shall see in Section 2.12.

2.7.4. Appendix: Sequential compactness of Borel probability measures. We now recall some standard facts from measure theory about Borel probability measures on a compact metrisable space X. Recall that a sequence of such measures μ_n converges in the *vague topology* to another μ if we have $\int_X f\, d\mu_n \to \int_X f\, d\mu$ for all $f \in C(X)$.

Lemma 2.7.16 (Vague sequential compactness). *The space* $\mathrm{Pr}(X)$ *of Borel probability measures on X is sequentially compact in the vague topology.*

Proof. The Riesz representation theorem identifies $\Pr(X)$ with the dual of $C(X)$. From the Stone-Weierstrass theorem we know that $C(X)$ is separable. The claim then follows from the usual Arzelà-Ascoli diagonalisation argument. □

Corollary 2.7.17 (Krylov-Bogolubov theorem). *Let $(X, \mathcal{F}, T)$ be a topological dynamical system. Then there exists a T-invariant probability measure μ on X.*

Proof. Pick any point $x_0 \in X$ and consider the finite probability measures

$$(2.55) \qquad \mu_N := \frac{1}{N} \sum_{n=1}^{N} \delta_{T^n x_0},$$

where δ_x is the Dirac mass at x. By Lemma 2.7.16, some subsequence μ_{N_j} converges in the vague topology to another Borel probability measure μ. Since we have

$$(2.56) \qquad \int Tf \, d\mu_N = \int f \, d\mu_N + O_f(1/N)$$

for all bounded continuous f, we conclude on taking vague limits and using the Riesz representation theorem that μ is T-invariant as required. □

Remark 2.7.18. Note that Corollary 2.7.17, like many other results obtained via compactness methods, guarantees existence of an invariant measure but not uniqueness (this latter property is known as *unique ergodicity*). Even for minimal systems, it is possible for uniqueness to fail, although actually constructing an example is tricky (see for instance [**Fu1961**]). However, as already observed in the proof of Theorem 2.7.12, any invariant measure on a minimal topological dynamical system must be *full* (i.e., its support must be the whole space).

Exercise 2.7.14. Show that any topological dynamical system which is uniquely ergodic is necessarily minimal.

Notes. This lecture first appeared at

terrytao.wordpress.com/2008/01/28.

2.8. The mean ergodic theorem

We now leave topological dynamics, and begin our study of *measure-preserving systems* $(X, \mathcal{X}, \mu, T)$, i.e., a probability space $(X, \mathcal{X}, \mu)$ together with a probability space isomorphism $T : (X, \mathcal{X}, \mu) \to (X, \mathcal{X}, \mu)$ (thus $T : X \to X$ is invertible, with T and T^{-1} both being measurable, and $\mu(T^n E) = \mu(E)$ for all $E \in \mathcal{X}$ and all n). For various technical reasons it is convenient to restrict to the case when the σ-algebra $\mathcal{X}$ is separable, i.e., countably generated. One reason for this is as follows.

Exercise 2.8.1. Let $(X, \mathcal{X}, \mu)$ be a probability space with $\mathcal{X}$ separable. Then the Banach spaces $L^p(X, \mathcal{X}, \mu)$ are separable (i.e., have a countable dense subset) for every $1 \leq p < \infty$; in particular, the Hilbert space $L^2(X, \mathcal{X}, \mu)$ is separable. Show that the claim can fail for $p = \infty$. (We allow the L^p spaces to be either real- or complex-valued, unless otherwise specified.)

Remark 2.8.1. In practice, the requirement that $\mathcal{X}$ be separable is not particularly onerous. For instance, if one is studying the recurrence properties of a function $f : X \to \mathbf{R}$ on a non-separable measure-preserving system $(X, \mathcal{X}, \mu, T)$, one can restrict $\mathcal{X}$ to the separable sub-σ-algebra $\mathcal{X}'$ generated by the level sets $\{x \in X : T^n f(x) > q\}$ for integer n and rational q, thus passing to a separable measure-preserving system $(X, \mathcal{X}', \mu, T)$ on which f is still measurable. Thus we see that in many cases of interest, we can immediately reduce to the separable case. (In particular, for many of the theorems in this course, the hypothesis of separability can be dropped, though we will not bother to specify for which ones this is the case.)

We are interested in the recurrence properties of sets $E \in \mathcal{X}$ or functions $f \in L^p(X, \mathcal{X}, \mu)$. The simplest such recurrence theorem is

Theorem 2.8.2 (Poincaré recurrence theorem). *Let $(X, \mathcal{X}, \mu, T)$ be a measure-preserving system, and let $E \in \mathcal{X}$ be a set of positive measure. Then $\limsup_{n \to +\infty} \mu(E \cap T^n E) \geq \mu(E)^2$. In particular, $E \cap T^n E$ has positive measure (and is thus non-empty) for infinitely many n.*

Remark 2.8.3. This theorem should be compared with Theorem 2.3.1.

Proof. For any integer $N > 1$, observe that $\int_X \sum_{n=1}^N 1_{T^n E} \, d\mu = N\mu(E)$, and thus by the Cauchy-Schwarz inequality,

$$(2.57) \qquad \int_X \left(\sum_{n=1}^N 1_{T^n E} \right)^2 d\mu \geq N^2 \mu(E)^2.$$

The left-hand side of (2.57) can be rearranged as

$$(2.58) \qquad \sum_{n=1}^N \sum_{m=1}^N \mu(T^n E \cap T^m E).$$

On the other hand, $\mu(T^n E \cap T^m E) = \mu(E \cap T^{m-n} E)$. From this one easily obtains the asymptotic

$$(2.59) \qquad (2.58) \leq \left(\limsup_{n \to \infty} \mu(E \cap T^n E) + o(1) \right) N^2,$$

where $o(1)$ denotes an expression which goes to zero as N goes to infinity. Combining (2.57), (2.58), (2.59) and taking limits as $N \to +\infty$ we obtain

$$(2.60) \qquad \limsup_{n \to \infty} \mu(E \cap T^n E) \geq \mu(E)^2.$$

By shift-invariance we have $\mu(E \cap T^{-n}E) = \mu(E \cap T^n E)$, and the claim follows. $\square$

Remark 2.8.4. In classical physics, the evolution of a physical system in a compact phase space is given by a (continuous-time) measure-preserving system (this is Hamilton's equations of motion combined with Liouville's theorem). The Poincaré recurrence theorem then has the following unintuitive consequence: Every collection E of states of positive measure, no matter how small, must eventually return to overlap itself given sufficient time. For instance, if one were to burn a piece of paper in a closed system, then there exist arbitrarily small perturbations of the initial conditions such that, if one waits long enough, the piece of paper will eventually reassemble (modulo arbitrarily small error)! This seems to contradict the second law of thermodynamics, but the reason for the discrepancy is because the time required for the recurrence theorem to take effect is inversely proportional to the measure of the set E, which in physical situations is exponentially small in the number of degrees of freedom (which is already typically quite large, e.g. of the order of the Avogadro constant). This gives more than enough[32] opportunity for *Maxwell's demon* to come into play to reverse the increase of entropy. The more sophisticated recurrence theorems we will see later have much poorer quantitative bounds still, so much so that they basically have no direct significance for any physical dynamical system with many relevant degrees of freedom.

Exercise 2.8.2. Prove the following generalisation of the Poincaré recurrence theorem: If $(X, \mathcal{X}, \mu, T)$ is a measure-preserving system and $f \in L^1(X, \mathcal{X}, \mu)$ is non-negative, then $\limsup_{n \to +\infty} \int_X f T^n f \geq (\int_X f \, d\mu)^2$.

Exercise 2.8.3. Give examples to show that the quantity $\mu(X)^2$ in the conclusion of Theorem 2.8.2 cannot be replaced by any smaller quantity in general, regardless of the actual value of $\mu(X)$. *Hint*: Use a Bernoulli system example.

Exercise 2.8.4. Using the pigeonhole principle instead of the Cauchy-Schwarz inequality (and in particular, the statement that if $\mu(E_1) + \ldots + \mu(E_n) > 1$, then the sets $E_1, \ldots, E_n$ cannot all be disjoint), prove the weaker statement that for any set E of positive measure in a measure-preserving

---

[32]This can be viewed as a manifestation of the *curse of dimensionality*.

system, the set $E \cap T^n E$ is non-empty for infinitely many n. (This exercise illustrates the general point that the Cauchy-Schwarz inequality can be viewed as a quantitative strengthening of the pigeonhole principle.)

For this section and the next we shall study several variants of the Poincaré recurrence theorem. We begin by looking at the mean ergodic theorem, which studies the limiting behaviour of the ergodic averages $\frac{1}{N} \sum_{n=1}^{N} T^n f$ in various L^p spaces, and in particular in L^2.

2.8.1. Hilbert space formulation.
We begin with the Hilbert space formulation of the mean ergodic theorem, due to von Neumann.

Theorem 2.8.5 (Von Neumann ergodic theorem). *Let $U : H \to H$ be a unitary operator on a separable Hilbert space H. Then for every $v \in H$ we have*

$$(2.61) \qquad \lim_{N \to +\infty} \frac{1}{N} \sum_{n=0}^{N-1} U^n v = \pi(v),$$

where $\pi : H \to H^U$ is the orthogonal projection from H to the closed subspace $H^U := \{v \in H : Uv = v\}$ consisting of the U-invariant vectors.

Proof. We give the slick (but not particularly illuminating) proof of von Neumann's theorem. It is clear that (2.61) holds if v is already invariant (i.e., $v \in H^U$). Next, let W denote the (possibly non-closed) space $W := \{Uw - w : w \in H\}$. If $Uw - w$ lies in W and v lies in H^U, then by unitarity

$$(2.62) \qquad \langle Uw - w, v \rangle = \langle w, U^{-1}v \rangle - \langle w, v \rangle = \langle w, v \rangle - \langle w, v \rangle = 0,$$

and thus W is orthogonal to H^U. In particular $\pi(Uw - w) = 0$. From the telescoping identity

$$(2.63) \qquad \frac{1}{N} \sum_{n=0}^{N-1} U^n (Uw - w) = \frac{1}{N}(U^N w - w)$$

we conclude that (2.61) also holds if $v \in W$; by linearity we conclude that (2.61) holds for all v in $H^U + W$. A standard limiting argument (using the fact that the linear transformations $v \mapsto \pi(v)$ and $v \mapsto \frac{1}{N} \sum_{n=0}^{N-1} U^n v$ are bounded on H, uniformly in n) then shows that (2.61) holds for v in the closure $\overline{H^U + W}$.

To conclude, it suffices to show that the closed space $\overline{H^U + W}$ is all of H. Suppose for contradiction that this is not the case. Then there exists a non-zero vector w which is orthogonal to all of $\overline{H^U + W}$. In particular, w is orthogonal to $Uw - w$. Applying the easily verified identity $\|Uw - w\|^2 = -2\mathrm{Re}\langle Uw - w, w \rangle$ (related to the parallelogram law) we conclude that

$Uw = w$, thus w lies in H^U. This implies that w is orthogonal to itself and is thus zero, a contradiction. $\qquad\square$

On a measure-preserving system $(X, \mathcal{X}, \mu, T)$, the shift map $f \mapsto Tf$ is a unitary transformation on the separable Hilbert space $L^2(X, \mathcal{X}, \mu)$. We conclude

Corollary 2.8.6 (Mean ergodic theorem). *Let $(X, \mathcal{X}, \mu, T)$ be a measure-preserving system, and let $f \in L^2(X, \mathcal{X}, \mu)$. Then $\frac{1}{N} \sum_{n=1}^{N} T^n f$ converges in the $L^2(X, \mathcal{X}, \mu)$ norm to $\pi(f)$, where $\pi(f) : L^2(X, \mathcal{X}, \mu) \to L^2(X, \mathcal{X}, \mu)^T$ is the orthogonal projection to the space $\{f \in L^2(X, \mathcal{X}, \mu) : Tf = f\}$ that consists of the shift-invariant functions in $L^2(X, \mathcal{X}, \mu)$.*

Example 2.8.7 (Finite case). Suppose that $(X, \mathcal{X}, \mu, T)$ is a finite measure-preserving system, with $\mathcal{X}$ discrete and μ the uniform probability measure. Then T is a permutation on X and thus decomposes into the direct sum of disjoint cycles (possibly including trivial cycles of length 1). Therefore, the shift-invariant functions are precisely those functions which are constant on each of these cycles, and the map $f \mapsto \pi(f)$ replaces a function $f : X \to \mathbf{C}$ with its average value on each of these cycles. It is then an instructive exercise to verify the mean ergodic theorem by hand in this case.

Exercise 2.8.5. With the notation and assumptions of Corollary 2.8.6, show that the limit $\lim_{N\to\infty} \frac{1}{N} \sum_{n=0}^{N-1} \int_X T^n f \overline{f} \, d\mu$ exists, is real, and is greater than or equal to $|\int_X f|^2$. (*Hint*: The constant function 1 lies in $L^2(X, \mathcal{X}, \mu)^T$.) Note that this is stronger than the conclusion of Exercise 2.8.2.

Let us now give some other proofs of the von Neumann ergodic theorem. We first give a proof using the spectral theorem for unitary operators. This theorem asserts (among other things) that a unitary operator $U : H \to H$ can be expressed in the form $U = \int_{S^1} \lambda \, d\mu(\lambda)$, where $S^1 := \{z \in \mathbf{C} : |z| = 1\}$ is the unit circle and μ is a projection-valued Borel measure on the circle. More generally, we have

$$(2.64) \qquad U^n = \int_{S^1} \lambda^n \, d\mu(\lambda)$$

and so for any vector v in H and any positive integer N,

$$(2.65) \qquad \frac{1}{N} \sum_{n=0}^{N-1} U^n v = \int_{S^1} \frac{1}{N} \sum_{n=0}^{N-1} \lambda^n \, d\mu(\lambda) v.$$

We separate off the $\lambda = 1$ portion of this integral. For $\lambda \neq 1$, we have the geometric series formula

$$(2.66) \qquad \frac{1}{N} \sum_{n=0}^{N-1} \lambda^n = \frac{1}{N} \frac{\lambda^N - 1}{\lambda - 1}$$

(compare with (2.63)), thus we can rewrite (2.65) as

$$(2.67) \qquad \mu(\{1\})v + \int_{S^1 \setminus \{1\}} \frac{1}{N} \frac{\lambda^N - 1}{\lambda - 1} \, d\mu(\lambda)v.$$

Now observe (using (2.66)) that $\frac{1}{N} \frac{\lambda^N - \lambda}{\lambda - 1}$ is bounded in magnitude by 1 and converges to zero as $N \to \infty$ for any fixed $\lambda \neq 1$. Applying the dominated convergence theorem (which requires a little bit of justification in this vector-valued case), we see that the second term in (2.67) goes to zero as $N \to \infty$. So we see that (2.65) converges to $\mu(\{1\})v$. But $\mu(\{1\})$ is just the orthogonal projection to the eigenspace of U with eigenvalue 1, i.e., the space H^U, thus recovering the von Neumann ergodic theorem.[33]

Remark 2.8.8. The above argument in fact shows that the rate of convergence in the von Neumann ergodic theorem is controlled by the spectral gap of U—i.e., how well is the trivial component $\{1\}$ of the spectrum separated from the rest of the spectrum. This is one of the reasons why results on spectral gaps of various operators are highly prized.

We now give another proof of Theorem 2.8.5, based on the *energy decrement method*; this proof is significantly lengthier, but is particularly well suited for conversion to finitary quantitative settings. For any positive integer N, define the averaging operators

$$A_N := \frac{1}{N} \sum_{n=0}^{N-1} U^n;$$

by the triangle inequality we see that $\|A_N v\| \leq \|v\|$ for all v. Now we observe

Lemma 2.8.9 (Lack of uniformity implies energy decrement). *Suppose* $\|A_N v\| \geq \varepsilon$. *Then* $\|v - A_N^* A_N v\|^2 \leq \|v\|^2 - \varepsilon^2$.

Proof. This follows from the identity

$$(2.68) \qquad \|v - A_N^* A_N v\|^2 = \|v\|^2 - 2\|A_N v\|^2 + \|A_N^* A_N v\|^2$$

and the fact that A_N^* has operator norm at most 1. $\square$

We now iterate this to obtain

---

[33]It is instructive to use spectral theory to interpret von Neumann's proof of this theorem and see how it relates to the argument just given.

Proposition 2.8.10 (Koopman-von Neumann type theorem). *Let v be a unit vector, let $\varepsilon > 0$, and let $1 < N_1 < N_2 < \ldots < N_J$ be a sequence of integers with $J > 1/\varepsilon^2 + 2$. Then there exists $1 \le j < J$ and a decomposition $v = s + r$ where $\|Us - s\| = O(J\frac{1}{N_{j+1}})$ and $\|A_N r\| \le \varepsilon$ for all $N \ge N_j$.*

Remark 2.8.11. The letters s, r stand for "structured" and "random" (or "residual"), respectively. For more on decompositions into structured and random components, see [**Ta2007b**].

Proof. We perform the following algorithm:

(1) Initialise $j := J - 1$, $s := 0$, and $r := v$.

(2) If $\|A_N r\| \le \varepsilon$ for all $N \ge N_j$ then STOP. If instead $\|A_N r\| > \varepsilon$ for some $N \ge N_j$, observe from Lemma 2.8.9 that $\|r - A_N^* A_N r\|^2 \le \|r\|^2 - \varepsilon^2$.

(3) Replace r with $r - A_N^* A_N r$, replace s with $s + A_N^* A_N r$, and replace j with $j - 1$. Then return to step (2).

Observe that this procedure must terminate in at most $1/\varepsilon^2$ steps (since the energy $\|r\|^2$ starts at 1, drops by at least ε^2 at each stage, and cannot go below zero). In particular, j stays positive. Observe also that r always has norm at most 1, and thus $\|(U - I)A_N^* A_N r\| = O(1/N)$ at any given stage of the algorithm. From this and the triangle inequality one easily verifies the required claims. $\qquad\square$

Corollary 2.8.12 (Partial von Neumann ergodic theorem). *For any vector v, the averages $A_N v$ form a Cauchy sequence in H.*

Proof. Without loss of generality we can take v to be a unit vector. Suppose for contradiction that $A_N v$ was not Cauchy. Then one could find $\varepsilon > 0$ and $1 < N_1 < M_1 < N_2 < M_2 < \ldots$ such that $\|A_{N_j} v - A_{M_j} v\| \ge 5\varepsilon$ (say) for all j. By sparsifying the sequence if necessary we can assume that N_{j+1} is large compared to N_j, M_j, and ε. Now we apply Proposition 2.8.10 to find $j = O_\varepsilon(1)$ and a decomposition $v = s + r$ such that $\|Us - s\| = O_\varepsilon(1/N_{j+1})$ and $\|A_{N_j} r\|, \|A_{M_j} r\| \le \varepsilon$. If N_{j+1} is large enough depending on N_j, M_j, ε, we thus have $\|A_{N_j} s - s\|, \|A_{M_j} s - s\| \le \varepsilon$, and thus by the triangle inequality, $\|A_{N_j} v - A_{M_j} v\| \le 4\varepsilon$, a contradiction. $\qquad\square$

Remark 2.8.13. This result looks weaker than Theorem 2.8.5, but the argument is much more robust; for instance, one can modify it to establish convergence of multiple averages such as $\frac{1}{N}\sum_{n=1}^{N} T_1^n f_1 T_2^n f_2 T_3^n f_3$ in L^p norms for commuting shifts T_1, T_2, T_3; see [**Ta2008**]. Further quantitative analysis of the mean ergodic theorem can be found in [**AvGeTo2008**].

Corollary 2.8.12 can be used to recover Theorem 2.8.5 in its full strength, by combining it with a weak form of Theorem 2.8.5.

Proposition 2.8.14 (Weak von Neumann ergodic theorem). *The conclusion (2.61) of Theorem 2.8.5 holds in the weak topology.*

Proof. The averages $A_N v$ lie in a bounded subset of the separable Hilbert space H, and are thus sequentially precompact in the weak topology by the Banach-Alaoglu theorem. Thus, if (2.61) fails, then there exists a subsequence $A_{N_j} v$ which converges in the weak topology to some limit w other than $\pi(v)$. By telescoping series we see that $\|U A_{N_j} v - A_{N_j} v\| \leq 2\|v\|/N_j$, and so on taking limits we see that $\|Uw - w\| = 0$, i.e., $w \in H^U$. On the other hand, if y is any vector in H^U, then $A_{N_j}^* y = y$, and thus on taking inner products with v we obtain $\langle y, A_{N_j} v \rangle = \langle y, v \rangle$. Taking limits we obtain $\langle y, w \rangle = \langle y, v \rangle$, i.e., $v - w$ is orthogonal to H^U. These facts imply that $w = \pi(v)$, giving the desired contradiction. $\square$

2.8.2. Conditional expectation. We now turn away from the abstract Hilbert approach to the ergodic theorem (which is excellent for proving the mean ergodic theorem, but not flexible enough to handle more general ergodic theorems) and turn to a more measure-theoretic dynamics approach, based on manipulating the four components $X, \mathcal{X}, \mu, T$ of the underlying system separately, rather than working with the single object $L^2(X, \mathcal{X}, \mu)$ (with the unitary shift T). In particular, it is useful to replace the σ-algebra $\mathcal{X}$ by a sub-σ-algebra $\mathcal{X}' \subset \mathcal{X}$, thus reducing the number of measurable functions. This creates an isometric embedding of Hilbert spaces,

$$(2.69) \qquad L^2(X, \mathcal{X}', \mu) \subset L^2(X, \mathcal{X}, \mu),$$

and so the former space is a closed subspace of the latter. In particular, we have an orthogonal projection $\mathbf{E}(\cdot | \mathcal{X}') : L^2(X, \mathcal{X}, \mu) \to L^2(X, \mathcal{X}', \mu)$, which can be viewed as the adjoint of the inclusion (2.69). In other words, for any $f \in L^2(X, \mathcal{X}, \mu)$, $\mathbf{E}(f | \mathcal{X}')$ is the unique[34] element of $L^2(X, \mathcal{X}', \mu)$ such that

$$(2.70) \qquad \int_X \mathbf{E}(f | \mathcal{X}') \overline{g} \, d\mu = \int_X f \overline{g} \, d\mu$$

for all $g \in L^2(X, \mathcal{X}', \mu)$.

Example 2.8.15 (Finite case). Let X be a finite set; thus $\mathcal{X}$ can be viewed as a partition of X, and $\mathcal{X}' \subset \mathcal{X}$ is a coarser partition of X. To avoid degeneracies, assume that every point in X has positive measure with respect to μ. Then an element f of $L^2(X, \mathcal{X}, \mu)$ is just a function $f : X \to \mathbf{C}$ which is constant on each atom of $\mathcal{X}$. Similarly for $L^2(X, \mathcal{X}', \mu)$. The conditional expectation $\mathbf{E}(f | \mathcal{X}')$ is then the function whose value on each atom A of $\mathcal{X}'$

---

[34] A reminder: when dealing with L^p spaces, we identify any two functions which agree μ-almost everywhere. Thus, technically speaking, elements of L^p spaces are not actually functions, but rather equivalence classes of functions.

is equal to the average value $\frac{1}{\mu(A)} \int_A f \, d\mu$ on that atom. (What needs to be changed here if some points have zero measure?)

We leave the following standard properties of conditional expectation as an exercise.

Exercise 2.8.6. Let $(X, \mathcal{X}, \mu)$ be a probability space, and let $\mathcal{X}'$ be a sub-σ-algebra. Let $f \in L^2(X, \mathcal{X}, \mu)$.

(1) The operator $f \mapsto \mathbf{E}(f|\mathcal{X}')$ is a bounded self-adjoint projection on $L^2(X, \mathcal{X}, \mu)$. It maps real functions to real functions, it preserves constant functions (and more generally preserves $\mathcal{X}'$-valued functions), and commutes with complex conjugation.

(2) If f is non-negative, then $\mathbf{E}(f|\mathcal{X}')$ is non-negative (up to sets of measure zero, of course). More generally, we have a *comparison principle*: If f, g are real-valued and $f \leq g$ pointwise a.e., then $\mathbf{E}(f|\mathcal{X}') \leq \mathbf{E}(g|\mathcal{X}')$ a.e. Similarly, we have the *triangle inequality* $|\mathbf{E}(f|\mathcal{X}')| \leq \mathbf{E}(|f||\mathcal{X}')$ a.e.

(3) (Module property) If $g \in L^\infty(X, \mathcal{X}', \mu)$, then $\mathbf{E}(fg|\mathcal{X}') = \mathbf{E}(f|\mathcal{X}')g$ a.e.

(4) (Contraction) If $f \in L^2(X, \mathcal{X}, \mu) \cap L^p(X, \mathcal{X}, \mu)$ for some $1 \leq p \leq \infty$, then $\|\mathbf{E}(f|\mathcal{X}')\|_{L^p} \leq \|f\|_{L^p}$. *Hint:* Do the $p = 1$ and $p = \infty$ cases first. (This implies in particular that conditional expectation has a unique continuous extension to $L^p(X, \mathcal{X}, \mu)$ for $1 \leq p \leq \infty$; the $p = \infty$ case is exceptional, but note that L^∞ is contained in L^2 since μ is finite.)

For applications to ergodic theory, we will only be interested in taking conditional expectations with respect to a *shift-invariant* sub-σ-algebra $\mathcal{X}'$, thus T and T^{-1} preserve $\mathcal{X}'$. In that case T preserves $L^2(X, \mathcal{X}', \mu)$, and thus T commutes with conditional expectation, or in other words,

$$(2.71) \qquad \mathbf{E}(T^n f|\mathcal{X}') = T^n \mathbf{E}(f|\mathcal{X}')$$

a.e. for all $f \in L^2(X, \mathcal{X}, \mu)$ and all n.

Now we connect conditional expectation to the mean ergodic theorem. Let $\mathcal{X}^T := \{E \in \mathcal{X} : TE = E \text{ a.e.}\}$ be the set of essentially shift-invariant sets. One easily verifies that this is a shift-invariant sub-σ-algebra of $\mathcal{X}$.

Exercise 2.8.7. Show that if E lies in $\mathcal{X}^T$, then there exists a set $F \in \mathcal{X}$ which is genuinely invariant ($TF = F$) and which differs from E only by a set of measure zero. Thus it does not matter whether we deal with shift-invariance or essential shift-invariance here. (More generally, it will not make any significant difference if we modify any of the sets in our σ-algebras by null sets.)

The relevance of this algebra to the mean ergodic theorem arises from the following identity:

Exercise 2.8.8. Show that $L^2(X, \mathcal{X}, \mu)^T = L^2(X, \mathcal{X}^T, \mu)$.

As a corollary of this and Corollary 2.8.6, we have

Corollary 2.8.16 (Mean ergodic theorem, again)**.** *Let $(X, \mathcal{X}, \mu, T)$ be a measure-preserving system. Then for any $f \in L^2(X, \mathcal{X}, \mu)$, the averages $\frac{1}{N} \sum_{n=0}^{N-1} T^n f$ converge in L^2 norm to $\mathbf{E}(f|\mathcal{X}^T)$.*

Exercise 2.8.9. Show that Corollary 2.8.12 continues to hold if L^2 is replaced throughout by L^p for any $1 \leq p < \infty$. *Hint*: For the case $p < 2$, use that L^2 is dense in L^p. For the case $p > 2$, use that L^∞ is dense in L^p. What happens when $p = \infty$?

Let us now give another proof of Corollary 2.8.16 (leading to a fourth proof of the mean ergodic theorem). The key here will be the decomposition[35] $f = f_{U^\perp} + f_U$, where $f_{U^\perp} := \mathbf{E}(f|\mathcal{X}^T)$ is the "structured" part of f (at least as far as the mean ergodic theorem is concerned) and $f_U := f - f_{U^\perp}$ is the "random" part. As $f_{U^\perp}$ is shift-invariant, we clearly have

$$(2.72) \qquad \frac{1}{N} \sum_{n=0}^{N-1} T^n f_{U^\perp} = f_{U^\perp}$$

so it suffices to show that

$$(2.73) \qquad \left\| \frac{1}{N} \sum_{n=0}^{N-1} T^n f_U \right\|_{L^2}^2 \to 0$$

as $N \to \infty$. But we can expand out the left-hand side (using the unitarity of T) as

$$(2.74) \qquad \langle F_N, f_U \rangle := \int_X F_N \overline{f_U} \, d\mu,$$

where F_N is the *dual function* of f_U, defined as

$$(2.75) \qquad F_N := \frac{1}{N^2} \sum_{n=0}^{N-1} \sum_{m=0}^{N-1} T^{n-m} f_U.$$

Now, from the triangle inequality we know that the sequence of dual functions F_N is uniformly bounded in the L^2 norm, and so by the Cauchy-Schwarz inequality we know that the inner products $\langle F_N, f_U \rangle$ are bounded. If they converge to zero, we are done; otherwise, by the Bolzano-Weierstrass

---

[35]The subscripts $U^\perp, U$ stand for "anti-uniform" and "uniform", respectively; this notation is not standard.

theorem, we have $\langle F_{N_j}, f_U \rangle \to c$ for some subsequence N_j and some non-zero c. (One could also use ultrafilters instead of subsequences here if desired; it makes little difference to the argument.) By the *Banach-Alaoglu theorem* (or more precisely, the sequential version of this in the separable case), there is a further subsequence $F_{N'_j}$ which converges *weakly* (or equivalently in this Hilbert space case, in the *weak** sense) to some limit $F_\infty \in L^2(X, \mathcal{X}, \mu)$. Since c is non-zero, F_∞ must also be non-zero. On the other hand, from telescoping series one easily computes that $\|TF_N - F_N\|_{L^2}$ decays like $O(1/N)$ as $N \to \infty$, so on taking limits we have $TF_\infty - F_\infty = 0$. In other words, F_∞ lies in $L^2(X, \mathcal{X}^T, \mu)$.

On the other hand, by construction of f_U we have $\mathbf{E}(f_U|\mathcal{X}^T) = 0$. From (2.71) and linearity we conclude that $\mathbf{E}(F_N|\mathcal{X}^T) = 0$ for all N, so on taking limits we have $\mathbf{E}(F_\infty|\mathcal{X}^T) = 0$. But since F_∞ is already in $L^2(X, \mathcal{X}^T, \mu)$, we conclude that $F_\infty = 0$, a contradiction.

Remark 2.8.17. The above argument is lengthier than some of the other proofs of the mean ergodic theorem, but it turns out to be fairly robust; it demonstrates (using the compactness properties of certain "dual functions") that a function f_U with sufficiently strong "mixing" properties (in this case, we require that $\mathbf{E}(f_U|\mathcal{X}^T) = 0$) will cancel itself out when taking suitable ergodic averages, thus reducing the study of averages of f to the study of averages of $f_U = \mathbf{E}(f|\mathcal{X}^T)$. In the modern jargon, this means that $\mathcal{X}^T$ is (the σ-algebra induced by) a characteristic factor of the ergodic average $f \mapsto \lim_{N \to \infty} \frac{1}{N} \sum_{n=1}^{N} T^n f$. We will see further examples of characteristic factors for other averages later in this course.

Exercise 2.8.10. Let $(\Gamma, \cdot)$ be a countably infinite discrete group. A *Følner sequence* is a sequence of increasing finite non-empty sets F_n in Γ with $\bigcup_n F_n = \Gamma$ with the property that for any given finite set $S \subset \Gamma$, we have $|(F_n \cdot S)\Delta F_n|/|F_n| \to 0$ as $n \to \infty$, where $F_n \cdot S := \{fs : f \in F_n, s \in S\}$ is the product set of F_n and S, $|F_n|$ denotes the cardinality of F_n, and Δ denotes *symmetric difference*. (For instance, in the case $\Gamma = \mathbf{Z}$, the sequence $F_n := \{-n, \ldots, n\}$ is a Følner sequence.) Show that if Γ acts (on the left) in a measure-preserving manner on a probability space $(X, \mathcal{X}, \mu)$, and $f \in L^2(X, \mathcal{X}, \mu)$, then $\frac{1}{|F_n|} \sum_{\gamma \in F_n} f \circ \gamma^{-1}$ converges in L^2 to $\mathbf{E}(f|\mathcal{X}^\Gamma)$, where $\mathcal{X}^\Gamma$ is the collection of all measurable sets which are Γ-invariant modulo null sets, and $f \circ \gamma^{-1}$ is the function $x \mapsto f(\gamma^{-1}x)$.

Notes. This lecture first appeared at

terrytao.wordpress.com/2008/01/30.

Thanks to Lior Silberman, Pedro Lauridsen Ribeiro, Orr, mmailliw/william, Sugata, and Liu Xiao Chuan for corrections.

2.9. Ergodicity

We continue our study of basic ergodic theorems, establishing the maximal and pointwise ergodic theorems of Birkhoff. Using these theorems, we can then give several equivalent notions of the fundamental concept of *ergodicity*, which (roughly speaking) plays the role in measure-preserving dynamics that minimality plays in topological dynamics. A general measure-preserving system is not necessarily ergodic, but we shall introduce the *ergodic decomposition*, which allows one to express any non-ergodic measure as an average of ergodic measures (generalising the decomposition of a permutation into disjoint cycles).

2.9.1. The maximal ergodic theorem. Just as we derived the mean ergodic theorem from the more abstract von Neumann ergodic theorem in Section 2.8, we shall derive the maximal ergodic theorem from the following abstract maximal inequality.

Theorem 2.9.1 (Dunford-Schwartz maximal inequality). *Let $(X, \mathcal{X}, \mu)$ be a probability space, and let $P : L^1(X, \mathcal{X}, \mu) \to L^1(X, \mathcal{X}, \mu)$ be a linear operator with $P1 = 1$ and $P^*1 = 1$ (i.e., $\int_X Pf\, d\mu = \int_X f\, d\mu$ for all $f \in L^1(X, \mathcal{X}, \mu)$). Assume also that P maps non-negative functions to non-negative functions. Then the maximal function $Mf := \sup_{N>0} \frac{1}{N} \sum_{n=1}^{N} P^n f$ obeys the inequality*

$$(2.76) \qquad\qquad \lambda \mu(\{Mf > \lambda\}) \leq \int_{Mf > \lambda} f\, d\mu$$

for any $\lambda \in \mathbf{R}$.

Proof. We can rewrite (2.76) as

$$(2.77) \qquad\qquad \int_{Mf - \lambda > 0} (f - \lambda)\, d\mu \geq 0.$$

Since $Mf - \lambda = M(f - \lambda)$, we thus see (by replacing f with $f - \lambda$) that we can reduce to proving (2.77) in the case $\lambda = 0$.

For every $m \geq 1$, consider the modified maximal function

$$F_m := \sup_{0 \leq N \leq m} \sum_{n=0}^{N-1} P^n f.$$

Observe that $Mf(x) > 0$ if and only if $F_m(x) > 0$ for all sufficiently large m. By the dominated convergence theorem, it thus suffices to show that

$$(2.78) \qquad\qquad \int_{F_m > 0} f\, d\mu \geq 0$$

for all m. But observe from definition of F_m (and the positivity-preserving nature of P) that we have the pointwise recursive inequality

$$(2.79) \qquad F_m(x) \leq F_{m+1}(x) = \max(0, f + PF_m(x)).$$

Integrating this over the region $F_m > 0$ and using the non-negativity of F_m, we obtain

$$(2.80) \qquad \int_X F_m \, d\mu \leq \int_{F_m > 0} f + \int_X PF_m \, d\mu.$$

Since $F_m \in L^1(X, \mathcal{X}, \mu)$ and $P^*1 = 1$, the claim follows. $\qquad \square$

Applying this in the case when P is a shift operator, and replacing f by $|f|$, we obtain

Corollary 2.9.2 (Maximal ergodic theorem). *Let* $(X, \mathcal{X}, \mu, T)$ *be a measure-preserving system. Then for any* $f \in L^1(X, \mathcal{X}, \mu)$ *and* $\lambda > 0$ *one has*

$$(2.81) \qquad \mu\left(\left\{\sup_N \frac{1}{N} \sum_{n=0}^{N-1} |T^n f| > \lambda\right\}\right) \leq \frac{1}{\lambda} \|f\|_{L^1(X, \mathcal{X}, \mu)}.$$

Note that this inequality implies Markov's inequality

$$(2.82) \qquad \mu(\{|f| > \lambda\}) \leq \frac{1}{\lambda} \int_X |f| \, d\mu$$

as a special case. Applying the real interpolation method, one also easily deduces the maximal inequality

$$(2.83) \qquad \left\| \sup_N \frac{1}{N} \sum_{n=0}^{N-1} |T^n f| \right\|_{L^p(X, \mathcal{X}, \mu)} \leq C_p \|f\|_{L^p(X, \mathcal{X}, \mu)}$$

for all $1 < p \leq \infty$, where the constant C_p depends on p (it blows up like $O(1/(p-1))$ in the limit $p \to 1$).

Exercise 2.9.1 (Rising sun inequality). If $f \in l^1(\mathbf{Z})$, and

$$f^*(m) := \sup_N \frac{1}{N} \sum_{n=0}^{N-1} f(m + n),$$

establish the *rising sun inequality*

$$(2.84) \qquad \lambda |\{m \in \mathbf{Z} : f^*(m) > \lambda\}| \leq \sum_{m \in \mathbf{Z}} f(m)$$

for any $\lambda > 0$. *Hint*: One can either adapt the proof of Theorem 2.9.1, or else partition the set appearing in (2.84) into disjoint intervals. The latter proof also leads to a proof of Corollary 2.9.2 which avoids the Dunford-Schwartz trick of introducing the functions F_m. The terminology "rising sun" comes from seeing how these intervals interact with the graph of the partial sums of f, which resembles the shadows cast on a hilly terrain by a rising sun.

Exercise 2.9.2 (Transference principle). Show that Corollary 2.9.2 can be deduced directly from (2.84). *Hint*: Given $f \in L^1(X, \mathcal{X}, \mu)$, apply (2.84) to the functions $f_x(n) := T^n f(x)$ for each $x \in X$ (truncating the integers to a finite set if necessary), and then integrate in x using Fubini's theorem. (This is an example of a *transference principle* between maximal inequalities on $\mathbf{Z}$ and maximal inequalities on measure-preserving systems.)

Exercise 2.9.3 (Stein-Stromberg maximal inequality [**StSt1983**]). Derive a continuous version of the Dunford-Schwartz maximal inequality, in which the operators P^n are replaced by a semigroup P_t acting on both L^1 and L^∞, in which the underlying measure space is only assumed to be σ-finite rather than a probability space, and the averages $\frac{1}{N} \sum_{n=0}^{N-1} P^n$ are replaced by $\frac{1}{T} \int_0^T P^t \, dt$. Apply this continuous version with $P_t := e^{t\Delta}$ equal to the heat operator on $\mathbf{R}^d$ for $d \geq 1$ to deduce the *Stein-Stromberg maximal inequality*[36]

$$(2.85) \quad m\left(\left\{x \in \mathbf{R}^d : \sup_{R>0} \frac{1}{m(B(x,R))} \int_{B(x,r)} |f| \, dm > \lambda\right\}\right) \leq \frac{Cd}{\lambda} \|f\|_{L^1(\mathbf{R}^d, dm)}$$

for all $\lambda > 0$ and $f \in L^1(\mathbf{R}^d, dm)$, where m is Lebesgue measure, $B(x, R)$ is the Euclidean ball of radius R centred at x, and the constant C is absolute (independent of d).

Remark 2.9.3. The study of maximal inequalities in ergodic theory is, of course, a subject in itself; a classical reference is [**St1970**].

2.9.2. The pointwise ergodic theorem. Using the maximal ergodic theorem and a standard limiting argument we can now deduce

Theorem 2.9.4 (Pointwise ergodic theorem). *Let $(X, \mathcal{X}, \mu, T)$ be a measure-preserving system, and let $f \in L^1(X, \mathcal{X}, \mu)$. Then for μ-almost every $x \in X$, $\frac{1}{N} \sum_{n=0}^{N-1} T^n f(x)$ converges to $\mathbf{E}(f|\mathcal{X}^T)(x)$.*

Proof. By subtracting $\mathbf{E}(f|\mathcal{X}^T)$ from f if necessary, it suffices to show that

$$(2.86) \quad \limsup_{N \to \infty} \left| \frac{1}{N} \sum_{n=0}^{N-1} T^n f(x) \right| = 0$$

a.e. whenever $\mathbf{E}(f|\mathcal{X}^T) = 0$. By telescoping series, (2.86) is already true when f takes the form $f = Tg - g$ for some $g \in L^\infty(X, \mathcal{X}, \mu)$. So by the arguments used to prove Theorem 2.8.5, we have already established the

---

[36]This improves upon the *Hardy-Littlewood maximal inequality*, which gives the same estimate but with Cd replaced by C^d. It is an open question whether the dependence on d can be removed entirely; the estimate (2.85) is still the best known in high dimension. For $d = 1$, the best constant C is known to be $\frac{11+\sqrt{61}}{12} = 1.567\ldots$, a result of Melas [**Me2003**].

claim for a dense class of functions f in $L^2(X, \mathcal{X}, \mu)$ with $\mathbf{E}(f|\mathcal{X}^T) = 0$, and thus also for a dense class of functions in $L^1(X, \mathcal{X}, \mu)$ with $\mathbf{E}(f|\mathcal{X}^T) = 0$ (since the latter space is dense in the former, and the L^2 norm controls the L^1 norm by the Cauchy-Schwarz inequality).

Now we use a standard limiting argument. Let $f \in L^1(X, \mathcal{X}, \mu)$ with $\mathbf{E}(f|\mathcal{X}^T) = 0$. Then we can find a sequence f_j in the above dense class which converges in L^1 to f. For almost every x, we thus have

$$(2.87) \qquad \lim_{N \to \infty} \left| \frac{1}{N} \sum_{n=0}^{N-1} T^n f_j(x) \right| = 0$$

for all j, and so by the triangle inequality we have

$$(2.88) \qquad \limsup_{N \to \infty} \left| \frac{1}{N} \sum_{n=0}^{N-1} T^n f(x) \right| \le \sup_N \frac{1}{N} \sum_{n=0}^{N-1} T^n |f - f_j|(x).$$

But by Corollary 2.9.2 we see that the right-hand side of (2.88) converges to zero in measure as $j \to \infty$. Since the left-hand side does not depend on j, it must vanish almost everywhere, as required. $\qquad \square$

Remark 2.9.5. More generally, one can derive a pointwise convergence result on a class of rough functions by first establishing convergence for a dense subclass of functions, and then establishing a maximal inequality which is strong enough to allow one to take limits and establish pointwise convergence for all functions in the larger class. Conversely, principles such as Stein's maximal principle [**St1961**] indicate that in many cases this is in some sense the *only* way to establish such pointwise convergence results for rough functions.

Remark 2.9.6. Using the dominated convergence theorem (starting first with bounded functions f in order to get the domination), one can deduce the mean ergodic theorem from the pointwise ergodic theorem. But the converse is significantly more difficult; pointwise convergence for various ergodic averages is often a much harder result to establish than the corresponding norm convergence result (in particular, many of the techniques discussed in this course appear to be of sharply limited utility for pointwise convergence problems), and many questions in this area remain open.

Exercise 2.9.4 (Lebesgue differentiation theorem). Let $f \in L^1(\mathbf{R}^d, dm)$ with Lebesgue measure dm. Show that for almost every $x \in \mathbf{R}^d$, we have that $\lim_{r \to 0^+} \frac{1}{m(B(x,r))} \int_{B(x,r)} |f(y) - f(x)| \, dx = 0$, and in particular that $\lim_{r \to 0^+} \frac{1}{m(B(x,r))} \int_{B(x,r)} f(y) \, dx = f(x)$.

2.9.3. Ergodicity. Combining the mean ergodic theorem with the pointwise ergodic theorem (and with Exercises 2.8.7, 2.8.8) we obtain

Theorem 2.9.7 (Characterisations of ergodicity). *Let $(X, \mathcal{X}, \mu, T)$ be a measure-preserving system. Then the following are equivalent:*

(1) *Any set $E \in \mathcal{X}$ which is invariant (thus $TE = E$) has either full measure $\mu(E) = 1$ or zero measure $\mu(E) = 0$.*

(2) *Any set $E \in \mathcal{X}$ which is almost invariant (thus TE differs from E by a null set) has either full measure or zero measure.*

(3) *Any measurable function f with $Tf = f$ a.e. is constant a.e.*

(4) *For any $1 < p < \infty$ and $f \in L^p(X, \mathcal{X}, \mu)$, the averages $\frac{1}{N} \sum_{n=0}^{N} T^n f$ converge in the L^p norm to $\int_X f$.*

(5) *For any two $f, g \in L^\infty(X, \mathcal{X}, \mu)$, we have*

$$\lim_{N \to \infty} \frac{1}{N} \sum_{n=1}^{N} \int_X (T^n f) g \, d\mu = \left(\int_X f \, d\mu \right) \left(\int_X g \, d\mu \right).$$

(6) *For any two measurable sets E and F, we have*

$$\lim_{N \to \infty} \frac{1}{N} \sum_{n=1}^{N} \mu(T^n E \cap F) = \mu(E)\mu(F).$$

(7) *For any $f \in L^1(X, \mathcal{X}, \mu)$, the averages $\frac{1}{N} \sum_{n=0}^{N} T^n f$ converge pointwise almost everywhere to $\int_X f \, d\mu$.*

A measure-preserving system with any (and hence all) of the above properties is said to be *ergodic*.

Remark 2.9.8. Strictly speaking, ergodicity is a property that applies to a measure-preserving system $(X, \mathcal{X}, \mu, T)$. However, we shall sometimes abuse notation and apply the adjective "ergodic" to a single component of a system, such as the measure μ or the shift T, when the other three components of the system are clear from the context.

Here are some simple examples of ergodicity:

Example 2.9.9. If X is finite with uniform measure, then a shift map $T : X \to X$ is ergodic if and only if it is a cycle.

Example 2.9.10. If a shift T is ergodic, then so is T^{-1}. However, from Example 2.9.9 we see that it is not necessarily true that T^n is ergodic for all n (this latter property is also known as *total ergodicity*).

Exercise 2.9.5. Show that the circle shift $(\mathbf{R}/\mathbf{Z}, x \mapsto x+\alpha)$ (with the usual Lebesgue measure) is ergodic if and only if α is irrational. *Hint*: Analyse the equation $Tf = f$ for (say) $f \in L^2(X, \mathcal{X}, \mu)$ using Fourier analysis. Another way to proceed is to use the Lebesgue density theorem (or Lebesgue differentiation theorem) combined with Exercise 2.6.15.

Exercise 2.9.6. Let $(\Omega, \mathcal{B}, \mu)$ be a probability space. Show that the Bernoulli shift on the product system $(\Omega^{\mathbf{Z}}, \mathcal{B}^{\mathbf{Z}}, \mu^{\mathbf{Z}})$ is ergodic. *Hint*: First establish property (6) of Theorem 2.9.7 when E and F each depend on only finitely many of the coordinates of $\Omega^{\mathbf{Z}}$.

Exercise 2.9.7. Let $(X, \mathcal{X}, \mu, T)$ be an ergodic system. Show that if λ is an eigenvalue of $T : L^2(X, \mathcal{X}, \mu) \to L^2(X, \mathcal{X}, \mu)$, then $|\lambda| = 1$, the eigenspace $\{f \in L^2(X, \mathcal{X}, \mu) : Tf = \lambda f\}$ is one-dimensional, and every eigenfunction f has constant magnitude $|f|$ a.e. Show that the eigenspaces are orthogonal to each other in $L^2(X, \mathcal{X}, \mu)$, and the set of all eigenvalues of T forms an at most countable subgroup of the unit circle S^1.

Now we give a less trivial example of an ergodic system.

Proposition 2.9.11 (Ergodicity of skew shift). *Let $\alpha \in \mathbf{R}$ be irrational. Then the skew shift $((\mathbf{R}/\mathbf{Z})^2, (x, y) \mapsto (x + \alpha, y + x))$ is ergodic.*

Proof. Write the skew shift system as $(X, \mathcal{X}, \mu, T)$. To simplify the notation we shall omit the phrase "almost everywhere" in what follows.

We use an argument of Parry [**Pa1969**]. If the system is not ergodic, then we can find a non-constant $f \in L^2(X, \mathcal{X}, \mu)$ such that $Tf = f$. Next, we use Fourier analysis to write $f = \sum_m f_m$, where

$$f_m(x, y) := \int_{\mathbf{R}/\mathbf{Z}} f(x, y + \theta) e^{-2\pi i m \theta} \, d\theta.$$

Since f is T-invariant, and the vertical rotations $(x, y) \mapsto (x, y + \theta)$ commute with T, we see that the f_m are also T-invariant. The function f_0 depends only on the x variable, and so is constant by Exercise 2.9.5. So it suffices to show that f_m is zero for all non-zero m.

Fix m. We can factorise $f_m(x, y) = F_m(x) e^{2\pi i m y}$. The T-invariance of f_m now implies that $F_m(x + \alpha) = e^{-2\pi i m x} F_m(x)$. If we then define $F_{m,\theta} := F_m(x + \theta) \overline{F_m(x)}$ for $\theta \in \mathbf{R}$, we see that $F_{m,\theta}(x + \alpha) = e^{-2\pi i m \theta} F_{m,\theta}(x)$, thus $F_{m,\theta}$ is an eigenfunction of the circle shift with eigenvalue $e^{-2\pi i m \theta}$. But this implies (by Exercise 2.9.7) that $F_{m,\theta}$ is orthogonal to $F_{m,0}$ for θ close to zero. Taking limits we see that $F_{m,0}$ is orthogonal to itself and must vanish; this implies that F_m and hence f_m vanish as well, as desired. $\square$

Exercise 2.9.8. Show that for any irrational α and any $d \geq 1$, the iterated skew shift system $(\mathbf{R}/\mathbf{Z}^d, (x_1, \ldots, x_d) \to (x_1 + \alpha, x_2 + x_1, \ldots, x_d + x_{d-1}))$ is ergodic.

2.9.4. Generic points. Now let us suppose that we have a topological measure preserving system $(X, \mathcal{F}, \mu, T)$, i.e., a measure-preserving system $(X, \mathcal{X}, \mu, T)$ which is also a topological dynamical system $(X, \mathcal{F}, T)$, with $\mathcal{X}$ the Borel σ-algebra of T. Then we have the space $C(X)$ of continuous (real- or complex-valued) functions on X, which is dense inside $L^2(X)$. From the Stone-Weierstrass theorem we also see that $C(X)$ is separable.

Definition 2.9.12. Let $(X, \mathcal{X}, \mu)$ be a probability space. A sequence $x_1, x_2, x_3, \ldots$ in X is said to be *uniformly distributed* with respect to μ if we have

$$(2.89) \qquad \lim_{N \to \infty} \frac{1}{N} \sum_{n=1}^{N} f(x_n) = \int_X f \, d\mu$$

for all $f \in C(X)$. A point x in X is said to be *generic* if the forward orbit $x, Tx, T^2x, \ldots$ is uniformly distributed.

Exercise 2.9.9. Let $(X, \mathcal{F}, \mu)$ be a compact metrisable space with a Borel probability measure μ, and let $x_1, x_2, \ldots$ be a sequence in X. Show that this sequence is uniformly distributed if and only if

$$\lim_{N \to \infty} \frac{1}{N} |\{1 \leq i \leq N : x_i \in U\}| = \mu(U)$$

for all open sets U in X.

From Theorem 2.9.7 and the separability of $C(X)$ we obtain

Proposition 2.9.13. *A topological measure-preserving system is ergodic if and only if almost every point is generic.*

A topological measure-preserving system is said to be *uniquely ergodic* if *every* point is generic. The following exercise explains the terminology:

Exercise 2.9.10. Show that a topological measure-preserving system $(X, \mathcal{F}, \mu, T)$ is uniquely ergodic if and only if the only T-invariant Borel probability measure on T is μ. *Hint*: Use Lemma 2.7.16. Because of this fact, one can sensibly define what it means for a topological dynamical system $(X, \mathcal{F}, T)$ to be uniquely ergodic, namely that it has a unique T-invariant Borel probability measure.

It is not always the case that an ergodic system is uniquely ergodic. For instance, in the Bernoulli system $\{0, 1\}^{\mathbf{Z}}$ (with uniform measure on $\{0, 1\}$, say), the point $0^{\mathbf{Z}}$ is not generic. However, for more algebraic systems, it turns out that ergodicity and unique ergodicity are largely equivalent. We illustrate this with the circle and skew shifts:

Exercise 2.9.11. Show that the circle shift $(\mathbf{R}/\mathbf{Z}, x \mapsto x + \alpha)$ (with the usual Lebesgue measure) is uniquely ergodic if and only if α is irrational.

Hint: First show in the circle shift system that any translate of a generic point is generic.

Proposition 2.9.14 (Unique ergodicity of skew shift). *Let $\alpha \in \mathbf{R}$ be irrational. Then the skew shift $((\mathbf{R}/\mathbf{Z})^2, (x,y) \mapsto (x+\alpha, y+x))$ is uniquely ergodic.*

Proof. We use an argument of Furstenberg [**Fu1981**]. We again write the skew shift as $(X, \mathcal{X}, \mu, T)$. Suppose this system is not uniquely ergodic; then by Exercise 2.9.10 there is another shift-invariant Borel probability measure $\mu' \neq \mu$. If we push μ and μ' down to the circle shift system $(\mathbf{R}/\mathbf{Z}, x \mapsto x+\alpha)$ by the projection map $(x,y) \mapsto x$, then by Exercises 2.9.10, 2.9.11 we must get the same measure. Thus μ and μ' must agree on any set of the form $A \times (\mathbf{R}/\mathbf{Z})$.

Let E denote the points in X which are generic with respect to μ; note that this set is Borel measurable. By Proposition 2.9.13, this set has full measure in μ. Also, since the vertical rotations $(x,y) \mapsto (x, y+\theta)$ commute with T and preserve μ, we see that E must be invariant under such rotations; thus they are of the form $A \times (\mathbf{R}/\mathbf{Z})$ for some A. By the preceding discussion, we conclude that E also has full measure in μ'. But then (by the pointwise or mean ergodic theorem for $(X, \mathcal{X}, \mu', T)$) we conclude that $\mathbf{E}_{\mu'}(f|\mathcal{X}^T) = \int_X f \, d\mu$ μ'-almost everywhere for every continuous f, and thus on integrating with respect to μ' we obtain $\int_X f \, d\mu' = \int_X f \, d\mu$ for every continuous f. But then, by the Riesz representation theorem, we have $\mu = \mu'$, a contradiction. $\square$

Corollary 2.9.15. *If $\alpha \in \mathbf{R}$ is irrational, then the sequence $(\alpha n^2 \bmod 1)_{n \in \mathbf{N}}$ is uniformly distributed in $\mathbf{R}/\mathbf{Z}$ (with respect to uniform measure).*

Exercise 2.9.12. Show that the systems considered in Exercise 2.9.8 are uniquely ergodic. Conclude that the exponent 2 in Corollary 2.9.15 can be replaced by any positive integer d.

Note that the topological dynamics theory developed in Section 2.6 only establishes the weaker statement that the above sequence is dense in $\mathbf{R}/\mathbf{Z}$ rather than uniformly distributed. More generally, it seems that ergodic theory methods can prove topological dynamics results, but not vice versa. Here is another simple example of the same phenomenon:

Exercise 2.9.13. Show that a uniquely ergodic topological dynamical system is necessarily minimal. (The converse is not necessarily true, as already mentioned in Remark 2.7.18.)

2.9.5. The ergodic decomposition. Just as not every topological dynamical system is minimal, not every measure-preserving system is ergodic.

Nevertheless, there is an important decomposition that allows one to represent non-ergodic measures as averages of ergodic measures. One can already see this in the finite case, when X is a finite set with the discrete σ-algebra, and $T : X \to X$ is a permutation on X, which can be decomposed as the disjoint union of cycles on a partition $X = C_1 \cup \ldots \cup C_m$ of X. In this case, all shift-invariant probability measures take the form

$$(2.90) \qquad \qquad \mu = \sum_{j=1}^{m} \alpha_j \mu_j,$$

where μ_j is the uniform probability measure on the cycle C_j, and the α_j are non-negative constants adding up to 1. Each of the μ_j is ergodic, but no non-trivial linear combination of these measures is ergodic. Thus we see in the finite case that every shift-invariant measure can be uniquely expressed as a convex combination of ergodic measures.

It turns out that a similar decomposition is available in general, at least if the underlying measure space is a compact topological space (or more generally, a Radon space). This is because of the following general theorem from measure theory.

Definition 2.9.16 (Probability kernel). Let $(X, \mathcal{X})$ and $(Y, \mathcal{Y})$ be measurable spaces. A *probability kernel* $y \mapsto \mu_y$ is an assignment of a probability measure μ_y on X to each $y \in Y$ in such a way that the map $y \mapsto \int_X f \, d\mu_y$ is measurable for every bounded measurable $f : X \to \mathbf{C}$.

Example 2.9.17. Every measurable map $\phi : Y \to X$ induces a probability kernel $y \mapsto \delta_{\phi(y)}$. Every probability measure on X can be viewed as a probability kernel from a point to X. If $y \mapsto \mu_y$ and $x \mapsto \nu_x$ are two probability kernels from Y to X and from X to Z respectively, their composition $x \mapsto (\mu \circ \nu)_x := \int_X \mu_y \, d\nu_x(y)$ is also a probability kernel, where $\int_X \mu_y \, d\nu_x(y)$ is the measure that assigns $\int_X \mu_y(E) \, d\nu_x(y)$ to any measurable set E in Z. Thus one can view the class of measurable spaces and their probability kernels as a category, which includes the class of measurable spaces and their measurable maps as a subcategory.

Definition 2.9.18 (Regular space). A measurable space $(X, \mathcal{X})$ is said to be *regular* if there exists a compact metrisable topology $\mathcal{F}$ on X for which $\mathcal{X}$ is the Borel σ-algebra.

Example 2.9.19. Every topological measure-preserving system is regular.

Remark 2.9.20. Measurable spaces $(X, \mathcal{X})$ in which $\mathcal{X}$ is the Borel σ-algebra of a topological space generated by a separable complete metric space (i.e., a Polish space) are known as *standard Borel spaces*. It is a non-trivial theorem from descriptive set theory that, up to measurable isomorphism, there are only three types of standard Borel spaces: finite discrete spaces, countable discrete spaces, and the unit interval $[0,1]$ with the usual Borel σ-algebra. From this, one can see that regular spaces are the same as standard Borel spaces, though we will not need this fact here.

Theorem 2.9.21 (Disintegration theorem). *Let $(X, \mathcal{X}, \mu)$ and $(Y, \mathcal{Y}, \nu)$ be probability spaces, with $(X, \mathcal{X})$ regular. Let $\pi : X \to Y$ be a morphism (thus $\nu = \pi_{\#}\mu$). Then there exists a probability kernel $y \mapsto \mu_y$ such that*

$$(2.91) \qquad \int_X f(g \circ \pi) \, d\mu = \int_Y \left(\int_X f \, d\mu_y \right) g(y) \, d\nu(y)$$

for any bounded measurable $f : X \to \mathbf{C}$ and $g : Y \to \mathbf{C}$. Also, for any such g, we have

$$(2.92) \qquad\qquad g \circ \pi = g(y) \quad \mu_y\text{-a.e.}$$

for ν-a.e. y.

Furthermore, this probability kernel is unique up to ν-almost everywhere equivalence, in the sense that if $y \mapsto \mu'_y$ is another probability kernel with the same properties, then $\mu_y = \mu'_y$ for ν-almost every y.

We refer to the probability kernel $y \mapsto \mu_y$ generated by the above theorem as the *disintegration* of μ relative to the factor map π.

Proof. We begin by proving uniqueness. Suppose we have two probability kernels $y \mapsto \mu_y$, $y \mapsto \mu'_y$ with the above properties. Then on subtraction we have

$$(2.93) \qquad \int_Y \left(\int_X f \, d(\mu_y - \mu'_y) \right) g(y) \, d\nu(y) = 0$$

for all bounded measurable $f : X \to \mathbf{C}$, $g : Y \to \mathbf{C}$. Specialising to $f = 1_E$ for some measurable set $E \in \mathcal{X}$, we conclude that $\mu_y(E) = \mu'_y(E)$ for ν-almost every y. Since $\mathcal{X}$ is regular, it is separable and we conclude that $\mu_y = \mu'_y$ for ν-almost every y, as required.

Now we prove existence. The pullback map

$$\pi^{\#} : L^2(Y, \mathcal{Y}, \nu) \to L^2(X, \mathcal{X}, \mu)$$

defined by $g \mapsto g \circ \pi$ has an adjoint $\pi_{\#} : L^2(X, \mathcal{X}, \mu) \to L^2(Y, \mathcal{Y}, \nu)$, thus

$$(2.94) \qquad \int_X f(g \circ \pi) \, d\mu = \int_Y (\pi_{\#}f) g \, d\nu$$

for all $f \in L^2(X, \mathcal{X}, \mu)$ and $g \in L^2(Y, \mathcal{Y}, \nu)$. It is easy to see from duality that we have $\|\pi_{\#}f\|_{L^\infty(Y,\mathcal{Y},\nu)} \le \|f\|_{C(X)}$ for all $f \in C(X)$ (where we select a compact metrisable topology that generates the regular σ-algebra $\mathcal{X}$). Recall that $\pi_{\#}f$ is not quite a measurable function, but is instead an equivalence class of measurable functions modulo ν-almost everywhere equivalence. Since $C(X)$ is separable, we find a measurable representative

$\tilde{\pi}_\# f : Y \to \mathbf{C}$ of $\pi_\# f$ for every $f \in C(X)$, which varies linearly with f, and is such that $|\tilde{\pi}_\# f(y)| \leq \|f\|_{C(X)}$ for all y outside of a set E of ν-measure zero and for all $f \in C(X)$. For all such y, we can then apply the Riesz representation theorem to obtain a Radon probability measure μ_y with

$$(2.95) \qquad\qquad \tilde{\pi}_\# f(y) = \int_X f \, d\mu_y$$

for all such y. We set μ_y equal to some arbitrarily fixed Radon probability measure for $y \in E$. We then observe that the required properties (including the measurability of $y \mapsto \int_X f \, d\mu_y$) are already obeyed for $f \in C(X)$. To generalise this to bounded measurable f, observe that the class $\mathcal{C}$ of f obeying the required properties is closed under dominated pointwise convergence, and so contains the indicator functions of open or compact sets (by *Urysohn's lemma*). Applying dominated pointwise convergence again, together with inner and outer regularity, we see that the indicator functions of Borel sets lie in $\mathcal{C}$. Thus all simple measurable functions lie in $\mathcal{C}$, and on taking uniform limits we obtain the claim.

Finally, we prove (2.92). From two applications of (2.91) we have

$$(2.96) \quad \int_Y \left(\int_X f(g \circ \pi) \, d\mu_y \right) h(y) \, d\nu(y) = \int_Y \left(\int_X f g(y) \, d\mu_y \right) h(y) \, d\nu(y)$$

for all bounded measurable $f : X \to \mathbf{C}$ and $h : Y \to \mathbf{C}$. The claim follows (using the separability of the space of all f). $\qquad\qquad\qquad\qquad\qquad\square$

Exercise 2.9.14. Let the notation and assumptions be as in Theorem 2.9.21. Suppose that $\mathcal{Y}$ is also regular, and that the map $\pi : X \to Y$ is continuous with respect to some compact metrisable topologies that generate $\mathcal{X}$ and $\mathcal{Y}$, respectively. Then show that for ν-almost every y, the probability measure ν_y is supported in $\pi^{-1}(\{y\})$.

Proposition 2.9.22 (Ergodic decomposition). *Let $(X, \mathcal{X}, \mu, T)$ be a regular measure-preserving system. Let $(Y, \mathcal{Y}, \nu, S)$ be the system defined by $Y := X$, $\mathcal{Y} := \mathcal{X}^T$, $\nu := \mu \downharpoonright_{\mathcal{Y}}$, and $S := T$, and let $\pi : X \to Y$ be the identity map. Let $y \mapsto \mu_y$ be the disintegration of μ with respect to the factor map π. Then for ν-almost every y, the measure μ_y is T-invariant and ergodic.*

Proof. Observe from the T-invariance $\mu = T_\# \mu$ of μ (and of $\mathcal{X}^T$) that the probability kernel $y \mapsto T_\# \mu_y$ would also be a disintegration of μ. Thus we have $\mu_y = T_\# \mu_y$ for ν-almost every y.

Now we show the ergodicity. As the space of bounded measurable $f : X \to \mathbf{C}$ is separable, it suffices by Theorem 2.9.7 and a limiting argument to show that for any fixed such f, the averages $\frac{1}{N} \sum_{n=1}^N T^n f$ converge pointwise μ_y-a.e. to $\int_X f \, d\mu_y$ for ν-a.e. y.

From the pointwise ergodic theorem, we already know that $\frac{1}{N}\sum_{n=1}^{N} T^n f$ converges to $\mathbf{E}(f|\mathcal{X}^T)$ outside of a set of μ-measure zero. By (2.91), this set also has μ_y-measure zero for ν-almost every y. Thus it will suffice to show that $\mathbf{E}(f|\mathcal{X}^T)$ is μ_y-a.e. equal to $\int_X f \, d\mu_y$ for ν-a.e. y. Now observe that $\mathbf{E}(f|\mathcal{X}^T)(x) = \pi_\# f(\pi(x))$, so the claim follows from (2.92) and (2.95). $\qquad\square$

Exercise 2.9.15. Let $(X, \mathcal{X})$ be a separable measurable space, and let T be a bimeasurable bijection $T : X \to X$. Let $M(\mathcal{X})$ denote the Banach space of all finite measures on $\mathcal{X}$ with the total variation norm. Let $\mathrm{Pr}(\mathcal{X})^T \subset M(\mathcal{X})$ denote the collection of probability measures on $\mathcal{X}$ which are T-invariant. Show that this is a closed convex subset of $M(\mathcal{X})$, and the extreme points of $\mathrm{Pr}(\mathcal{X})^T$ are precisely the ergodic probability measures (which also form a closed subset of $M(\mathcal{X})$). (This allows one to prove a variant of Proposition 2.9.22 using *Choquet's theorem*.)

Exercise 2.9.16. Show that a topological measure-preserving system $(X, \mathcal{F}, T, \mu)$ is uniquely ergodic if and only if the only ergodic shift-invariant Borel probability measure on X is μ.

Notes. This lecture first appeared at

terrytao.wordpress.com/2008/02/04.

Thanks to mykola, Lior Silberman, and Liu Xiao Chuan for corrections.

2.10. The Furstenberg correspondence principle

In this lecture, we describe the simple but fundamental *Furstenberg correspondence principle*, which connects the "soft analysis" subject of ergodic theory (in particular, recurrence theorems) with the "hard analysis" subject of combinatorial number theory (or more generally with results of "density Ramsey theory" type).[37] Rather than try to set up the most general and abstract version of this principle, we shall instead study the canonical example of this principle in action, namely the equating of the *Furstenberg multiple recurrence theorem* with Szemerédi's theorem on arithmetic progressions.

In [**Sz1975**], Szemerédi established the following theorem, which had been conjectured by Erdős and Turán [**ErTu1936**]:

Theorem 2.10.1 (Szemerédi's theorem). *Let $k \geq 1$ be an integer, and let A be a set of integers of positive upper density, thus*

$$\limsup_{N \to \infty} \frac{1}{2N+1} |A \cap \{-N, \ldots, N\}| > 0.$$

---

[37]See Section 1.3 of *Structure and Randomness* for a discussion of the relationship between soft and hard analysis.

Then A contains a non-trivial arithmetic progression $n, n+r, \ldots, n+(k-1)r$ of length k. (By "non-trivial" we mean that $r \neq 0$.) Or more succinctly: Every set of integers of positive upper density contains arbitrarily long arithmetic progressions.

Remark 2.10.2. This theorem is trivial for $k = 1$ and $k = 2$. The first non-trivial case is $k = 3$, which was proven in [**Ro1953**] and will be discussed in Section 2.12.4. The $k = 4$ case was also established earlier in [**Sz1969**].

In [**Fu1977**], Furstenberg gave another proof of Szemerédi's theorem, by establishing the following equivalent statement:

Theorem 2.10.3 (Furstenberg multiple recurrence theorem). *Let $k \geq 1$ be an integer, let $(X, \mathcal{X}, \mu, T)$ be a measure-preserving system, and let E be a set of positive measure. Then there exists $r > 0$ such that $E \cap T^{-r}E \cap \ldots \cap T^{-(k-1)r}E$ is non-empty.*

Remark 2.10.4. The negative signs here can be easily removed because T is invertible, but I have placed them for consistency with some later results involving non-invertible transformations, in which the negative sign becomes important.

Exercise 2.10.1. Prove that Theorem 2.10.3 is equivalent to the apparently stronger theorem in which "is non-empty" is replaced by "has positive measure", and "there exists $r > 0$" is replaced by "there exist infinitely many $r > 0$".

Note that the $k = 1$ case of Theorem 2.10.3 is trivial, while the $k = 2$ case follows from the Poincaré recurrence theorem (Theorem 2.8.2). We will prove the higher k cases of this theorem in Sections 2.11–2.15. In this section, we explain why, for any fixed k, Theorem 2.10.1 and Theorem 2.10.3 are equivalent.

Let us first give the easy part that Theorem 2.10.1 implies Theorem 2.10.3. This follows immediately from

Lemma 2.10.5. *Let $(X, \mathcal{X}, \mu, T)$ be a measure-preserving system, and let E be a set of positive measure. Then there exists a point x in X such that the recurrence set $\{n \in \mathbf{Z} : T^n x \in E\}$ has positive upper density.*

Indeed, from Lemma 2.10.5 and Theorem 2.10.1, we obtain a point x for which the set $\{n \in \mathbf{Z} : T^n x \in E\}$ contains an arithmetic progression of length k and some step r, which implies that $E \cap T^r E \cap \ldots \cap T^{(k-1)r}E$ is non-empty.

Proof of Lemma 2.10.5. Observe (from the shift-invariance of μ) that

$$(2.97) \qquad \int_X \frac{1}{2N+1} \sum_{n=-N}^{N} 1_{T^n E} \, d\mu = \mu(E).$$

On the other hand, the integrand is at most 1. We conclude that for each N, the set $A_N := \{x : \frac{1}{2N+1} \sum_{n=-N}^{N} 1_{T^n E}(x) \geq \mu(E)/2\}$ must have measure at least $\mu(E)/2$. This implies that the function $\sum_N 1_{A_N}$ is not absolutely integrable even after excluding an arbitrary set of measure up to $\mu(E)/4$, which implies that $\sum_N 1_{A_N}$ is not finite a.e., and the claim follows (cf. the proof of the Borel-Cantelli lemma, Lemma 1.4.5). $\qquad \square$

Now we show how Theorem 2.10.3 implies Theorem 2.10.1. If we could pretend that "upper density" was a probability measure on the integers, then this implication would be immediate by applying Theorem 2.10.3 to the dynamical system $(\mathbf{Z}, n \mapsto n+1)$. Of course, we know that the integers do not admit a shift-invariant probability measure (and upper density is not even additive, let alone a probability measure). So this does not work directly. Instead, we need to first lift from the integers to a more abstract universal space and use a standard "compactness and contradiction" argument in order to be able to build the desired probability measure properly.

More precisely, let A be as in Theorem 2.10.1. Consider the topological boolean Bernoulli dynamical system $2^{\mathbf{Z}}$ with the product topology and the shift $T : B \mapsto B+1$. The set A can be viewed as a point in this system, and the orbit closure $X := \overline{\{A+n : n \in \mathbf{Z}\}}$ of that point becomes a subsystem of that Bernoulli system, with the relative topology.

Suppose for contradiction that A contains no non-trivial progressions of length k, thus $A \cap A + r \cap \ldots \cap A + (k-1)r = \emptyset$ for all $r > 0$. Then, if we define the *cylinder set* $E := \{B \in X : 0 \in B\}$ to be the collection of all points in X which (viewed as sets of integers) contain 0, we see (after unpacking all the definitions) that $E \cap T^r E \cap \ldots \cap T^{(k-1)r} E = \emptyset$ for all $r > 0$.

In order to apply Theorem 2.10.3 and obtain the desired contradiction, we need to find a shift-invariant Borel probability measure μ on X which assigns a positive measure to E.

For each integer N, consider the measure μ_N which assigns a mass of $\frac{1}{2N+1}$ to the points $T^{-n}A$ in X for $-N \leq n \leq N$, and no mass to the rest of X. Then we see that $\mu_N(E) = \frac{1}{2N+1}|A \cap \{-N, \ldots, N\}|$. Thus, since A has positive upper density, there exists some sequence N_j going to infinity such that $\liminf_{j \to \infty} \mu_{N_j}(E) > 0$. On the other hand, by vague sequential compactness (Lemma 2.7.16) we know that some subsequence of μ_{N_j} converges in the *vague topology* to a probability measure μ, which then assigns a positive measure to the (clopen) set E. As the μ_{N_j} are asymptotically shift invariant, we see that μ is invariant also (as in the proof of Corollary 2.7.17). As μ now has all the required properties, we have completed the deduction of Theorem 2.10.1 from Theorem 2.10.3.

Exercise 2.10.2. Show that Theorem 2.10.3 in fact implies a seemingly stronger version of Theorem 2.10.1, in which the conclusion becomes the assertion that the set $\{n : n, n+r, \ldots, n+(k-1)r \in A\}$ has positive upper density for infinitely many r.

Exercise 2.10.3. Show that Theorem 2.10.1 in fact implies a seemingly stronger version of Theorem 2.10.3: If $E_1, E_2, E_3, \ldots$ are sets in a probability space with uniformly positive measure (i.e., $\inf_n \mu(E_n) > 0$), then for any k there exists positive integers n, r such that $\mu(E_n \cap E_{n+r} \cap \ldots \cap E_{n+(k-1)r}) > 0$.

2.10.1. Varnavides type theorems. In [**BeHoMcCPa2000**], it was observed that a similar "compactness and contradiction" argument (combined with a preliminary averaging-over-dilations trick of Varnavides [**Va1959**]) allows us to use Theorem 2.10.3 to imply the following apparently stronger statement.

Theorem 2.10.6 (Uniform Furstenberg multiple recurrence theorem). *Let $k \geq 1$ be an integer and $\delta > 0$. Then for any measure-preserving system $(X, \mathcal{X}, \mu, T)$ and any measurable set E with $\mu(E) \geq \delta$ we have*

$$(2.98) \qquad \frac{1}{N} \sum_{r=0}^{N-1} \mu(E \cap T^r E \cap \ldots \cap T^{(k-1)r} E) \geq c(k, \delta)$$

for all $N \geq 1$, where $c(k, \delta) > 0$ is a positive quantity which depends only on k and δ (i.e., it is uniform over all choices of the system and of the set E with measure at least δ).

Exercise 2.10.4. Assuming Theorem 2.10.6, show that[38] if N is sufficiently large depending on k and δ, then any subset of $\{1, \ldots, N\}$ with cardinality at least δN will contain at least $c'(k, \delta)N^2$ non-trivial arithmetic progressions of length k, for some $c'(k, \delta) > 0$. Conclude in particular that Theorem 2.10.6 implies Theorem 2.10.1.

It is clear that Theorem 2.10.6 implies Theorem 2.10.3; let us now establish the converse. We first use an averaging argument of Varnavides to reduce Theorem 2.10.6 to a weaker statement, in which the conclusion (2.98) is not asserted to hold for all N, but instead one asserts that

$$(2.99) \qquad \frac{1}{N_0} \sum_{r=1}^{N_0-1} \mu(E \cap T^r E \cap \ldots \cap T^{(k-1)r} E) \geq c(k, \delta)$$

---

[38]This result for $k = 3$ was first established in [**Va1959**] via an averaging argument from Roth's theorem.

is true for some $N_0 = N_0(k, \delta) > 0$ depending only on k and δ (note that the $r = 0$ term in (2.99) has been dropped, otherwise the claim is trivial). To see why one can recover (2.98) from (2.99), observe by replacing the shift T with a power T^a that we can amplify (2.99) to

$$(2.100) \qquad \frac{1}{N_0} \sum_{r=1}^{N_0-1} \mu(E \cap T^{ar}E \cap \ldots \cap T^{(k-1)ar}E) \geq c(k, \delta)$$

for all a. Averaging (2.100) over $1 \leq a \leq N$ we easily conclude (2.98).

It remains to prove that (2.100) holds under the hypotheses of Theorem 2.10.6. Our next reduction is to observe that it suffices to perform this task for the boolean Bernoulli system $X_0 := 2^{\mathbf{Z}}$ with the cylinder set $E_0 := \{B \in X_0 : 0 \in B\}$ as before. To see this, recall from Example 2.2.6 that there is a morphism $\phi : X \to X_0$ from any measure-preserving system $(X, \mathcal{X}, \mu, T)$ with a distinguished set E to the system X_0 with the product σ-algebra $\mathcal{X}_0$, the usual shift T_0, and the set E_0, and with the push-forward measure $\mu_0 := \phi_\# \mu$. Specifically, ϕ sends any point x in X to its recurrence set $\phi(x) := \{n \in \mathbf{Z} : T^n x \in E\}$. Using this morphism it is not difficult to show that the claim (2.98) for $(X, \mathcal{X}, \mu, T)$ and E would follow from the same claim for $(X_0, \mathcal{X}_0, \mu_0, T_0)$ and E_0.

We still need to prove (2.99) for the boolean system. The point is that by lifting to this universal setting, the dynamical system $(X, \mathcal{X}, T)$ and the set E have been canonically fixed; the only remaining parameter is the probability measure μ. But now we can exploit vague sequential compactness again as follows.

Suppose for contradiction that Theorem 2.10.6 failed for the boolean system. Then by carefully negating all the quantifiers, we can find $\delta > 0$ such that for any N_0 there is a sequence of shift-invariant probability measures μ_j on X with $\mu_j(E) \geq \delta$,

$$(2.101) \qquad \frac{1}{N_0} \sum_{r=1}^{N_0-1} \mu_j(E \cap T^r E \cap \ldots \cap T^{(k-1)r}E) \to 0$$

as $j \to \infty$. Note that if (2.101) holds for one value of N_0, then it also holds for all smaller values of N_0. A standard diagonalisation argument then allows us to build a sequence μ_j as above, but which obeys (2.101) for *all* $N_0 \geq 1$.

Now we are finally in a good position to apply vague sequential compactness. By passing to a subsequence if necessary, we may assume that μ_j converges vaguely to a limit μ, which is a shift-invariant probability measure.

In particular we have $\mu(E) \geq \delta > 0$, while from (2.101) we see that

$$(2.102) \qquad \frac{1}{N_0} \sum_{r=1}^{N_0-1} \mu(E \cap T^r E \cap \ldots \cap T^{(k-1)r} E) = 0$$

for all $N_0 \geq 1$; thus the sets $E \cap T^r E \cap \ldots \cap T^{(k-1)r} E$ all have zero measure for $r > 0$. But this contradicts Theorem 2.10.3 (and Exercise 2.10.1). This completes the deduction of Theorem 2.10.6 from Theorem 2.10.3.

2.10.2. Other recurrence theorems and their combinatorial counterparts. The Furstenberg correspondence principle can be extended to relate several other recurrence theorems to their combinatorial analogues. We give some representative examples here (without proofs). Firstly, there is a multidimensional version of Szemerédi's theorem (compare with Exercise 2.4.8):

Theorem 2.10.7 (Multidimensional Szemerédi theorem [**FuKa1979**]). *Let $d \geq 1$, let $v_1, \ldots, v_k \in \mathbf{Z}^d$, and let $A \subset \mathbf{Z}^d$ be a set of positive upper Banach density (which means that $\limsup_{N \to \infty} |A \cap B_N|/|B_N| > 0$, where $B_N := \{-N, \ldots, N\}^d$). Then A contains a pattern of the form $n + rv_1, \ldots, n + rv_k$ for some $n \in \mathbf{Z}^d$ and $r > 0$.*

Note that Theorem 2.10.1 corresponds to the special case when $d = 1$ and $v_i = i - 1$.

This theorem was first proven by Furstenberg and Katznelson (see [**FuKa1979**]), who deduced it via the correspondence principle from the following generalisation of Theorem 2.10.3:

Theorem 2.10.8 (Recurrence for multiple commuting shifts). *Let $k \geq 1$ be an integer, let $(X, \mathcal{X}, \mu)$ be a probability space, let $T_1, \ldots, T_k : X \to X$ be measure-preserving bimeasurable maps which commute with each other, and let E be a set of positive measure. Then there exists $r > 0$ such that $T_1^r E \cap T_2^r E \cap \ldots \cap T_k^r E$ is non-empty.*

Exercise 2.10.5. Show that Theorem 2.10.7 and Theorem 2.10.8 are equivalent.

Exercise 2.10.6. State an analogue of Theorem 2.10.6 for multiple commuting shifts, and prove that it is equivalent to Theorem 2.10.8.

There is also a polynomial version of these theorems (cf. Theorem 2.5.1), which we will also state in general dimension:

Theorem 2.10.9 (Multidimensional polynomial Szemerédi theorem). *Let $d \geq 1$, let $P_1, \ldots, P_k : \mathbf{Z} \to \mathbf{Z}^d$ be polynomials with $P_1(0) = \ldots = P_k(0) = 0$, and let $A \subset \mathbf{Z}^d$ be a set of positive upper Banach density. Then A contains a pattern of the form $n + P_1(r), \ldots, n + P_k(r)$ for some $n \in \mathbf{Z}^d$ and $r > 0$.*

This theorem was established by Bergelson and Leibman [**BeLe1996**], who deduced it from

Theorem 2.10.10 (Polynomial recurrence for multiple commuting shifts). *Let k, $(X, \mathcal{X}, \mu)$, $T_1, \ldots, T_k : X \to X$, let E be as in Theorem 2.10.8, and let $P_1, \ldots, P_k$ be as in Theorem 2.10.9. Then there exists $r > 0$ such that $T^{-P_1(r)}E \cap T^{-P_2(r)}E \cap \ldots \cap T^{-P_k(r)}E$ is non-empty, where we adopt the convention $T^{(a_1, \ldots, a_k)} := T_1^{a_1} \ldots T_k^{a_k}$ (thus we are making the action of $\mathbf{Z}^d$ on X explicit).*

Exercise 2.10.7. Show that Theorem 2.10.9 and Theorem 2.10.10 are equivalent.

Exercise 2.10.8. State an analogue of Theorem 2.10.6 for polynomial recurrence for multiple commuting shifts, and prove that it is equivalent to Theorem 2.10.10. *Hint*: First establish this in the case that each of the P_j is a monomial, in which case there is enough dilation symmetry to use the Varnavides averaging trick. Interestingly, if one only restricts attention to one-dimensional systems $k = 1$, it does not seem possible to deduce the uniform polynomial recurrence theorem from the non-uniform polynomial recurrence theorem, thus indicating that the averaging trick is less universal in its applicability than the correspondence principle.

In the above theorems, the underlying action was given by either the integer group $\mathbf{Z}$ or the lattice group $\mathbf{Z}^d$. It is not too difficult to generalise these results to the semigroups $\mathbf{N}$ and $\mathbf{N}^d$ (thus dropping the assumption that the shift maps are invertible), by using a trick similar to that used in Exercise 2.4.10, or by using the correspondence principle back and forth a few times. A bit more surprisingly, it is possible to extend these results to even weaker objects than semigroups. To describe this we need some more notation.

Define a *partial semigroup* $(G, \cdot)$ to be a set G together with a partially defined multiplication operation $\cdot : \Omega \to G$ for a subset $\Omega \subset G \times G$, which is associative in the sense that if $(a \cdot b) \cdot c$ is defined, then $a \cdot (b \cdot c)$ is defined and equal to $(a \cdot b) \cdot c$, and vice versa. A good example of a partial semigroup is the finite subsets $\binom{S}{<\omega} := \{A \subset S : |A| < \infty\}$ of a fixed set S, where the multiplication operation $A \cdot B$ is disjoint union, or more precisely $A \cdot B := A \cup B$ when A and B are disjoint, and $A \cdot B$ is undefined otherwise.

Remark 2.10.11. One can extend a partial semigroup to be a genuine semigroup by adjoining a new element err to G, and redefining multiplication $a \cdot b$ to equal err if it was previously undefined (or if one of a or b was already equal to err). However, we will avoid using this trick here, as it tends to complicate the notation a little.

One can take Cartesian products of partial semigroups in the obvious manner to obtain more partial semigroups. In particular, we have the partial semigroup $\left(\begin{smallmatrix} \mathbf{N} \\ <\omega \end{smallmatrix}\right)^d$ for any $d \geq 1$, defined as the collection of d-tuples $(A_1, \ldots, A_d)$ of finite sets of natural numbers (not necessarily disjoint), with the partial semigroup law $(A_1, \ldots, A_d) \cdot (B_1, \ldots, B_d) := (A_1 \cup B_1, \ldots, A_d \cup B_d)$ whenever A_i and B_i are disjoint for each $1 \leq i \leq d$.

If $(X, \mathcal{X}, \mu)$ is a probability space and $(G, \cdot)$ is a partial semigroup, we define a *measure-preserving action* of G on X to be an assignment of a measure-preserving transformation $T^g : X \to X$ (not necessarily invertible) to each $g \in G$, such that $T^{g \cdot h} = T^g T^h$ whenever $g \cdot h$ is defined.

An action T of $\left(\begin{smallmatrix} \mathbf{N} \\ <\omega \end{smallmatrix}\right)$ on X is known as an *IP system* on X; it is generated by a countable number $T_1, T_2, \ldots$ of commuting measure-preserving transformations, with $T^A := \prod_{i \in A} T^i$. (Admittedly, it is possible that the action of the empty set is not necessarily the identity, but this turns out to have a negligible impact on matters.) An action T of $\left(\begin{smallmatrix} \mathbf{N} \\ <\omega \end{smallmatrix}\right)^d$ is then a collection of d simultaneously commuting IP systems.

In [**FuKa1985**], Furstenberg and Katznelson showed the following generalisation of Theorem 2.10.8:

Theorem 2.10.12 (IP multiple recurrence theorem). *Let T be an action of $\left(\begin{smallmatrix} \mathbf{N} \\ <\omega \end{smallmatrix}\right)^d$ on a probability space $(X, \mathcal{X}, \mu)$. Then there exists a non-empty set $A \in \left(\begin{smallmatrix} \mathbf{N} \\ <\omega \end{smallmatrix}\right)$ such that $E \cap (T^{A_1})^{-1}(E) \cap \ldots \cap (T^{A_d})^{-1}(E)$ is non-empty, where $A_i := (\emptyset, \ldots, \emptyset, A, \emptyset, \ldots, \emptyset)$ is the group element which equals A in the ith position and is the empty set otherwise.*

This theorem has a number of combinatorial consequences,[39] such as the following strengthening of Szemerédi's theorem:

Theorem 2.10.13 (IP Szemerédi theorem [**FuKa1985**]). *Let A be a set of integers of positive upper density, let $k \geq 1$, and let $B \subset \mathbf{N}$ be infinite. Then A contains an arithmetic progression $n, n + r, \ldots, n + (k - 1)r$ of length k in which r lies in $FS(B)$, the set of finite sums of B (cf. Theorem 2.5.18).*

Exercise 2.10.9. Deduce Theorem 2.10.13 from Theorem 2.10.12.

Exercise 2.10.10. Using Theorem 2.10.13, show that for any k, and any set of integers A of positive upper density, the set of steps r which occur in the arithmetic progressions in A of length k is syndetic.

Exercise 2.10.11. Using Theorem 2.10.12, show that if $\mathbf{F}$ is a finite field, and $\mathbf{F}^{<\omega} := \bigcup_{n=0}^{\infty} \mathbf{F}^n$ is the canonical vector space over $\mathbf{F}$ spanned (in the algebraic sense) by a countably infinite number of basis vectors, show that

---

[39]There is also a multidimensional version of this theorem, but it requires a fair amount of notation to state properly.

any subset A of $\mathbf{F}^{<\omega}$ of positive upper Banach density (which means that $\limsup_{n\to\infty} |A \cap \mathbf{F}^n|/|\mathbf{F}^n| > 0$) contains affine subspaces of arbitrarily high dimension.

The IP recurrence theorem is already very powerful, but even stronger theorems are known. For instance, in [**FuKa1991**], Furstenberg and Katznelson established the following deep strengthening of the Hales-Jewett theorem (Theorem 2.5.21), as well as of Exercise 2.10.11 above:

Theorem 2.10.14 (Density Hales-Jewett theorem). *Let A be a finite alphabet. If E is a subset of $A^{<\omega}$ of positive upper Banach density, then E contains a combinatorial line.*

This theorem was deduced (via an advanced form of the correspondence principle) by a somewhat complicated recurrence theorem which we will not state here; rather than the action of a group, semigroup, or partial semigroup, one works instead with an ensemble of sets (as in Exercise 2.10.3), and furthermore one regularises the system of the probability space and set ensemble (which can collectively be viewed as a random process) to be what Furstenberg and Katznelson call a *strongly stationary process*, which (very) roughly means that the statistics of this process look "the same" when restricted to any combinatorial subspace of a fixed dimension.

Remark 2.10.15. Similar correspondence principles can be established connecting property testing results for graphs and hypergraphs to the measure theory of exchangeable measures: see [**Ta2007c**], [**AuTa2008**], [**Ta2008**], [**AvGeTo2008**]. Finally, we have implicitly been using a similar correspondence principle between topological dynamics and colouring Ramsey theorems in Sections 2.3, 2.4, 2.5.

Remark 2.10.16. The Furstenberg correspondence principle also comes tantalisingly close to deducing my theorem with Ben Green [**GrTa2008**] that the primes contain arbitrarily long arithmetic progressions from Szemerédi's theorem. More precisely, it shows that any subset A of a *genuinely* random set of integers with logarithmic-type density B, with A having positive *relative* upper density with respect to B, contains arbitrarily long arithmetic progressions; see [**Ta**]. Unfortunately, the almost primes are not known to quite obey enough "correlation conditions" to behave sufficiently pseudorandomly that these arguments apply to the primes, though perhaps there is still a "softer" way to prove our theorem than the way we did it (see the recent papers [**Go2008**], [**ReTrTuVa2008**] for some progress in this direction).

Notes. This lecture first appeared at

terrytao.wordpress.com/2008/02/10.

Thanks to Liu Xiao Chuan for corrections.

2.11. Compact systems

The primary objective of this lecture and the next few will be to give a proof of the Furstenberg recurrence theorem (Theorem 2.10.3). Along the way we will develop a structural theory for measure-preserving systems.

The basic strategy of Furstenberg's proof is to first prove the recurrence theorems for very simple systems—either those with "almost periodic" (or *compact*) dynamics or with "weakly mixing" dynamics. These cases are quite easy, but do not cover all the cases. To go further, we need to consider various combinations of these systems. For instance, by viewing a general system as an extension of the maximal compact factor, we will be able to prove Roth's theorem (which is equivalent to the $k = 3$ form of the Fursten-berg recurrence theorem). To handle the general case, we need to consider compact extensions of compact factors, compact extensions of compact extensions of compact factors, etc., as well as weakly mixing extensions of all the previously mentioned factors.

In this section, we consider those measure-preserving systems $(X, \mathcal{X}, \mu, T)$ which are *compact* or *almost periodic*. These systems are analogous to the equicontinuous or isometric systems in topological dynamics discussed in Section 2.6, and as with those systems, we will be able to characterise such systems (or more precisely, the ergodic ones) algebraically as Kronecker systems, though this is not strictly necessary for the proof of the recurrence theorem.

2.11.1. Almost periodic functions. We begin with a basic definition.

Definition 2.11.1. Let $(X, \mathcal{X}, \mu, T)$ be a measure-preserving system. A function $f \in L^2(X, \mathcal{X}, \mu)$ is *almost periodic* if the orbit closure $\overline{\{T^n f : n \in \mathbf{Z}\}}$ is compact in $L^2(X, \mathcal{X}, \mu)$.

Example 2.11.2. If f is periodic (i.e., $T^n f = f$ for some $n > 0$) then it is clearly almost periodic. In particular, any shift-invariant function (such as a constant function) is almost periodic.

Example 2.11.3. In the circle shift system $(\mathbf{R}/\mathbf{Z}, x \mapsto x + \alpha)$, every function $f \in L^2(\mathbf{R}/\mathbf{Z})$ is almost periodic, because the orbit closure lies inside the set $\{f(\cdot + \theta) : \theta \in \mathbf{R}/\mathbf{Z}\}$, which is the continuous image of a circle $\mathbf{R}/\mathbf{Z}$ and therefore compact.

Exercise 2.11.1. Let $(X, \mathcal{X}, \mu, T)$ be a measure-preserving system, and let $f \in L^2(X, \mathcal{X}, \mu)$. Show that f is almost periodic in the ergodic theory sense (i.e., Definition 2.11.1 above) if and only if it is almost periodic in the topological dynamical systems sense (see Section 2.3), i.e., if the sets $\{n \in \mathbf{Z} : \|T^n f - f\|_{L^2(X,\mathcal{X},\mu)} \leq \varepsilon\}$ are syndetic for every $\varepsilon > 0$. *Hint*: If f is almost periodic in the ergodic theory sense, show that the orbit closure is an isometric system and thus a Kronecker system, at which point Theorem 2.3.5 can be applied. For the converse implication, use the *Heine-Borel theorem* and the isometric nature of T on L^2.

Exercise 2.11.2. Let $(X, \mathcal{X}, \mu, T)$ be a measure-preserving system. Show that the space of almost periodic functions in $L^2(X, \mathcal{X}, \mu)$ is a closed shift-invariant subspace which is also closed under the pointwise operations $f, g \mapsto \max(f, g)$ and $f, g \mapsto \min(f, g)$. Similarly, show that the space of almost periodic functions in $L^\infty(X, \mathcal{X}, \mu)$ is a closed subspace which is also an algebra (closed under products) as well as closed under max and min.

Exercise 2.11.3. Show that in any Bernoulli system $\Omega^{\mathbf{Z}}$, the only almost periodic functions are the constants. *Hint*: First show that if $f \in L^2(X, \mathcal{X}, \mu)$ has mean zero, then $\lim_{n \to \infty} \int_X f T^n f \, d\mu = 0$, by first considering elementary functions.

Let us recall the Furstenberg multiple recurrence theorem, which we now phrase in terms of functions rather than sets:

Theorem 2.11.4 (Furstenberg multiple recurrence theorem). *Suppose* $(X, \mathcal{X}, \mu, T)$ *is a measure-preserving system. Let* $k \geq 1$, *and let* $f \in L^\infty(X, \mathcal{X}, \mu)$ *be a non-negative function with* $\int_X f \, d\mu > 0$. *Then we have*

$$\liminf_{N \to \infty} \frac{1}{N} \sum_{r=0}^{N-1} \int_X f T^r f \dots T^{(k-1)r} f > 0.$$

Exercise 2.11.4. Show that Theorem 2.11.4 is equivalent to Theorem 2.10.3.

We can now quickly establish this theorem in the almost periodic case:

Proposition 2.11.5. *Theorem 2.11.4 is true whenever f is almost periodic.*

Proof. Without loss of generality we may assume that f is bounded a.e. by 1. Let $\varepsilon > 0$ be chosen later. Recall from Exercise 2.11.1 that $T^n f$ lies within ε of f in the L^2 topology for a syndetic set of n. For all such n, one also has $\|T^{(j+1)n} f - T^{jn} f\|_{L^2(X,\mathcal{X},\mu)} \leq \varepsilon$ for all j, since T acts isometrically. By the triangle inequality, we conclude that $T^{jn} f$ lies within $O_k(\varepsilon)$ of f in L^2 for $0 \leq j \leq k$. On the other hand, from Hölder's inequality we see that on the unit ball of $L^\infty(X, \mathcal{X}, \mu)$ with the L^2 topology, pointwise multiplication is Lipschitz. Applying this fact repeatedly, we conclude that for n in this syndetic set, $fT^n f \ldots T^{(k-1)n} f$ lies within $O_k(\varepsilon)$ in L^2 of f^k. In particular,

$$(2.103) \qquad \int_X fT^n f \ldots T^{(k-1)n} f \, d\mu = \int_X f^k \, d\mu + O_k(\varepsilon).$$

On the other hand, since $\int f \, d\mu > 0$, we must have $\int_X f^k \, d\mu > 0$. Choosing ε sufficiently small, we thus see that the left-hand side of (2.103) is uniformly bounded away from zero in a syndetic set, and the conclusion of Theorem 2.11.4 follows. $\qquad\qquad\qquad\qquad\qquad\qquad\qquad\qquad\qquad\square$

Remark 2.11.6. Because f lives in a Kronecker system, one can also obtain the above result using various multiple recurrence theorems from topological dynamics, such as Proposition 2.6.10 or the Birkhoff multiple recurrence theorem (Theorem 2.3.4), though to get the full strength of the results, one needs to use either syndetic van der Waerden theorem, see part (3) of Exercise 2.5.15, or the Varnavides averaging trick from Subsection 2.10.1. We leave the details to the reader.

2.11.2. Kronecker systems and Haar measure. We have seen how nice almost periodic functions are. Motivated by this, we give the following definition.

Definition 2.11.7. A measure-preserving system $(X, \mathcal{X}, \mu, T)$ is said to be *compact* if every function in $L^2(X, \mathcal{X}, \mu)$ is almost periodic.

Thus, for instance, by Example 2.11.3, the circle shift system is compact, but from Exercise 2.11.3, no non-trivial Bernoulli system is compact. From Proposition 2.11.5 we know that the Furstenberg recurrence theorem is true for compact systems.

One source of compact systems comes from *Kronecker systems*, as introduced in Definition 2.6.5. As such systems are topological rather than measure-theoretic, we will need to endow them with a canonical measure— *Haar measure*—first.

Let G be a compact metrisable topological group (not necessarily abelian). Without an ambient measure, we cannot yet define the convolution $f * g$ of two continuous functions $f, g \in C(G)$. However, we can define the convolution $\mu * f$ of a finite Borel measure μ on G and a continuous function $f \in C(G)$ to be the function

$$(2.104) \qquad\qquad \mu * f(x) := \int_G f(y^{-1}x) \, d\mu(y),$$

which (by the uniform continuity of f) is easily seen to be another continuous function. We similarly define

$$(2.105) \qquad\qquad f * \mu(x) := \int_G f(xy^{-1}) \, d\mu(y).$$

Also, one can introduce the convolution $\mu * \nu$ of two finite Borel measures to be the finite Borel measure defined as

$$(2.106) \qquad \mu * \nu(E) := \int_G \nu(y^{-1} \cdot E) \, d\mu(y)$$

for all Borel sets E. For instance, the convolution $\delta_x * \delta_y$ of two Dirac masses is another Dirac mass δ_{xy}. Fubini's theorem tells us that the convolution of two finite measures is another finite measure. Convolution is also bilinear and associative (thus $(\mu * \nu) * \rho = \mu * (\nu * \rho)$, $f * (\mu * \nu) = (f * \mu) * \nu$, $(\mu * f) * \nu = \mu * (f * \nu)$, and $(\mu * \nu) * f = \mu * (\nu * f)$ for measures μ, ν, ρ and continuous f); in particular, left convolution and right convolution commute. Also observe that the convolution of two Borel probability measures is again a Borel probability measure. Convolution also has a powerful *smoothing effect* that can upgrade weak convergence to strong convergence. Specifically, if μ_n converges in the vague sense to μ, and f is continuous, then an easy application of compactness of the underlying group G reveals that $\mu_n * f$ converges in the uniform sense to $\mu * f$.

Let us say that a number c is a *left mean* (resp. *right mean*) of a continuous function $f \in C(G)$ if there exists a probability measure μ such that $\mu * f$ (resp. $f * \mu_n$) is equal to a constant c. For compact metrisable groups G, this mean is well defined:

Lemma 2.11.8 (Existence and uniqueness of mean). *Let G be a compact metrisable topological group, and let $f \in C(G)$. Then there exists a unique constant c which is both a left mean and right mean of f.*

Proof. Without loss of generality we can take f to be real-valued. Let us first show that there exists a left mean. Define the oscillation of a real-valued continuous function to be the difference between its maximum and minimum. By the vague sequential compactness of probability measures (Lemma 2.7.16), one can find a probability measure μ which minimises the oscillation of $\mu * f$. If this oscillation is zero, we are done. If the oscillation is non-zero, then (using the compactness of the group and the transitivity of the group action) it is not hard to find a finite number of left rotations of $\mu * f$ whose average has strictly smaller oscillation than that of $\mu * f$ (basically by rotating the places where $\mu * f$ is near its maximum to cover where it is near its mimum). Thus we have a finitely supported probability measure ν with $\nu * \mu * f$ having smaller oscillation than $\mu * f$, a contradiction. We thus see that a left mean exists. Similarly, a right mean exists. But since left convolution commutes with right convolution, we see that all left means are equal to all right means, and the claim follows. $\square$

The map $f \mapsto c$ from a continuous function to its mean is a bounded non-negative linear functional on $C(G)$ which preserves constants, and thus by the Riesz representation theorem is given by a unique probability measure μ; since left and right convolutions commute, we see that this measure is both left- and right-invariant. Conversely, given any such measure μ we easily see (again using the commutativity of left and right convolutions) that $f * \mu = \mu * f = c$. We have thus shown

Corollary 2.11.9 (Existence and uniqueness of Haar measure). *If G is a compact metrisable topological group, then there exists a unique Borel probability measure μ on G which is both left- and right-invariant.*

In particular, every topological Kronecker system $(K, x \mapsto x + \alpha)$ can be canonically converted into a measure-preserving system, which is then compact by the same argument used to establish Example 2.11.3. (Actually this observation works for non-abelian Kronecker systems as well as abelian ones.)

Remark 2.11.10. One can also build left and right Haar measures for locally compact groups; these measures are locally finite Radon measures rather than Borel probability measures, and are unique up to constants; however it is no longer the case that such measures are necessarily equal to each other except in special cases, such as when the group is abelian or compact. These measures play an important role in the harmonic analysis and representation theory of such groups, but we will not discuss these topics further here.

2.11.3. Classification of compact systems. We have just seen that every Kronecker system is a compact system. The converse is not quite true; consider for instance the disjoint union of two Kronecker systems from different groups (with the probability measure being split, say, 50–50 between the two components). The situation is similar to that in Section 2.6, in which every Kronecker system was equicontinuous and isometric, but the converse only held under the additional assumption of minimality. There is a similar situation here, but first we need to define the notion of *equivalence* of two measure-preserving systems.

Define an *abstract measure-preserving system* $(\mathcal{X}, \mu, T)$ to be an abstract separable σ-algebra $\mathcal{X}$ (i.e., a Boolean algebra in which every countable sequence has both a supremum and an infimum), together with an abstract probability measure[40] $\mu : \mathcal{X} \to [0, 1]$ and an abstract invertible shift $T : \mathcal{X} \to \mathcal{X}$ which preserves the measure μ (but does not necessarily come from an invertible map $T : X \to X$ on some ambient space). There is an obvious notion of a morphism $\Phi : (\mathcal{X}, \mu, T) \to (\mathcal{Y}, \nu, S)$ between abstract measure-preserving systems, in which $\Phi : \mathcal{Y} \to \mathcal{X}$ (note the contravariance) is a σ-algebra homomorphism with $\nu = \mu \circ \Phi$ and $S \circ \Phi = \Phi \circ T$. This makes the class

---

[40] An abstract measure space $(\mathcal{X}, \mu)$ is sometimes also known as a *measure algebra*.

of abstract measure-preserving systems into a category. In particular we have a notion of two abstract measure-preserving systems being isomorphic.

Example 2.11.11. Let $(X, \mathcal{X}, \mu, T)$ be a skew shift $(y, z) \mapsto (y + \alpha, z + y)$ and let $(Y, \mathcal{Y}, \nu, S)$ be the underlying circle shift $y \mapsto y + \alpha$. These systems are of course non-isomorphic, although there is a factor map $\pi : X \to Y$ which is a morphism. If however we consider the σ-algebra $\pi^{\#}(\mathcal{Y}) \subset \mathcal{X}$ (which is formed by the Cartesian products of horizontal Borel sets with the vertical circle $\mathbf{R}/\mathbf{Z}$), we see that π induces an isomorphism between the abstract measure-preserving systems $(\pi^{\#}(\mathcal{Y}), \mu, T)$ and $(\mathcal{Y}, \nu, S)$.

Given a concrete measure-preserving system $(X, \mathcal{X}, \mu, T)$, we can define its *abstraction* $(\mathcal{X}/ \sim, \mu, T)$, where $\sim$ is the equivalence relation of almost everywhere equivalence modulo μ. In category-theoretic language, abstraction is a covariant functor from the category of concrete measure-preserving systems to the category of abstract measure-preserving systems. We say that two concrete measure-preserving systems are *equivalent* if their abstractions are isomorphic. Thus for instance, in Example 2.11.11 above, $(X, \pi^{\#}(\mathcal{Y}), \mu, T)$ and $(Y, \mathcal{Y}, \nu, S)$ are equivalent; there is no concrete isomorphism between these two systems, but once one abstracts away the underlying sets X and Y, we can recover an equivalence. As another example, we see that if we add or remove a null set to a measure-preserving system, we obtain an abstractly equivalent measure-preserving system.

Remark 2.11.12. Up to null sets, we can also identify an abstract measure-preserving system $(\mathcal{X}, \mu, T)$ with its commutative *von Neumann algebra* $L^{\infty}(\mathcal{X}, \mu)$ (which acts on the Hilbert space $L^2(\mathcal{X}, \mu)$ by pointwise multiplication), together with an automorphism T of that algebra; conversely, one can recover the algebra $\mathcal{X}$ as the idempotents 1_E of the von Neumann algebra, and the measure $\mu(E)$ of a set being the trace of the idempotent 1_E. A significant portion of ergodic theory can in fact be rephrased in terms of von Neumann algebras (which, in particular, naturally suggests a non-commutative generalisation of the subject), although we will not adopt this perspective here.

Many results and notions about concrete measure-preserving systems $(X, \mathcal{X}, \mu, T)$ can be rephrased to not require knowledge of the underlying space X (and to be stable under modification by null sets), and so can be converted to statements about abstract measure-preserving systems; for instance, the Furstenberg recurrence theorem is of this form once one replaces "non-empty" with "positive measure" (see Exercise 2.10.1). The notion of ergodicity is also of this form. In particular, such results and notions automatically become preserved under equivalence. In view of this, the following classification result is of interest.

Theorem 2.11.13 (Classification of ergodic compact systems). *Every ergodic compact system is equivalent to an (abelian) Kronecker system.*

To prove this theorem, it is convenient to use a harmonic analysis approach. Define an *eigenfunction* of a measure-preserving system $(X, \mathcal{X}, \mu, T)$ to be a bounded measurable function f, not a.e. zero, such that $Tf = \lambda f$ a.e.

Let $\mathcal{Z}_1 \subset \mathcal{X}$ denote the σ-algebra generated by all the eigenfunctions. Note that this contains $\mathcal{Z}_0 := \mathcal{X}^T$, which is the σ-algebra generated by the eigenfunctions with eigenvalue 1. We have the following fundamental result:

Proposition 2.11.14 (Description of the almost periodic functions). *Let $(X, \mathcal{X}, \mu, T)$ be an ergodic measure-preserving system, and $f \in L^2(X, \mathcal{X}, \mu)$. Then f is almost periodic if and only if it lies in $L^2(X, \mathcal{Z}_1, \mu)$, i.e., if it is $\mathcal{Z}_1$-measurable (note that $\mathcal{Z}_1$ contains all null sets of $\mathcal{X}$).*

Remark 2.11.15. One can view $(X, \mathcal{Z}_1, \mu, T)$ as the maximal compact factor of $(X, \mathcal{X}, \mu, T)$, in much the same way that $(X, \mathcal{Z}_0, \mu, T)$ is the maximal factor on which the system is essentially trivial (every function is essentially invariant).

Proof. It is clear that every eigenfunction is almost periodic. From repeated application of Exercise 2.11.2 we conclude that the indicator of any set in $\mathcal{Z}_1$ is also almost periodic, and thus (by more applications of Exercise 2.11.2) every function in $L^2(X, \mathcal{Z}_1, \mu)$ is almost periodic.

Conversely, suppose $f \in L^2(X, \mathcal{X}, \mu)$ is almost periodic. Then the orbit closure $Y_f \subset L^2(X, \mathcal{X}, \mu)$ of f is an isometric system; the orbit of f is clearly dense in Y_f, and thus by isometry the orbit of every other point is also dense. Thus Y_f is minimal, and therefore Kronecker by Proposition 2.6.7; thus we have an isomorphism $\phi : K \to Y_f$ from a group rotation $(K, x \mapsto x + \alpha)$ to Y_f. By rotating if necessary we may assume that $\phi(0) = f$.

By Corollary 2.11.9, K comes with an invariant probability measure ν. The theory of Fourier analysis on compact abelian groups then says that $L^2(K, \nu)$ is spanned by an (orthonormal) basis of characters χ. In particular, the Dirac mass at 0 (the group identity of K) can be expressed as the weak limit of finite linear combinations of such characters.

Now we need to move this information back to X. For this we use the operator $S : L^2(K, \nu) \to L^2(X, \mathcal{X}, \mu)$ defined by $Sh := \int_K \phi(y)h(y)\,d\nu(y)$; one checks from Minkowski's integral inequality that this is a bounded linear map. Because ϕ is a morphism, and each character is an eigenfunction of the group rotation $x \mapsto x + \alpha$, one easily checks that the image $S\chi$ of a character χ is an eigenfunction. Since the image of the Dirac mass is (formally) just f, we thus conclude that f is the weak limit[41] of finite linear combinations of characters. In particular, f is equivalent a.e. to a $\mathcal{Z}_1$-measurable function, as desired. $\qquad\square$

Exercise 2.11.5 (Spectral description of Kronecker factor). Show that the product of two eigenfunctions is again an eigenfunction. Using this and Proposition 2.11.14, conclude that $L^2(X, \mathcal{Z}_1, \mu)$ is in fact equal to $\mathbf{H}_{pp}$, the closed subspace of the Hilbert space $\mathbf{H} := L^2(X, \mathcal{X}, \mu)$ generated by the eigenfunctions of the shift operator T.

Exercise 2.11.6. Let $(X, \mathcal{X}, \mu, T)$ be a measure-preserving system, and let $f \in L^2(X, \mathcal{X}, \mu)$. We say that f is quasiperiodic if the orbit $\{T^n f : n \in \mathbf{Z}\}$ lies in a finite-dimensional space. Show that a function is quasiperiodic if and only if it is a finite linear combination of eigenfunctions. Deduce that a function is almost periodic if and only if it is the limit in L^2 of quasiperiodic functions.

Exercise 2.11.7. The purpose of this exercise is to show how abstract measure-preserving systems, and the morphisms between them, can be satisfactorily modeled by concrete systems and morphisms.

(1) Let $(\mathcal{X}, \mu, T)$ be an abstract measure-preserving system. Show that there exists a concrete regular measure-preserving system $(X', \mathcal{X}', \mu', T')$ which is equivalent to $(\mathcal{X}, \mu, T)$ (thus after omitting X' and quotienting out both σ-algebras by null sets, the two resulting abstract measure-preserving systems are isomorphic); the notion of regularity was introduced in Definition 2.9.18. *Hint*: Take a countable shift-invariant family of sets that generate $\mathcal{X}$ (thus T acts on this space by permutation), and use this to create a σ-algebra morphism from $\mathcal{X}$ to $\mathcal{X}'$, the product σ-algebra of some boolean space $X' := 2^{\mathbf{Z}}$, endowed with a permutation action T'.

(2) Let $\phi : (\mathcal{X}, \mu, T) \to (\mathcal{Y}, \nu, S)$ be an abstract morphism. Show that there exist regular measure-preserving systems $(X', \mathcal{X}', \mu', T')$ and $(Y', \mathcal{Y}', \nu', S')$ equivalent to $(\mathcal{X}, \mu, T)$ and $(\mathcal{Y}, \nu, S)$, together with a concrete morphism $\phi' : X' \to Y'$, such that obvious commuting square connecting the abstract σ-algebras $\mathcal{X}, \mathcal{Y}, \mathcal{X}', \mathcal{Y}'$ quotiented out by null sets does indeed commute.

Remark 2.11.16. Exercise 2.11.7 (and various related results) show that the distinction between concrete and abstract measure-preserving systems is very minor in practice. There are however other areas of mathematics in which taking an abstract or "point-less" approach by deleting (or at least downplaying) the underlying space can lead to non-trivial generalisations or

---

[41]One can in fact use compactness and continuity to make this a strong limit, but this is not necessary here.

refinements of the original concrete concept, for instance when moving from varieties to schemes.

Proof of Theorem 2.11.13. Note that if f is an eigenfunction then $T|f| = |f|$, and so (if the system is ergodic) $|f|$ is a.e. constant (which implies also that the eigenvalue lies on the unit circle). In particular, any eigenfunction is invertible. The quotient of two eigenfunctions of the same eigenvalue is then T-invariant and thus constant a.e. by ergodicity, which shows that all eigenspaces have geometric multiplicity 1 modulo null sets. As T is unitary, any eigenfunctions of different eigenvalues are orthogonal to each other; as $L^2(X, \mathcal{X}, \mu)$ is separable, we conclude that the number of eigenfunctions (up to constants and a.e. equivalence) is at most countable.

Let $(\phi_n)_{n \in A}$ be a collection of representative eigenfunctions for some at most countable index set A with eigenvalues λ_n; we can normalise $|\phi_n| = 1$ a.e. By modifying each eigenfunction on a set of measure zero (cf. Exercise 2.8.7) we can assume that $T\phi_n = \lambda_n \phi_n$ and $|\phi_n| = 1$ *everywhere* rather than just almost everywhere. Then the map $\Phi : x \mapsto (\log \phi_n(x))_{n \in A}$ is a morphism from $(X, \mathcal{X}, \mu, T)$ to the torus $(\mathbf{R}/\mathbf{Z})^A$ with the product σ-algebra $\mathcal{B}$, the push-forward measure $\Phi_\# \mu$, and the shift $x \mapsto x + \alpha$, where $\alpha := (\log \lambda_n)_{n \in A}$. From Proposition 2.11.14 we see that every measurable set in $\mathcal{X}$ differs by a null set from a set in the pullback σ-algebra $\Phi^\#(\mathcal{B})$. From this it is not hard to see that $(X, \mathcal{X}, \mu, T)$ is equivalent to the system $((\mathbf{R}/\mathbf{Z})^A, \mathcal{B}, \Phi_\# \mu, x \mapsto x + \alpha)$.

Now, $(\mathbf{R}/\mathbf{Z})^A$ is a compact metrisable space. The orbit closure K of α inside this space is thus also compact metrisable. The support of $\Phi_\# \mu$ is shift-invariant and thus K-invariant; but from the ergodicity of μ we conclude that the support must in fact be a single translate of K. In particular, $\Phi_\# \mu$ is just a translate of Haar measure on K. From this one easily concludes that $((\mathbf{R}/\mathbf{Z})^A, \mathcal{B}, \Phi_\# \mu, x \mapsto x + \alpha)$ is equivalent to the Kronecker system $(K, x \mapsto x + \alpha)$ with the Borel σ-algebra and Haar measure, and the claim follows. $\square$

Exercise 2.11.8. Let $(X, \mathcal{X}, \mu, T)$ be a compact system which is not necessarily ergodic, and let $y \mapsto \mu_y$ be the ergodic decomposition of μ relative to the projection $\pi : (X, \mathcal{X}, \mu, T) \to (Y, \mathcal{Y}, \nu, S)$ given by Proposition 2.9.22. Show that $(X, \mathcal{X}, \mu_y, T)$ is a compact ergodic system for ν-almost every y. From this and Theorem 2.11.13, we conclude that every compact system can be disintegrated into ergodic Kronecker systems (cf. the discussion after Proposition 2.6.9).

Remark 2.11.17. We comment here on finitary versions of the above concepts. Consider the cyclic group system $(\mathbf{Z}/N\mathbf{Z}, x \mapsto x+1)$ with the discrete

σ-algebra and uniform probability measure. Strictly speaking, every function on this system is periodic with period N and thus almost periodic, and so this is a compact system. But suppose we consider N as a large parameter going to infinity (in which case one can view these systems, together with some function $f = f_N$ on them "converging" to some infinite system with some limit function f, as in the derivation of Theorem 2.10.6 from Theorem 2.10.3). Then we would be interested in *uniform* control on the almost periodicity of the function or the compactness of the system, i.e., quantitative bounds involving expressions such as $O(1)$ which are bounded uniformly in N. With such a perspective, the analogue of a quasiperiodic function (see Exercise 2.11.6) is a function $f : \mathbf{Z}/N\mathbf{Z} \to \mathbf{C}$ which is a linear combination of at most $O(1)$ characters (i.e., its Fourier transform is non-zero at only $O(1)$ frequencies), whilst an almost periodic function f is one which is approximable in L^2 by quasiperiodic functions, thus for every $\varepsilon > 0$ one can find a function with only $O_\varepsilon(1)$ frequencies which lies within ε of f in the L^2 norm. Most functions on $\mathbf{Z}/N\mathbf{Z}$ for large N are not like this, and so the cyclic shift system is not compact in the asymptotic limit $N \to \infty$; however if one coarsens the underlying σ-algebra significantly one can recover compactness, though unfortunately one has to replace exact shift-invariance by approximate shift-invariance when one does so. For instance if one considers a σ-algebra $\mathcal{B}$ generated by a bounded ($O(1)$) number of Bohr sets $\{n \in \mathbf{Z}/N\mathbf{Z} : \|\frac{\xi n}{N} - a\|_{\mathbf{R}/\mathbf{Z}} \le \varepsilon\}$, then $\mathcal{B}$ is no longer shift-invariant in general, but all the functions which are measurable with respect to this algebra are uniformly almost periodic in the above sense. For some further developments of these sorts of "quantitative ergodic theory" ideas, see [**GrTa2008**], [**GrTa2009a**], [**GrTa2006**], [**Ta2006**], [**Ta2006b**], [**GrTa2009b**], [**Ta2008**].

Notes. This lecture first appeared at

terrytao.wordpress.com/2008/02/11.

Thanks to Emmanuel Kowalski and Liu Xiao Chuan for corrections.

As was pointed out to me anonymously, Theorem 2.12.26 was essentially established by von Neumann and Halmos (more precisely, they showed that any ergodic system in which the spectrum of the shift map is purely discrete is equivalent to a Kronecker system). It is also possible to construct the Kronecker system explicitly via Pontryagin duality.

2.12. Weakly mixing systems

In Section 2.11, we studied the recurrence properties of compact systems, which are systems in which all measurable functions exhibit almost periodicity—they almost return completely to themselves after repeated shifting.

Now, we consider the opposite extreme of *mixing systems*—those in which all measurable functions (of mean zero) exhibit *mixing*[42]—they become orthogonal to themselves after repeated shifting.

We shall see that for weakly mixing systems, averages such as

$$\frac{1}{N} \sum_{n=0}^{N-1} T^n f \ldots T^{(k-1)n} f$$

can be computed very explicitly (in fact, this average converges to the constant $(\int_X f \, d\mu)^{k-1}$). More generally, we shall see that weakly mixing components of a system tend to average themselves out and thus become irrelevant when studying many types of ergodic averages. Our main tool here will be the humble Cauchy-Schwarz inequality, and in particular a certain consequence of it, known as the *van der Corput lemma*.

As one application of this theory, we will be able to establish Roth's theorem [**Ro1953**] (the $k = 3$ case of Szemerédi's theorem).

2.12.1. Mixing functions. Much as compact systems were characterised by their abundance of almost periodic functions, we will characterise mixing systems by their abundance of mixing functions (this is not standard terminology). To define and motivate this concept, it will be convenient to introduce a weak notion of convergence (this notation is also not standard):

Definition 2.12.1 (Cesàro convergence). A sequence c_n in a normed vector space is said to *converge in the Cesàro sense* to a limit c if the averages $\frac{1}{N} \sum_{n=0}^{N-1} c_n$ converge strongly to c, in which case we write C-$\lim_{n \to \infty} c_n = c$. We also write C-$\sup_{n \to \infty} c_n := \limsup_{N \to \infty} \|\frac{1}{N} \sum_{n=0}^{N-1} c_n\|$ (therefore, C-$\lim_{n \to \infty} c_n = 0$ if and only if C-$\sup_{n \to \infty} c_n = 0$).

Example 2.12.2. The sequence $0, 1, 0, 1, \ldots$ has a Cesàro limit of $1/2$.

Exercise 2.12.1. Let c_n be a bounded sequence of *non-negative* numbers. Show that the following three statements are equivalent:

(1) C-$\lim_{n \to \infty} c_n = 0$.

(2) C-$\lim_{n \to \infty} |c_n|^2 = 0$.

(3) c_n converges to zero in density.[43]

Which of the implications between (1), (2), and (3) remain valid if c_n is not bounded?

---

[42] Actually, there are two different types of mixing, *strong mixing* and *weak mixing*, depending on whether the orthogonality occurs individually or on the average; it is the latter concept which is of more importance to the task of establishing the Furstenberg recurrence theorem.

[43] We say that c_n *converges in density* to c if for any $\varepsilon > 0$, the set $\{n \in \mathbf{N} : |c_n - c| > \varepsilon\}$ has upper density zero.

Let $(X, \mathcal{X}, \mu, T)$ be a measure-preserving system, and let $f \in L^2(X, \mathcal{X}, \mu)$ be a function. We consider the *correlation coefficients* $\langle T^n f, f \rangle :=$ $\int_X T^n f \overline{f} \, d\mu$ as n goes to infinity. Note that we have the symmetry $\langle T^n f, f \rangle = \overline{\langle T^{-n} f, f \rangle}$, so we only need to consider the case when n is positive. The mean ergodic theorem (Corollary 2.8.16) shows us the Cesàro behaviour of these coefficients. Indeed, we have

$$(2.107) \qquad \text{C-lim}_{n \to \infty} \langle T^n f, f \rangle = \langle \mathbf{E}(f | \mathcal{X}^T), f \rangle = \| \mathbf{E}(f | \mathcal{X}^T) \|^2_{L^2(X, \mathcal{X}, \mu)},$$

where $\mathcal{X}^T$ is the σ-algebra of essentially shift-invariant sets. In particular, if the system is ergodic, and f has mean zero (i.e., $\int_X f \, d\mu = 0$), then

$$(2.108) \qquad \text{C-lim}_{n \to \infty} \langle T^n f, f \rangle = 0,$$

thus the correlation coefficients go to zero in the Cesàro sense. However, this does not necessarily imply that these coefficients go to zero pointwise. For instance, consider a circle shift system $(\mathbf{R}/\mathbf{Z}, x \mapsto x + \alpha)$ with α irrational (and with uniform measure), thus this system is ergodic by Exercise 2.9.5. Then the function $f(x) := e^{2\pi i x}$ has mean zero, but one easily computes that $\langle T^n f, f \rangle = e^{2\pi i n \alpha}$. The coefficients $e^{2\pi i n \alpha}$ converge in the Cesàro sense to zero, but have magnitude 1 and thus do not converge to zero pointwise.

Definition 2.12.3 (Mixing). Let $(X, \mathcal{X}, \mu, T)$ be a measure-preserving system. A function $f \in L^2(X, \mathcal{X}, \mu)$ is *strongly mixing* if $\lim_{n \to \infty} \langle T^n f, f \rangle = 0$, and *weakly mixing* if $\text{C-lim}_{n \to \infty} |\langle T^n f, f \rangle| = 0$.

Remark 2.12.4. Clearly strong mixing implies weak mixing. From (2.107) we also see that if f is weakly mixing, then $\mathbf{E}(f | \mathcal{X}^T)$ must vanish a.e.

Exercise 2.12.2. Show that if f is both almost periodic and weakly mixing, then it must be 0 almost everywhere. In particular, in a compact system, the only weakly mixing function is 0 (up to a.e. equivalence).

Exercise 2.12.3. In any Bernoulli system $\Omega^{\mathbf{Z}}$ with the product σ-algebra and a product measure, and the standard shift, show that any function of mean zero is strongly mixing. *Hint*: First do this for functions that depend on only finitely many of the variables.

Exercise 2.12.4. Consider a skew shift system $((\mathbf{R}/\mathbf{Z})^2, (x, y) \mapsto (x + \alpha, y + x))$ with the usual Lebesgue measure and Borel σ-algebra, and with α irrational. Show that the function $f(x, y) := e^{2\pi i x}$ is neither strongly mixing nor weakly mixing, but that the function $g(x, y) := e^{2\pi i y}$ is both strongly mixing and weakly mixing.

Exercise 2.12.5. Let $X := \mathbf{C}^{\mathbf{Z}}$ be given the product Borel σ-algebra $\mathcal{X}$ and the shift $T : (z_n)_{n \in \mathbf{Z}} \to (z_{n+1})_{n \in \mathbf{Z}}$. For each $d \geq 1$, let μ_d be the probability distribution in X of the random sequence $(z_n)_{n\in\mathbf{Z}}$ given by the rule

$$(2.109) \qquad z_n := \frac{1}{2^{d/2}} \sum_{\omega_1,\dots,\omega_d\in\{0,1\}} w_{\omega_1,\dots,\omega_d} e^{2\pi i \sum_{j=1}^{d} \omega_j n/100^j},$$

where the $w_{\omega_1,\dots,\omega_d}$ are iid standard complex Gaussians (thus each w has probability distribution $e^{-\pi|w|^2}\,dw$). Show that each μ_d is shift invariant. If μ is a vague limit point of the sequence μ_d, and $f : X \to \mathbf{C}$ is the function defined as $f((z_n)_{n\in\mathbf{Z}}) := \operatorname{sgn}(\operatorname{Re} z_0)$, show that f is weakly mixing but not strongly mixing (and more specifically, that $\langle T^{100^j} f, f\rangle$ stays bounded away from zero) with respect to the system $(X, \mathcal{X}, \mu, T)$.

Remark 2.12.5. Exercise 2.12.5 illustrates an important point, namely that *stationary processes* yield a rich source of measure-preserving systems (indeed the two notions are almost equivalent in some sense, especially after one distinguishes a specific function f on the measure-preserving system). However, we will not adopt this more probabilistic perspective to ergodic theory here.

Remark 2.12.6. We briefly discuss the finitary analogue of the weak mixing concept in the context of functions $f : \mathbf{Z}/N\mathbf{Z} \to \mathbf{C}$ on a large cyclic group $\mathbf{Z}/N\mathbf{Z}$ with the usual shift $x \mapsto x + 1$. Then one can compute

$$(2.110) \qquad \text{C-}\lim_{n\to\infty} |\langle T^n f, f\rangle|^2 = \sum_{\xi\in\mathbf{Z}/N\mathbf{Z}} |\hat{f}(\xi)|^4,$$

where $\hat{f}(\xi) := \frac{1}{N}\sum_{x\in\mathbf{Z}/N\mathbf{Z}} f(x)e^{-2\pi i x\xi/N}$ are the Fourier coefficients of f. Comparing this against the Plancherel identity $\|f\|_{L^2}^2 = \sum_{\xi\in\mathbf{Z}/N\mathbf{Z}} |\hat{f}(\xi)|^2$ we thus see that a function f bounded in the L^2 norm should be considered "weakly mixing" if it has no large Fourier coefficients. Contrast this with Remark 2.11.17.

Now let us see some consequences of the weak mixing property. We need the following lemma, which gives a useful criterion as to whether a sequence of bounded vectors in a Hilbert space converges in the Cesàro sense to zero.

Lemma 2.12.7 (van der Corput lemma). *Let $v_1, v_2, v_3, \dots$ be a bounded sequence of vectors in a Hilbert space H. If*

$$(2.111) \qquad \text{C-}\lim_{h\to\infty} \text{C-}\sup_{n\to\infty}\langle v_n, v_{n+h}\rangle = 0,$$

then $\text{C-}\lim_{n\to\infty} v_n = 0$.

Informally, this lemma asserts that if each vector in a bounded sequence tends to be orthogonal to nearby elements in that sequence, then the vectors will converge to zero in the Cesàro sense. This formulation of the lemma is essentially the version in [**Be1987**], except that we have made the minor change of replacing one of the Cesàro limits with a Cesàro supremum.

Proof. We can normalise so that $\|v_n\| \leq 1$ for all n. In particular, we have $v_n = O(1)$, where $O(1)$ denotes a vector of bounded magnitude. For any h and $N \geq 1$, we thus have the telescoping identity

$$(2.112) \qquad \frac{1}{N} \sum_{n=0}^{N-1} v_{n+h} = \frac{1}{N} \sum_{n=0}^{N-1} v_n + O(|h|/N);$$

averaging this over all h from 0 to $H-1$ for some $H \geq 1$, we obtain

$$(2.113) \qquad \frac{1}{N} \sum_{n=0}^{N-1} \frac{1}{H} \sum_{h=0}^{H-1} v_{n+h} = \frac{1}{N} \sum_{n=0}^{N-1} v_n + O(H/N);$$

by the triangle inequality we thus have

$$(2.114) \qquad \left\| \frac{1}{N} \sum_{n=0}^{N-1} v_n \right\| \leq \frac{1}{N} \sum_{n=0}^{N-1} \left\| \frac{1}{H} \sum_{h=0}^{H-1} v_{n+h} \right\| + O(H/N),$$

where the $O(\ )$ terms are now scalars rather than vectors. We square this (using the crude inequality $(a+b)^2 \leq 2a^2 + 2b^2$) and apply the Cauchy-Schwarz inequality to obtain

$$(2.115) \qquad \left\| \frac{1}{N} \sum_{n=0}^{N-1} v_n \right\|^2 \leq O\left(\frac{1}{N} \sum_{n=0}^{N-1} \left\| \frac{1}{H} \sum_{h=0}^{H-1} v_{n+h} \right\|^2 \right) + O(H^2/N^2),$$

which we rearrange as
(2.116)

$$\left\| \frac{1}{N} \sum_{n=0}^{N-1} v_n \right\|^2 \leq O\left(\frac{1}{H^2} \sum_{0 \leq h,h' < H} \frac{1}{N} \sum_{n=0}^{N-1} \langle v_{n+h}, v_{n+h'} \rangle \right) + O(H^2/N^2).$$

We take limits as $N \to \infty$ (keeping H fixed for now) to conclude that

$$(2.117) \quad \limsup_{N \to \infty} \left\| \frac{1}{N} \sum_{n=0}^{N-1} v_n \right\|^2 \leq O\left(\frac{1}{H^2} \sum_{0 \leq h,h' < H} \text{C-sup}_{n \to \infty} \langle v_{n+h}, v_{n+h'} \rangle \right).$$

Another telescoping argument (and symmetry) gives us

$$(2.118) \qquad \text{C-sup}_{n \to \infty} \langle v_{n+h}, v_{n+h'} \rangle = \text{C-sup}_{n \to \infty} \langle v_{n+|h-h'|}, v_n \rangle$$

and so

$$(2.119) \qquad \limsup_{N \to \infty} \left\| \frac{1}{N} \sum_{n=0}^{N-1} v_n \right\|^2 \leq O\left(\frac{1}{H} \sum_{0 \leq h < H} \text{C-sup}_{n \to \infty} \langle v_{n+h}, v_n \rangle \right).$$

Taking limits as $H \to \infty$ and using (2.111) we obtain the claim. $\qquad\square$

Exercise 2.12.6. Let $P : \mathbf{Z} \to \mathbf{R}/\mathbf{Z}$ be a polynomial with at least one irrational non-constant coefficient. Using Lemma 2.12.7 (in the scalar case $H = \mathbf{C}$) and an induction on degree, show that C-$\lim_{n\to\infty} e^{2\pi i P(n)} = 0$. Conclude that the sequence $(P(n))_{n\in\mathbf{N}}$ is uniformly distributed with respect to uniform measure (see Definition 2.9.12 for a definition of uniform distribution).

Exercise 2.12.7. Using Exercise 2.12.6, give another proof of Theorem 2.6.26.

We now apply the van der Corput lemma to weakly mixing functions.

Corollary 2.12.8. *Let $(X, \mathcal{X}, \mu, T)$ be a measure-preserving system, and let $f \in L^2(X, \mathcal{X}, \mu)$ be weakly mixing. Then for any $g \in L^2(X, \mathcal{X}, \mu)$ we have* C-$\lim_{n\to\infty} |\langle T^n f, g\rangle| = 0$ *and* C-$\lim_{n\to\infty} |\langle f, T^n g\rangle| = 0$.

Proof. We just prove the first claim, as the second claim is similar. By Exercise 2.12.1, it suffices to show that

$$(2.120) \qquad \frac{1}{N} \sum_{n=0}^{N-1} |\langle T^n f, g\rangle|^2 \to 0$$

as $N \to \infty$. The left-hand side can be rewritten as

$$(2.121) \qquad \Big\langle \frac{1}{N} \sum_{n=0}^{N-1} \langle g, T^n f\rangle T^n f, g \Big\rangle,$$

so by the Cauchy-Schwarz inequality it suffices to show that

$$(2.122) \qquad \text{C-}\lim_{N\to\infty} \langle g, T^n f\rangle T^n f = 0.$$

Applying the van der Corput lemma and discarding the bounded coefficients $\langle g, T^n f\rangle$, it suffices to show that

$$(2.123) \qquad \text{C-}\lim_{H\to\infty} \text{C-}\sup_{n\to\infty} |\langle T^{n+h} f, T^n f\rangle| = 0.$$

But $\langle T^{n+h} f, T^n f\rangle = \langle T^h f, f\rangle$, and the claim now follows from the weakly mixing nature of f. $\qquad\qquad\qquad\qquad\qquad\qquad\qquad\qquad\square$

2.12.2. Weakly mixing systems. Now we consider systems which are full of mixing functions.

Definition 2.12.9 (Mixing systems). A measure-preserving system $(X, \mathcal{X}, \mu, T)$ is *weakly mixing* (resp. *strongly mixing*) if every function $f \in L^2(X, \mathcal{X}, \mu)$ with mean zero is weakly mixing (resp. strongly mixing).

Example 2.12.10. From Exercise 2.12.2, we know that any system with a non-trivial Kronecker factor is not weakly mixing (and thus not strongly mixing). On the other hand, from Exercise 2.12.3, we know that any Bernoulli system is strongly mixing (and thus weakly mixing also). From Remark 2.12.4, we see that any strongly or weakly mixing system must be ergodic.

Exercise 2.12.8. Show that the system in Exercise 2.12.5 is weakly mixing but not strongly mixing.

Here is another characterisation of weak mixing:

Exercise 2.12.9. Let $(X, \mathcal{X}, \mu, T)$ be a measure-preserving system. Show that the following assertions are equivalent:

  (1) $(X, \mathcal{X}, \mu, T)$ is weakly mixing.
  (2) For every $f, g \in L^2(X, \mathcal{X}, \mu)$, $\langle T^n f, g \rangle$ converges in density to $(\int_X f \, d\mu)(\int_X \overline{g} \, d\mu)$. (See Exercise 2.12.1 for a definition of convergence in density.)
  (3) For any measurable E, F, $\mu(T^n E \cap F)$ converges in density to $\mu(E)\mu(F)$.
  (4) The product system $(X \times X, \mathcal{X} \times \mathcal{X}, \mu \times \mu, T \times T)$ is ergodic.

Hint: To equate (1) and (2), use the decomposition $f = (f - \int_X f \, d\mu) + \int_X f \, d\mu$ of a function into its mean and mean-free components. To equate (2) and (4), use the fact that the space $L^2(X \times X, \mathcal{X} \times \mathcal{X}, \mu \times \mu)$ is spanned (in the topological vector space sense) by tensor products $(x, y) \mapsto f(x)g(y)$ with $f, g \in L^2(X, \mathcal{X}, \mu)$.

Exercise 2.12.10. Show that the equivalences between (1), (2), and (3) in Exercise 2.12.9 remain if "weak mixing" and "converges in density" are replaced by "strong mixing" and "converges", respectively.

Exercise 2.12.11. Let $(X, \mathcal{F}, T)$ be any minimal topological system with Borel σ-algebra $\mathcal{B}$, and let μ be a shift-invariant Borel probability measure. Show that if $(X, \mathcal{B}, \mu, T)$ is weakly mixing (resp. strongly mixing), then $(X, \mathcal{F}, T)$ is topologically weakly mixing (resp. topologically mixing), as defined in Definition 2.12.9 and Exercise 2.7.12.

Exercise 2.12.12. If $(X, \mathcal{X}, \mu, T)$ is weakly mixing, show that $(X, \mathcal{X}, \mu, T^n)$ is weakly mixing for any non-zero n.

Exercise 2.12.13. Let $(X, \mathcal{X}, \mu, T)$ be a measure-preserving system. Show that the following are equivalent:

  (1) $(X, \mathcal{X}, \mu, T)$ is weakly mixing.
  (2) Whenever $(Y, \mathcal{Y}, \nu, S)$ is ergodic, the product system $(X \times Y, \mathcal{X} \times \mathcal{Y}, \mu \times \nu, T \times S)$ is ergodic.

Hint: To obtain (1) from (2), use Exercise 2.12.9. To obtain (2) from (1), repeat the *methods* used to prove Exercise 2.12.9.

Exercise 2.12.14. Show that the product of two weakly mixing systems is again weakly mixing. *Hint*: Use Exercises 2.12.9 and 2.12.13.

Now we come to an important type of observation for the purposes of establishing the Furstenberg recurrence theorem: in weakly mixing systems, functions of mean zero are negligible as far as multiple averages are concerned.

Proposition 2.12.11. *Let $a_1, \ldots, a_k \in \mathbf{Z}$ be distinct non-zero integers for some $k \geq 1$. Let $(X, \mathcal{X}, \mu, T)$ be weakly mixing, and let $f_1, \ldots, f_k \in L^\infty(X, \mathcal{X}, \mu)$ be such that at least one of $f_1, \ldots, f_k$ has mean zero. Then we have*

$$(2.124) \qquad \text{C-}\lim_{n \to \infty} T^{a_1 n} f_1 \ldots T^{a_k n} f_k = 0$$

in $L^2(X, \mathcal{X}, \mu)$.

Proof. We induct on k. When $k = 1$, the claim follows from the mean ergodic theorem and Exercise 2.12.12 (recall from Example 2.12.10 that all weakly mixing systems are ergodic).

Now let $k \geq 2$ and suppose that the claim has already been proven for $k - 1$. Without loss of generality we may assume that it is f_1 which has mean zero. Applying the van der Corput lemma (Lemma 2.12.7), it suffices to show that

$$(2.125) \qquad \text{C-sup}_{n \to \infty} \langle T^{a_1(n+h)} f_1 \ldots T^{a_k(n+h)} f_k, T^{a_1 n} f_1 \ldots T^{a_k n} f_k \rangle$$

converges in density to zero as $h \to \infty$. But the left-hand side can be rearranged as

$$(2.126) \qquad \text{C-sup}_{n \to \infty} \int_X T^{(a_1 - a_k)n} f_{1,h} \ldots T^{(a_{k-1} - a_k)n} f_{k-1,h} f_{k,h} \, d\mu$$

where $f_{j,h} := T^{a_j h} f_j \overline{f_j}$. Applying the Cauchy-Schwarz inequality, it suffices to show that

$$(2.127) \qquad \text{C-sup}_{n \to \infty} T^{(a_1 - a_k)n} f_{1,h} \ldots T^{(a_{k-1} - a_k)n} f_{k-1,h}$$

converges in density to zero as $h \to \infty$.

Since $(X, \mathcal{X}, \mu, T)$ is weakly mixing, the mean-zero function f_1 is weakly mixing, and so the mean of $f_{1,h}$ goes to zero in density as $h \to \infty$. As all functions are assumed to be bounded, we can thus subtract the mean from $f_{1,h}$ in (2.127) without affecting the desired conclusion, leaving behind the mean-zero component $f_{1,h} - \int_X f_{1,h} \, d\mu$. But then the contribution of this expression to (2.127) vanishes by the induction hypothesis. $\qquad \square$

Remark 2.12.12. The key point here was that the functions f of mean zero were weakly mixing and thus had the property that $T^h f \overline{f}$ almost had mean zero, and hence were almost weakly mixing. One could iterate this further to investigate the behaviour of "higher derivatives" of f such as $T^{h+h'} f \overline{T^h f T^{h'} f} f$. Pursuing this analysis further leads to the Gowers-Host-Kra seminorms [**HoKr2005**], which are closely related to the Gowers uniformity norms [**Go2001**] in additive combinatorics.

Corollary 2.12.13. *Let* $a_1, \dots, a_k \in \mathbf{Z}$ *be distinct integers for some* $k \geq 1$, *let* $(X, \mathcal{X}, \mu, T)$ *be a weakly mixing system, and let* $f_1, \dots, f_k \in L^\infty(X, \mathcal{X}, \mu)$. *Then* $\int_X T^{a_1 n} f_1 \dots T^{a_k n} f_k \, d\mu$ *converges in the Cesàro sense to*

$$\left(\int_X f_1 \, d\mu \right) \dots \left(\int_X f_k \, d\mu \right).$$

Note in particular that this establishes the Furstenberg recurrence theorem (Theorem 2.11.4) in the case of weakly mixing systems.

Proof. We again induct on k. The $k = 1$ case is trivial, so suppose $k > 1$ and the claim has already been proven for $k - 1$. If any of the functions f_j is constant, then the claim follows from the induction hypothesis, so we may subtract the mean from each function and assume that all functions have mean zero. By shift-invariance we may also fix a_k (say) to be zero. The claim now follows from Proposition 2.12.11 and the Cauchy-Schwarz inequality. $\qquad \square$

Exercise 2.12.15. Show that the Cesàro convergence in Corollary 2.12.8 can be strengthened to convergence in density. *Hint*: First reduce to the mean zero case, then apply Exercise 2.12.14 to work with the product system instead.

Exercise 2.12.16. Let $(X, \mathcal{X}, \mu, T)$ be a weakly mixing system, and let $f \in L^\infty(X, \mathcal{X}, \mu)$ have mean zero. Show that $T^{n^2} f$ converges in the Cesàro sense in $L^2(X, \mathcal{X}, \mu)$ to zero. *Hint*: Use van der Corput's lemma and Proposition 2.12.11 or Corollary 2.12.13.

Exercise 2.12.17. Show that Corollary 2.12.13 continues to hold if the linear polynomials $a_1 n, \dots, a_k n$ are replaced by arbitrary polynomials $P_1(n)$, $\dots, P_k(n)$ from the integers to the integers, so long as the difference between any two of these polynomials is non-constant. *Hint*: You will need the "PET induction" machinery from Exercise 2.5.3. This result was first established in [**Be1987**].

2.12.3. Hilbert-Schmidt operators. We have now established the Furstenberg recurrence theorem for two distinct types of systems: compact systems and weakly mixing systems. From Example 2.12.10 we know that these systems are indeed quite distinct from each other. Here is another indication of "distinctness":

Exercise 2.12.18. In any measure-preserving system $(X, \mathcal{X}, \mu, T)$, show that almost periodic functions and weakly mixing functions are always orthogonal to each other.

On the other hand, there are certainly systems which are neither weakly mixing nor compact (e.g. the skew shift). But we have the following important dichotomy (cf. Theorem 2.7.12):

Theorem 2.12.14. *Suppose that* $(X, \mathcal{X}, \mu, T)$ *is a measure-preserving system. Then exactly one of the following statements is true:*

(1) *(Structure)* $(X, \mathcal{X}, \mu, T)$ *has a non-trivial compact factor.*[44]

(2) *(Randomness)* $(X, \mathcal{X}, \mu, T)$ *is weakly mixing.*

In Example 2.12.10 we have already shown that (1) and (2) cannot be both true; the tricky part is to show that lack of weak mixing implies a non-trivial compact factor.

In order to prove this result, we recall some standard results about Hilbert-Schmidt operators on a separable[45] Hilbert space. We begin by recalling the notion of tensor product of two Hilbert spaces:

Proposition 2.12.15. *Let* H, H' *be two separable Hilbert spaces. Then there exists another separable Hilbert space* $H \otimes H'$ *and a bilinear tensor product map* $\otimes : H \times H' \to H \otimes H'$ *such that*

$$(2.128) \qquad \langle v \otimes v', w \otimes w' \rangle_{H \otimes H'} = \langle v, w \rangle_H \langle v', w' \rangle_{H'}$$

for all $v, w \in H$ *and* $v', w' \in H'$. *Furthermore, the tensor products* $(e_n \otimes e'_{n'})_{n \in A, n' \in A'}$ *between any orthonormal bases* $(e_n)_{n \in A}$, $(e'_{n'})_{n' \in A'}$ *of* H *and* H' *respectively, form an orthonormal basis of* $H \otimes H'$.

It is easy to see that $H \otimes H'$ is unique up to isomorphism, and so we shall abuse notation slightly and refer to $H \otimes H'$ as **the** tensor product of H and H'.

Proof. Take any orthonormal bases $(e_n)_{n \in A}$ and $(e'_{n'})_{n' \in A'}$ of H and H' respectively, and let $H \otimes H'$ be the Hilbert space generated by declaring the

---

[44]In ergodic theory, a *factor* of a measure-preserving system is simply a morphism from that system to some other measure-preserving system. Unlike the case with topological dynamics, we do not need to assume surjectivity of the morphism, since in the measure-theoretic setting, the image of a morphism always has full measure.

[45]As usual, the hypothesis of separability is not absolutely essential, but is convenient to assume throughout; for instance, it assures that orthonormal bases always exist and are at most countable.

formal quantities $e_n \otimes e'_{n'}$ to form an orthonormal basis. If one then defines

$$(2.129) \qquad \left(\sum_n c_n e_n \right) \otimes \left(\sum_{n'} c'_{n'} e_{n'} \right) := \sum_n \sum_{n'} c_n c'_{n'} e_n \otimes e_{n'}$$

for all square-summable sequences c_n and $c'_{n'}$, one easily verifies that $\otimes$ is indeed a bilinear map that obeys (2.128). In particular, if $(f_m)_{m \in B}$ and $(f'_{m'})_{m' \in B'}$ are some other orthonormal bases of H, H' respectively, then, from (2.128), $(f_m \otimes f'_{m'})_{m \in B, m' \in B'}$ is an orthonormal set, and one can approximate any element $e_n \otimes e'_{n'}$ in the original orthonormal basis to arbitrary accuracy by linear combinations from this orthonormal set, and so this set is in fact an orthonormal basis as required. $\qquad \square$

Example 2.12.16. The tensor product of $L^2(X, \mathcal{X}, \mu)$ and $L^2(Y, \mathcal{Y}, \nu)$ is $L^2(X \times Y, \mathcal{X} \times \mathcal{Y}, \mu \times \nu)$, with the tensor product operation $f \otimes g(x, y) := f(x)g(y)$. The tensor product of $\mathbf{C}^m$ and $\mathbf{C}^n$ is $\mathbf{C}^{n \times m}$, which can be thought of as the Hilbert space of $n \times m$ (or $m \times n$) matrices, with the inner product $\langle A, B \rangle := \text{tr}(AB^\dagger) = \text{tr}(A^\dagger B)$.

Given a Hilbert space H, define its *complex conjugate* $\overline{H}$ to be the same set as H, but with the conjugated scalar multiplication structure $z, v \mapsto \bar{z}v$ and the conjugated inner product $\langle z, w \rangle_{\overline{H}} := \overline{\langle z, w \rangle_H} = \langle w, z \rangle_H$, but with all other structures unchanged. This is also a Hilbert space.[46]

Example 2.12.17. The conjugation map $f \mapsto \overline{f}$ is a Hilbert space isometry between the Hilbert space $L^2(X, \mathcal{X}, \mu)$ and its complex conjugate.

Every element $K \in \overline{H} \otimes H'$ induces a bounded linear operator $T_K : H \to H'$, defined via duality by the formula

$$(2.130) \qquad \langle T_K v, v' \rangle_{H'} := \langle K, v \otimes v' \rangle$$

for all $v \in H, v' \in H'$. We refer to K as the *kernel* of T_K. Any operator $T = T_K$ that arises in this manner is called a *Hilbert-Schmidt operator* from H to H'. The Hilbert space structure on the space $\overline{H} \otimes H'$ of kernels induces an analogous Hilbert space structure on the Hilbert-Schmidt operators, leading to the Hilbert-Schmidt norm $\|T\|_{HS}$ and inner product $\langle S, T \rangle_{HS}$ for such operators. Here are some other characterisations of this concept:

Exercise 2.12.19. Let H, H' be Hilbert spaces with orthonormal bases $(e_n)_{n \in A}$ and $(e'_{n'})_{n' \in A'}$ respectively, and let $T : H \to H'$ be a bounded linear operator. Show that the following are equivalent:

(1) T is a Hilbert-Schmidt operator.

(2) $\sum_{n \in A} \|T e_n\|_{H'}^2 < \infty$.

---

[46] Of course, for real Hilbert spaces rather than complex, the notion of complex conjugation is trivial.

(3) $\sum_{n \in A} \sum_{n' \in A'} |\langle Te_n, e'_{n'} \rangle_{H'}|^2 < \infty$.

Also, show that if $T, S : H \to H'$ are Hilbert-Schmidt operators, then

$$(2.131) \qquad \langle T, S \rangle_{HS} = \sum_{n \in A} \langle Te_n, Se_n \rangle_{H'}$$

and

$$(2.132) \qquad \|T\|_{HS}^2 = \sum_{n \in A} \|Te_n\|_{H'}^2 = \sum_{n \in A} \sum_{n' \in A'} |\langle Te_n, e'_{n'} \rangle_{H'}|^2.$$

As one consequence of the above exercise, we see that the Hilbert-Schmidt norm controls the operator norm, thus $\|Tv\| \leq \|T\|_{HS}\|v\|$ for all vectors v.

Remark 2.12.18. From this exercise and Fatou's lemma, we see in particular that the limit (in either the norm topology or strong or weak operator topologies) of a sequence of Hilbert-Schmidt operators with uniformly bounded Hilbert-Schmidt norm is still Hilbert-Schmidt. We also see that the composition of a Hilbert-Schmidt operator with a bounded operator is still Hilbert-Schmidt (thus the Hilbert-Schmidt operators can be viewed as a closed two-sided ideal in the space of bounded operators).

Example 2.12.19. An operator $T : L^2(X, \mathcal{X}, \mu) \to L^2(Y, \mathcal{Y}, \nu)$ is Hilbert-Schmidt if and only if it takes the form $Tf(y) := \int_X K(x, y)f(x)\,d\mu(x)$ for some kernel $K \in L^2(X \times Y, \mathcal{X} \times \mathcal{Y}, \mu \times \nu)$, in which case the Hilbert-Schmidt norm is $\|K\|_{L^2(X \times Y, \mathcal{X} \times \mathcal{Y}, \mu \times \nu)}$. The Hilbert-Schmidt inner product is defined similarly.

Example 2.12.20. The identity operator on an infinite-dimensional Hilbert space is never Hilbert-Schmidt, despite being bounded. On the other hand, every finite rank operator is Hilbert-Schmidt.

One of the key properties of Hilbert-Schmidt operators which will be relevant to us is the following.

Lemma 2.12.21. *If $T : H \to H'$ is Hilbert-Schmidt, then it is compact (i.e., the image of any bounded set is precompact).*

Proof. Let $\varepsilon > 0$ be arbitrary. By Exercise 2.12.19 and monotone convergence, we can find a finite orthonormal set $e_1, \ldots, e_N$ such that

$$\sum_{n=1}^{N} \|Te_n\|_{H'}^2 \geq \|T\|_{HS}^2 - \varepsilon^2,$$

and in particular that $\|Te_{n+1}\|_{H'} \leq \varepsilon$ for any e_{n+1} orthogonal to $e_1, \ldots, e_n$. As a consequence, the image of the unit ball of H under T lies within ε of the image of the unit ball of the finite-dimensional space span$(e_1, \ldots, e_N)$. This image is therefore totally bounded and thus precompact. $\square$

The following exercise may help illuminate the distinction between bounded operators, Hilbert-Schmidt operators, and compact operators:

Exercise 2.12.20. Let λ_n be a sequence of complex numbers, and consider the diagonal operator $T : (z_n)_{n \in \mathbf{N}} \mapsto (\lambda_n z_n)_{n \in \mathbf{N}}$ on $l^2(\mathbf{N})$.

(1) Show that T is a well-defined bounded linear operator on $l^2(\mathbf{N})$ if and only if the sequence (λ_n) is bounded.

(2) Show that T is Hilbert-Schmidt if and only if the sequence (λ_n) is square-summable.

(3) Show that T is compact if and only if the sequence (λ_n) goes to zero as $n \to \infty$.

Now we apply the above theory to establish Theorem 2.12.14. Let $(X, \mathcal{X}, \mu, T)$ be a measure-preserving system, and let $f \in L^2(X, \mathcal{X}, \mu)$. The rank one operators $g \mapsto \langle g, T^n f \rangle T^n f$ can easily be verified to have a Hilbert-Schmidt norm of $\|f\|_{L^2}^2$, and so by the triangle inequality, their averages $S_{f,N} : g \mapsto \frac{1}{N} \sum_{n=0}^{N-1} \langle g, T^n f \rangle T^n f$ have a Hilbert-Schmidt norm of at most $\|f\|_{L^2}^2$. On the other hand, from the identity

$$(2.133) \qquad \langle S_{f,N} g, h \rangle = \frac{1}{N} \sum_{n=0}^{N-1} \langle g \otimes \overline{h}, (T \otimes T)^n (f \otimes \overline{f}) \rangle$$

and the mean ergodic theorem (applied to the product space) we see that $S_{f,N}$ converges in the weak operator topology[47] to some limit S_f, which is then also Hilbert-Schmidt by Remark 2.12.18, and thus compact by Lemma 2.12.21. Also, it is easy to see that S_f is self-adjoint and commutes with T. As a consequence, we conclude that for any $g \in L^2(X, \mathcal{X}, \mu)$, the image $S_f g$ is almost periodic (since $\{T^n S_f g : n \in \mathbf{Z}\} = S_f \{T^n g : n \in \mathbf{Z}\}$ is the image of a bounded set by the compact operator S_f and therefore precompact).

On the other hand, observe that

$$(2.134) \qquad \langle S_f f, f \rangle = \text{C-lim}_{n \to \infty} |\langle T^n f, f \rangle|^2.$$

Thus by Definition 2.12.3 (and Exercise 2.12.1), we see that $\langle S_f f, f \rangle \neq 0$ whenever f is not weakly mixing. In particular, f is not orthogonal to the almost periodic function $S_f f$. From this and Exercise 2.12.18, we have thus shown

Proposition 2.12.22 (Dichotomy between structure and randomness). *Let* $(X, \mathcal{X}, \mu, T)$ *be a measure-preserving system. A function* $f \in L^2(X, \mathcal{X}, \mu)$ *is*

---

[47]Actually, $S_{f,N}$ converges to S_f in the Hilbert-Schmidt norm, and thus also in the operator norm and in the strong topology: this is another application of the mean ergodic theorem, which we leave as an exercise. Since each of the $S_{f,N}$ is clearly finite rank, this gives a direct proof of the compactness of S_f.

weakly mixing if and only if it is orthogonal to all almost periodic functions (or equivalently, orthogonal to all eigenfunctions).

Remark 2.12.23. It is interesting that essentially the same result appears in the spectral and scattering theory of linear Schrödinger equations, which in that context is known as the "RAGE theorem" [**Ru1969**], [**AmGe1973**], [**En1978**].

Remark 2.12.24. The finitary analogue of the expression $S_f f$ is the *dual function* (of order 2) of f (the dual function of order 1 was briefly discussed in Section 2.8). If we are working on $\mathbf{Z}/N\mathbf{Z}$ with the usual shift, then S_f can be viewed as a Fourier multiplier which multiplies the Fourier coefficient at ξ by $|\hat{f}(\xi)|^2$; informally, S_f filters out all the low amplitude frequencies of f, leaving only a handful of high-amplitude frequencies.

Recall from Proposition 2.12.15 and Exercise 2.12.5 of Section 2.11 that a function $f \in L^2(X, \mathcal{X}, \mu)$ is almost periodic if and only if it is $\mathcal{Z}_1$-measurable, or if it lies in the pure point component $\mathbf{H}_{pp}$ of the shift operator T. We thus have

Corollary 2.12.25 (Koopman-von Neumann theorem). *Let $(X, \mathcal{X}, \mu, T)$ be a measure-preserving system, and let $f \in L^2(X, \mathcal{X}, \mu)$. Let $\mathcal{Z}_1$ be the σ-algebra generated by the eigenfunctions of T.*

(1) *f is almost periodic if and only if $f \in L^2(X, \mathcal{Z}_1, \mu)$ if and only if $f \in \mathbf{H}_{pp}$.*

(2) *f is weakly mixing if and only if $\mathbf{E}(f|\mathcal{Z}_1) = 0$ a.e. if and only if $f \in \mathbf{H}_c = \mathbf{H}_{sc} + \mathbf{H}_{ac}$ (corresponding to the continuous spectrum of T).*

(3) *In general, f has a unique decomposition $f = f_{U^\perp} + f_U$ into an almost periodic function $f_{U^\perp}$ and a weakly mixing function f_U. Indeed, $f_{U^\perp} = \mathbf{E}(f|\mathcal{Z}_1)$ and $f_U = f - \mathbf{E}(f|\mathcal{Z}_1)$.*

Theorem 2.12.14 follows immediately from Corollary 2.12.25. Indeed, if a system is not weakly mixing, then by the above corollary we see that $\mathcal{Z}_1$ is non-trivial, and the identity map from $(X, \mathcal{X}, \mu, T)$ to $(X, \mathcal{Z}_1, \mu, T)$ yields a non-trivial compact factor.

2.12.4. Roth's theorem. As a quick application of the above machinery we give a proof of Roth's theorem. We first need a variant of Corollary 2.12.8, which is proven by much the same means:

Exercise 2.12.21. Let $(X, \mathcal{X}, \mu, T)$ be an ergodic measure-preserving system, let a_1, a_2, a_3 be distinct integers, and let $f_1, f_2, f_3 \in L^\infty(X, \mathcal{X}, \mu)$ with at least one of f_1, f_2, f_3 weakly mixing. Show that

$$\text{C-}\lim_{n\to\infty} \int_X T^{a_1 n} f_1 T^{a_2 n} f_2 T^{a_3 n} f_3 \, d\mu = 0.$$

Theorem 2.12.26 (Roth's theorem). *Let $(X, \mathcal{X}, \mu, T)$ be an ergodic measure-preserving system, and let $f \in L^\infty(X, \mathcal{X}, \mu)$ be non-negative with $\int_X f \, d\mu > 0$. Then*

$$(2.135) \qquad \liminf_{N\to\infty} \frac{1}{N} \sum_{n=0}^{N-1} \int_X f T^n f T^{2n} f \, d\mu > 0.$$

Proof. We decompose $f = f_{U^\perp} + f_U$ as in Corollary 2.12.25. The contribution of f_U is negligible by Exercise 2.12.21, so it suffices to show that

$$(2.136) \qquad \liminf_{N\to\infty} \frac{1}{N} \sum_{n=0}^{N-1} \int_X f_{U^\perp} T^n f_{U^\perp} T^{2n} f_{U^\perp} \, d\mu > 0.$$

Since $f_{U^\perp}$ is almost periodic, the claim follows from Proposition 2.11.5. $\qquad\square$

One can then immediately establish the $k = 3$ case of Furstenberg's theorem (Theorem 2.10.3) by combining the above result with the ergodic decomposition (Proposition 2.9.22). The $k = 3$ case of Szemerédi's theorem (i.e., Roth's theorem) then follows from the Furstenberg correspondence principle (see Section 2.10).

Exercise 2.12.22. Let $(X, \mathcal{X}, \mu, T)$ be a measure-preserving system, and let $f \in L^2(X, \mathcal{X}, \mu)$ be non-negative. Show that for every $\varepsilon > 0$, one has $\langle T^n f, f \rangle \geq \int_X (f \, d\mu)^2 - \varepsilon$ for infinitely many n. *Hint*: First show this when f is almost periodic, and then use Corollaries 2.12.8 and 2.12.25 to prove the general case. This is a simplified version of the *Khintchine recurrence theorem*, which asserts that the set of such n is not only infinite but also syndetic. Analogues of the Khintchine recurrence theorem hold for double recurrence but not for triple recurrence; see [**BeHoKr2005**] for details.

Notes. This lecture first appeared at

terrytao.wordpress.com/2008/02/21.

Thanks to Liu Xiao Chuan for corrections.

2.13. Compact extensions

In Section 2.11, we studied *compact* measure-preserving systems—those systems $(X, \mathcal{X}, \mu, T)$ in which every function $f \in L^2(X, \mathcal{X}, \mu)$ was almost periodic, which meant that their orbit $\{T^n f : n \in \mathbf{Z}\}$ was *precompact* in the

$L^2(X, \mathcal{X}, \mu)$ topology. Among other things, we were able to easily establish the Furstenberg recurrence theorem (Theorem 2.11.4) for such systems.

In this section, we generalise these results to a "relative" or "conditional" setting, in which we study systems which are compact relative to some factor $(Y, \mathcal{Y}, \nu, S)$ of $(X, \mathcal{X}, \mu, T)$. Such systems are to compact systems as isometric extensions are to isometric systems in topological dynamics. The main result we establish here is that the Furstenberg recurrence theorem holds for such compact extensions whenever the theorem holds for the base. The proof is essentially the same as in the compact case; the main new trick is to work not in the Hilbert spaces $L^2(X, \mathcal{X}, \mu)$ over the complex numbers, but rather in the *Hilbert module*[48] $L^2(X, \mathcal{X}, \mu | Y, \mathcal{Y}, \nu)$ over the (commutative) *von Neumann algebra* $L^\infty(Y, \mathcal{Y}, \nu)$. Because of the compact nature of the extension, it turns out that results from topological dynamics (and in particular, van der Waerden's theorem) can be exploited to good effect in this argument.[49]

2.13.1. Hilbert modules. Let $X = (X, \mathcal{X}, \mu, T)$ be a measure-preserving system, and let $\pi : X \to Y$ be a factor map, i.e., a morphism from X to another system $Y = (Y, \mathcal{Y}, \nu, S)$. The algebra $L^\infty(Y)$ can be viewed (using π) as a subalgebra of $L^\infty(X)$; indeed, it is isomorphic to $L^\infty(X, \pi^\#(\mathcal{Y}), \mu)$, where $\pi^\#(\mathcal{Y}) := \{\pi^{-1}(E) : E \in \mathcal{Y}\}$ is the pullback of $\mathcal{Y}$ by π.

Example 2.13.1. Throughout these notes we shall use the *skew shift* as our running example. Thus, in this example, $X = (\mathbf{R}/\mathbf{Z})^2$ with shift $T : (y, z) \mapsto (y + \alpha, z + y)$ for some fixed α (which can be either rational or irrational), and $Y = \mathbf{R}/\mathbf{Z}$ with shift $S : y \mapsto y + \alpha$, with factor map $\pi : (y, z) \mapsto y$. In this case, $L^\infty(Y)$ can be thought of (modulo equivalence on null sets, of course) as the space of bounded functions on $(\mathbf{R}/\mathbf{Z})^2$ which depend only on the first variable.

Example 2.13.2. Another (rather trivial) example is when the factor system Y is simply a point. In this case, $L^\infty(Y)$ is the space of constants and can be identified with $\mathbf{C}$. At the opposite extreme, another example is when Y is equal to X. It is instructive to see how all of the concepts behave in each of these two extreme cases, as well as the typical intermediate case presented in Example 2.13.1.

The idea here will be to try to "relativise" the machinery of Hilbert spaces over $\mathbf{C}$ to that of Hilbert modules over $L^\infty(Y)$. Roughly speaking, all concepts which used to be complex- or real-valued (e.g. inner products,

---

[48]Modules are to rings as vector spaces are to fields.

[49]Note that this operator-algebraic approach is not the only way to understand these extensions; one can also proceed by disintegrating μ into fibre measures μ_y for almost every $y \in Y$ and working fibre by fibre. We will discuss the connection between the two approaches below.

norms, coefficients, etc.) will now take values in the algebra $L^\infty(Y)$. The following table depicts the various concepts that will be relativised:

Absolute / unconditional	Relative / conditional
Constants $\mathbf{C}$	Factor-measurable functions $L^\infty(Y)$
Expectation $\mathbf{E}f = \int_X f\, d\mu \in \mathbf{C}$	Conditional expectation $\mathbf{E}(f\|Y) \in L^\infty(Y)$
Inner product $\langle f, g \rangle_{L^2(X)} = \mathbf{E}f\bar{g}$	Conditional inner product $\langle f, g \rangle_{L^2(X\|Y)} = \mathbf{E}(f\bar{g}\|Y)$
Hilbert space $L^2(X)$	Hilbert module $L^2(X\|Y)$
Finite-dimensional subspace $\{\sum_{j=1}^d c_j f_j : c_1, \ldots, c_d \in \mathbf{C}\}$	Finitely generated module $\{\sum_{j=1}^d c_j f_j : c_1, \ldots, c_d \in L^\infty(Y)\}$
Almost periodic function	Conditionally almost periodic function
Compact system	Compact extension
Hilbert-Schmidt operator	Conditionally Hilbert-Schmidt operator
Weakly mixing function	Conditionally weakly mixing function
Weakly mixing system	Weakly mixing extension

Remark 2.13.3. In information-theoretic terms, one can view Y as representing all the observables in the system X that have already been "measured" in some sense, so that it is now permissible to allow one's "constants" to depend on that data, and only study the remaining information present in X conditioning on the observed values in Y. Note though that once we activate the shift map T, the data in Y will similarly shift (by S), and so the various fibres of π can interact with each other in a non-trivial manner, so one should take some caution in applying information-theoretic intuition to this setting.

We have already seen that the factor Y induces a sub-σ-algebra $\pi^{\#}(\mathcal{Y})$ of $\mathcal{X}$. We therefore have a conditional expectation map $f \mapsto \mathbf{E}(f|Y)$ defined for all absolutely integrable f by the formula

$$(2.137) \qquad \mathbf{E}(f|Y) := \mathbf{E}(f|\pi^{\#}(\mathcal{Y})).$$

In general, this expectation only lies in $L^1(Y)$, though for the functions we shall eventually study, the expectation will always lie in $L^\infty(Y)$ when needed.

As stated in the table, conditional expectation will play the role in the conditional setting that the unconditional expectation $\mathbf{E}f = \int_X f\, d\mu$ plays in the unconditional setting. Note though that the conditional expectation takes values in the algebra $L^\infty(Y)$ rather than in the complex numbers. We recall that conditional expectation is linear over this algebra, thus

$$(2.138) \qquad \mathbf{E}(cf + dg|Y) = c\mathbf{E}(f|Y) + d\mathbf{E}(g|Y)$$

for all absolutely integrable f, g and all $c, d \in L^\infty(Y)$.

Example 2.13.4. Continuing Example 2.13.1, we see that for any absolutely integrable f on $(\mathbf{R}/\mathbf{Z})^2$, we have $\mathbf{E}(f|Y)(y, z) = \int_{\mathbf{R}/\mathbf{Z}} f(y, z')\, dz'$ almost everywhere.

Let $L^2(X|Y)$ be the space of all $f \in L^2(X, \mathcal{X}, \mu)$ such that the conditional norm

$$(2.139) \qquad \|f\|_{L^2(X|Y)} := \mathbf{E}(|f|^2|Y)^{1/2}$$

lies in $L^\infty(Y)$ (rather than merely in $L^2(Y)$, which it does automatically). Thus for instance we have the inclusions

$$(2.140) \qquad L^\infty(X) \subset L^2(X|Y) \subset L^2(X).$$

The space $L^2(X|Y)$ is easily seen to be a vector space over $\mathbf{C}$, and moreover (thanks to (2.138)) is a module over $L^\infty(Y)$.

Exercise 2.13.1. For the inner product defined by

$$(2.141) \qquad \langle f, g \rangle_{L^2(X|Y)} := \mathbf{E}(f\overline{g}|Y)$$

(which, initially, is only in $L^1(Y)$), establish the pointwise Cauchy-Schwarz inequality

$$(2.142) \qquad |\langle f, g \rangle_{L^2(X|Y)}| \leq \|f\|_{L^2(X|Y)}\|g\|_{L^2(X|Y)}$$

almost everywhere. In particular, the inner product lies in $L^\infty(Y)$. *Hint:* Repeat the standard proof of the Cauchy-Schwarz inequality verbatim, but with $L^\infty(Y)$ playing the role of the constants $\mathbf{C}$.

Example 2.13.5. Continuing Examples 2.13.1 and 2.13.4, $L^2(X|Y)$ consists (modulo null set equivalence) of all measurable functions $f(y, z)$ such that $\|f\|_{L^2(X|Y)} = (\int_{\mathbf{R}/\mathbf{Z}} |f(y, z)|^2\, dz)^{1/2}$ is bounded a.e. in y, with the relative inner product

$$(2.143) \qquad \langle f, g \rangle_{L^2(X|Y)}(y) := \int_{\mathbf{R}/\mathbf{Z}} f(y, z)\overline{g(y, z)}\, dz$$

defined a.e. in y. Observe in this case that the relative Cauchy-Schwarz inequality (2.142) follows easily from the standard Cauchy-Schwarz inequality.

Exercise 2.13.2. Show that the function $f \mapsto \|\|f\|_{L^2(X|Y)}\|_{L^\infty(Y)}$ is a norm on $L^2(X|Y)$, which turns that space into a Banach space.[50] *Hint*: You may need to "relativise" one of the standard proofs that $L^2(X)$ is complete. You may also want to start with the skew shift example to build some intuition.

As π is a morphism, one can easily check the intertwining relationship

$$(2.144) \qquad\qquad \mathbf{E}(T^n f|Y) = S^n \mathbf{E}(f|Y)$$

for all $f \in L^1(X)$ and integers n. As a consequence we see that the map T (and all of its powers) preserves the space $L^2(X|Y)$, and furthermore is conditionally unitary in the sense that

$$(2.145) \qquad\qquad \langle T^n f, T^n g \rangle_{L^2(X|Y)} = S^n \langle f, g \rangle_{L^2(X|Y)}$$

for all $f, g \in L^2(X|Y)$ and integers n.

In the Hilbert space $L^2(X)$ one can create finite-dimensional subspaces $\{c_1 f_1 + \ldots + c_d f_d : c_1, \ldots, c_d \in \mathbf{C}\}$ for any $f_1, \ldots, f_d \in L^2(X)$. Inside such subspaces we can create the bounded finite-dimensional *zonotopes*

$$\{c_1 f_1 + \ldots + c_d f_d : c_1, \ldots, c_d \in \mathbf{C}, |c_1|, \ldots, |c_d| \leq 1\}.$$

Observe (from the Heine-Borel theorem) that a subset E of $L^2(X)$ is precompact if and only if it can be approximated by finite-dimensional zonotopes in the sense that for every $\varepsilon > 0$, there exists a finite-dimensional zonotope Z of $L^2(X)$ such that E lies within the ε-neighbourhood of Z.

Remark 2.13.6. There is nothing special about zonotopes here; just about any family of bounded finite-dimensional objects would suffice for this purpose. In fact, it seems to be slightly better (for the purposes of quantitative analysis, and in particular in controlling the dependence on dimension d) to work instead with octahedra, in which the constraint $|c_1|, \ldots, |c_d| \leq 1$ is replaced by $|c_1| + \ldots + |c_d| = 1$; this perspective is used for instance in [**Ta2006**].

Inspired by this, let us make some definitions. A *finitely generated module* of $L^2(X|Y)$ is any submodule of $L^2(X|Y)$ of the form $\{c_1 f_1 + \ldots + c_d f_d : c_1, \ldots, c_d \in L^\infty(Y)\}$, where $f_1, \ldots, f_d \in L^2(X|Y)$. Inside such a module we can define a *finitely generated module zonotope* $\{c_1 f_1 + \ldots + c_d f_d : c_1, \ldots, c_d \in L^\infty(Y); \|c_1\|_{L^\infty(Y)}, \ldots, \|c_d\|_{L^\infty(Y)} \leq 1\}$.

Definition 2.13.7.

- A subset E of $L^2(X|Y)$ is said to be *conditionally precompact* if for every $\varepsilon > 0$, there exists a finitely generated module zonotope Z of $L^2(X|Y)$ such that E lies within the ε-neighbourhood of Z (using the norm from Exercise 2.13.2).

---

[50]Because of this completeness, we refer to $L^2(X|Y)$ as a *Hilbert module* over $L^\infty(Y)$.

- A function $f \in L^2(X|Y)$ is said to be *conditionally almost periodic* if its orbit $\{T^n f : n \in \mathbf{Z}\}$ is conditionally precompact.
- A function $f \in L^2(X|Y)$ is said to be *conditionally almost periodic in measure* if for every $\varepsilon > 0$ there exists a set E in Y of measure at most ε such that $f1_{E^c}$ is conditionally almost periodic.
- The system X is said to be a *compact extension* of Y if every function in $L^2(X|Y)$ is conditionally almost periodic in measure.

Example 2.13.8. Any bounded subset of $L^\infty(Y)$ is conditionally precompact (though note that it need not be precompact in the topological sense, using the topology from Exercise 2.13.2). In particular, every function in $L^\infty(Y)$ is conditionally almost periodic.

Example 2.13.9. Every system is a compact extension of itself. A system is a compact extension of a point if and only if it is a compact system.

Example 2.13.10. Consider the skew shift (Examples 2.13.1, 2.13.4, and 2.13.5), and consider the orbit of the function $f(y, z) := e^{2\pi i z}$. A computation shows that

$$(2.146) \qquad T^n f(y, z) = e^{2\pi i \frac{-n(-n-1)}{2}\alpha} e^{-2\pi i n y} f,$$

which reveals (for α irrational) that f is not almost periodic in the unconditional sense. However, observe that all the shifts $T^n f$ lie in the zonotope $\{cf : c \in L^\infty(Y), \|c\|_{L^\infty(Y)} \leq 1\}$ generated by a single generator f, and so f is *conditionally* almost periodic.

Exercise 2.13.3. Consider the skew shift (Examples 2.13.1, 2.13.4, 2.13.5, and 2.13.10). Show that a sequence $f_n \in L^\infty(X)$ is conditionally precompact if and only if the sequences $f_n(y, \cdot) \in L^\infty(\mathbf{R}/\mathbf{Z})$ are precompact in $L^2(\mathbf{R}/\mathbf{Z})$ (with the usual Lebesgue measure) for almost every y.

Exercise 2.13.4. Show that the space of conditionally almost periodic functions in $L^2(X|Y)$ is a shift-invariant $L^\infty(Y)$ module, i.e., it is closed under addition, under multiplication by elements of $L^\infty(Y)$, and under powers T^n of the shift operator.

Exercise 2.13.5. Consider the skew shift (Examples 2.13.1, 2.13.4, 2.13.5, 2.13.10 and Exercise 2.13.3) with α irrational, and let $f \in L^2(X|Y)$ be the function defined by setting $f(y, z) := e^{2\pi i n z}$ whenever $n \geq 1$ and $y \in (1/(n+1), 1/n]$. Show that f is conditionally almost periodic in measure, but not conditionally almost periodic. Thus the two notions can be distinct even for bounded functions (a subtlety that does not arise in the unconditional setting).

Exercise 2.13.6. Let $\mathcal{Z}_{X|Y}$ denote the collection of all measurable sets E in X such that 1_E is conditionally almost periodic in measure. Show that $\mathcal{Z}_{X|Y}$ is a shift-invariant sub-σ-algebra of $\mathcal{X}$ that contains $\pi^{\#}\mathcal{Y}$, and that a function $f \in L^2(X|Y)$ is conditionally almost periodic in measure if and only if it is $\mathcal{Z}$-measurable. (In particular, $(X, \mathcal{Z}_{X|Y}, \mu, T)$ is the maximal compact extension of Y.) *Hint*: You may need to truncate the generators $f_1, \ldots, f_d$ of various module zonotopes to be in $L^{\infty}(X)$ rather than $L^2(X|Y)$.

Exercise 2.13.7. Show that the skew shift (Examples 2.13.1, 2.13.4, 2.13.5, 2.13.10 and Exercises 2.13.3, 2.13.5) is a compact extension of the circle shift. *Hint*: Use Example 2.13.10 and Exercise 2.13.6. Alternatively, approximate a function on the skew torus by its vertical Fourier expansions. For each fixed horizontal coordinate y, the partial sums of these vertical Fourier series converge (in the vertical L^2 sense) to the original function, pointwise in y. Now apply *Egorov's theorem*.

Exercise 2.13.8. Show that each of the iterated skew shifts (Exercise 2.9.8) are compact extensions of the preceding skew shift.

Exercise 2.13.9. Let $(Y, \mathcal{Y}, \nu, S)$ be a measure-preserving system, let G be a compact metrisable group with a closed subgroup H, let $\sigma : Y \to G$ be measurable, and let $Y \times_{\sigma} G/H$ be the extension of Y with underlying space $Y \times G/H$, with measure equal to the product of ν and Haar measure, and shift map $T : (y, \zeta) \mapsto (Sy, \sigma(y)\zeta)$. Show that $Y \times_{\sigma} G/H$ is a compact extension of Y.

2.13.2. Multiple recurrence for compact extensions. Let us say that a measure-preserving system $(X, \mathcal{X}, \mu, T)$ obeys the *uniform multiple recurrence* (UMR) property if the conclusion of the Furstenberg multiple recurrence theorem holds for this system, thus for all $k \geq 1$ and all non-negative $f \in L^{\infty}(X)$ with $\int_X f \, d\mu > 0$, we have

$$(2.147) \qquad \liminf_{N \to \infty} \frac{1}{N} \sum_{n=0}^{N-1} \int_X f T^n f \ldots T^{(k-1)n} f \, d\mu > 0.$$

Thus in Section 2.11 we showed that all compact systems obey UMR, and in Section 2.12 we showed that all weakly mixing systems obey UMR. The Furstenberg multiple recurrence theorem asserts, of course, that *all* measure-preserving systems obey UMR.

We now establish an important further step (and, in many ways, the *key* step) towards proving that theorem:

Theorem 2.13.11. *Suppose that $X = (X, \mathcal{X}, \mu, T)$ is a compact extension of $Y = (Y, \mathcal{Y}, \nu, S)$. If Y obeys UMR, then so does X.*

Note that the converse implication is trivial: if a system obeys UMR, then all of its factors automatically do also.

Proof. Fix $k \geq 1$, and fix a non-negative function $f \in L^\infty(X)$ with $\int_X f \, d\mu > 0$. Our objective is to show that (2.147) holds. As X is a compact extension, f is conditionally almost periodic in measure; by definition (and uniform integrability), this implies that f can be bounded from below by another conditionally almost periodic function which is non-negative with positive mean. Thus we may assume without loss of generality that f is conditionally almost periodic.

We may normalise $\|f\|_{L^\infty(X)} = 1$ and $\int_X f \, d\mu = \delta$ for some $0 < \delta < 1$. The reader may wish to follow this proof using the skew shift example as a guiding model.

Let $\varepsilon > 0$ be a small number (depending on k and δ) to be chosen later. If we set $E := \{y \in Y : \mathbf{E}(f|Y) > \delta/2\}$, then E must have measure at least $\delta/2$.

Since f is almost periodic, we can find a finitely generated module zonotope $\{c_1 f_1 + \ldots + c_d f_d : \|c_1\|_{L^\infty(Y)}, \ldots, \|c_d\|_{L^\infty(Y)} \leq 1\}$ whose ε-neighbourhood contains the orbit of f. In other words, we have an identity of the form

$$(2.148) \qquad T^n f = c_{1,n} f_1 + \ldots + c_{d,n} f_d + e_n$$

for all n, where $c_{1,n}, \ldots, c_{d,n} \in L^\infty(Y)$ with norm at most 1, and $e_n \in L^2(X, Y)$ is an error with $\|e_n\|_{L^2(X|Y)} = O(\varepsilon)$ almost everywhere.

By splitting into real and imaginary parts (and doubling d if necessary) we may assume that the $c_{j,n}$ are real-valued. By further duplication we can also assume that $\|f_i\|_{L^2(X|Y)} \leq 1$ for each i. By rounding off $c_{j,n}(y)$ to the nearest multiple of ε/d for each y (and absorbing the error into the e_n term) we may assume that $c_{j,n}(y)$ is always a multiple of ε/d. Thus each $c_{j,n}$ only takes on $O_{\varepsilon,d}(1)$ values.

Let K be a large integer (depending on k, d, δ, ε) to be chosen later. Since the factor space Y obeys UMR, and E has positive measure in Y, we know that

$$(2.149) \qquad \liminf_{N \to \infty} \frac{1}{N} \sum_{n=0}^{N-1} \int_Y 1_E T^n 1_E \ldots T^{(K-1)n} 1_E \, d\nu > 0.$$

In other words, there exists a constant $c > 0$ such that

$$(2.150) \qquad \nu(\Omega_n) > c$$

for a set of n of positive lower density, where Ω_n is the set

$$(2.151) \qquad \Omega_n := E \cap T^n E \cap \ldots \cap T^{(K-1)n} E.$$

Let n be as above. By definition of Ω_n and E (and (2.145)), we see that

(2.152) $$\mathbf{E}(T^{an}f|Y)(y) \geq \delta/2$$

for all $y \in \Omega_n$ and $0 \leq a < K$. Meanwhile, from (2.148) we have

(2.153) $$\|T^{an}f - c_{1,an}f_1 - \ldots - c_{d,an}f_d\|_{L^2(X|Y)}(y) = O(\varepsilon)$$

for all $y \in \Omega_n$ and $0 \leq a < K$.

Fix y. For each $0 \leq a < K$, the d-tuple $\vec{c}_{an}(y) := (c_{1,an}(y), \ldots, c_{d,an}(y))$ ranges over a set of cardinality $O_{d,\varepsilon}(1)$. One can view this as a colouring of $\{0, \ldots, K-1\}$ into $O_{d,\varepsilon}(1)$ colours. Applying van der Waerden's theorem (Exercise 2.4.3), we can thus find (if K is sufficiently large depending on d, ε, k) an arithmetic progression $a(y), a(y) + r(y), \ldots, a(y) + (k-1)r(y)$ in $\{0, \ldots, K-1\}$ for each y such that

(2.154) $$\vec{c}_{a(y)n}(y) = \vec{c}_{(a(y)+r(y))n}(y) = \ldots = \vec{c}_{(a(y)+(k-1)r(y))n}(y).$$

The quantities $a(y)$ and $r(y)$ can of course be chosen to be measurable in y. By the pigeonhole principle, we can thus find a subset Ω_n' of Ω_n of measure at least $\sigma > 0$ for some σ depending on c, K, d, ε but independent of n, and an arithmetic progression $a, a+r, \ldots, a+(k-1)r$ in $\{0, \ldots, K-1\}$ such that

(2.155) $$\vec{c}_{an}(y) = \vec{c}_{(a+r)n}(y) = \ldots = \vec{c}_{(a+(k-1)r)n}(y)$$

for all $y \in \Omega_n'$. (The quantities a and r can still depend on n, but this will not be of concern to us.)

Fix these values of a, r. From (2.153), (2.155), and the triangle inequality we see that

(2.156) $$\|T^{(a+jr)n}f - T^{an}f\|_{L^2(X|Y)}(y) = O(\varepsilon)$$

for all $1 \leq j \leq k$ and $y \in \Omega_n'$. Recalling that f was normalised to have the $L^\infty(X)$ norm 1, it is then not hard to conclude (by induction on k and the relative Cauchy-Schwarz inequality) that

(2.157) $$\|T^{an}f T^{(a+r)n}f \ldots T^{(a+(k-1)r)n}f - (T^{an}f)^k\|_{L^2(X|Y)}(y) = O_k(\varepsilon)$$

and thus (by another application of the relative Cauchy-Schwarz inequality)

(2.158) $$\mathbf{E}(T^{an}f T^{(a+r)n}f \ldots T^{(a+(k-1)r)n}f)(y) \geq \mathbf{E}((T^{an}f)^k|Y)(y) - O_k(\varepsilon).$$

But from (2.151), (2.152) and the relative Cauchy-Schwarz inequality again we have

(2.159) $$\mathbf{E}(T^{an}f|Y)(y) \geq \delta/2 - O(\varepsilon),$$

and so by several more applications of the same inequality we have

(2.160) $$\mathbf{E}((T^{an}f)^k|Y)(y) \geq c(k,\delta) > 0$$

for some positive quantity $c(k, \delta)$ (if ε is sufficiently small depending on k and δ). From (2.158), (2.160) we conclude that

$$(2.161) \qquad \mathbf{E}(T^{an}fT^{(a+r)n}f\ldots T^{(a+(k-1)r)n}f)(y) \geq c(k,\delta)/2$$

for $y \in \Omega'_n$, again if ε is small enough. Integrating this in y and using the shift-invariance we conclude that

$$(2.162) \qquad \int_X fT^{nr}f\ldots T^{(k-1)nr}f\,d\mu \geq c(k,\delta)\sigma/2.$$

The quantity r depends on n, but ranges between 1 and $K - 1$, and so (by the non-negativity of f)

$$(2.163) \qquad \sum_{s=1}^{K-1}\int_X fT^{ns}f\ldots T^{(k-1)ns}f\,d\mu \geq c(k,\delta)\sigma/2$$

for a set of n of positive lower density. Averaging this for n from 1 to N (say) one obtains (2.147) as desired. $\square$

Thus for instance we have now established UMR for the skew shift as well as higher iterates of that shift, thanks to Exercises 2.13.7 and 2.13.8.

Remark 2.13.12. One can avoid the use of Hilbert modules, etc. by instead appealing to the theory of disintegration of measures (Theorem 2.9.21). We sketch the details as follows. First, one has to restrict attention to those spaces X which are regular, though an inspection of the Furstenberg correspondence principle (Section 2.10) shows that this is in fact automatic for the purposes of such tasks as proving Szemerédi's theorem. Once one disintegrates μ with respect to ν, the situation now resembles the concrete example of the skew shift, with the fibre measures μ_y playing the role of integration along vertical fibres $\{(y, z) : z \in \mathbf{R}/\mathbf{Z}\}$. It is then not difficult (and somewhat instructive) to convert the above proof to one using norms such as $L^2(X, \mathcal{X}, mu_y)$ rather than the module norm $L^2(X|Y)$. We leave the details to the reader (who can also get them from [**Fu1981**]).

Remark 2.13.13. It is an intriguing question as to whether there is any interesting non-commutative extension of the above theory, in which the underlying von Neumann algebra $L^\infty(Y, \mathcal{Y}, \nu)$ is replaced by a non-commutative von Neumann algebra. While some of the theory seems to extend relatively easily, there does appear to be some genuine difficulties with other parts of the theory, particularly those involving multiple products such as $fT^nfT^{2n}f$.

Remark 2.13.14. Just as ergodic compact systems can be described as group rotation systems (Kronecker systems), it turns out that ergodic compact extensions can be described as (inverse limits of) group quotient extensions, somewhat analogously to Lemma 2.6.22. Roughly speaking, the idea is to first use some spectral theory to approximate conditionally almost periodic functions by conditionally *quasiperiodic* functions—those functions whose orbit lies on a finitely generated module zonotope (as opposed to merely being close to one). One can then use the generators of that zonotope as a basis from which to build the group quotient extension, and then use some further trickery to make the group consistent across all fibres. The precise machinery for this is known as *Mackey theory*; it is of particular importance in the deeper structural theory of dynamical systems, but we will not describe it in detail here, instead referring the reader to the papers of Furstenberg [**Fu1977**] and Zimmer [**Zi1976**].

Notes. This lecture first appeared at

terrytao.wordpress.com/2008/02/27.

Thanks to Liu Xiao Chuan for corrections.

2.14. Weakly mixing extensions

Having studied compact extensions in Section 2.13, we now consider the opposite type of extension, namely that of a *weakly mixing extension*. Just as compact extensions are "relative" versions of compact systems (see Section 2.11), weakly mixing extensions are "relative" versions of weakly mixing systems (see Section 2.12), in which the underlying algebra of scalars $\mathbf{C}$ is replaced by $L^\infty(Y)$. As in the case of unconditionally weakly mixing systems, we will be able to use the van der Corput lemma to neglect "conditionally weakly mixing" functions, thus allowing us to lift the uniform multiple recurrence property (UMR) from a system to any weakly mixing extension of that system.

Finishing the proof of the Furstenberg recurrence theorem requires two more steps. One is a relative version of the dichotomy between mixing and compactness: If a system is not weakly mixing relative to some factor, then that factor has a non-trivial compact extension. This will be accomplished using the theory of conditional Hilbert-Schmidt operators in this lecture. Finally, we need the (easy) result that the UMR property is preserved under limits of chains; this will be accomplished in the next section.

2.14.1. Conditionally weakly mixing functions. Recall that in a measure-preserving system $X = (X, \mathcal{X}, \mu, T)$, we call a function $f \in L^2(X) = L^2(X, \mathcal{X}, \mu)$ *weakly mixing* if the squared inner products $|\langle T^n f, f \rangle_X|^2 := (\int_X T^n f \bar{f} \, d\mu)^2$ converge in the Cesàro sense, thus

$$(2.164) \qquad \lim_{N \to \infty} \frac{1}{N} \sum_{n=0}^{N-1} \left| \int_X T^n f \bar{f} \, d\mu \right|^2 = 0.$$

Now let $Y = (Y, \mathcal{Y}, \nu, S)$ be a factor of X, so that $L^\infty(Y)$ can be viewed as a subspace of $L^\infty(X)$. Recall that we have the conditional inner product $\langle f, g \rangle_{X|Y} := \mathbf{E}(f\overline{g}|Y)$ and the Hilbert module $L^2(X|Y)$ of functions f for which $\langle f, f \rangle_{X|Y}$ lies in $L^\infty(Y)$. We shall say that a function $f \in L^2(X|Y)$ is *conditionally weakly mixing* relative to Y if the L^2 norms $\|\langle T^n f, f \rangle_{X|Y}\|_{L^2(Y)}^2$ converge to zero in the Cesàro sense, thus

$$(2.165) \qquad \lim_{N \to \infty} \frac{1}{N} \sum_{n=0}^{N-1} \int_X |\mathbf{E}(T^n f \overline{f}|Y)|^2 \, d\mu = 0.$$

Example 2.14.1. If $X = Y \times Z$ is a product system of the factor space $Y = (Y, \mathcal{Y}, \nu, S)$ and another system $Z = (Z, \mathcal{Z}, \rho, R)$, then a function $f(y, z) = f(z)$ of the vertical variable $z \in Z$ is weakly mixing relative to Y if and only if $f(z)$ is weakly mixing in Z.

Much of the theory of weakly mixing systems extends easily to the conditionally weakly mixing case. For instance:

Exercise 2.14.1. By adapting the proof of Corollary 2.12.13, show that if $f \in L^2(X|Y)$ is conditionally weakly mixing and $g \in L^2(X|Y)$, then $\|\langle T^n f, g \rangle_{X|Y}\|_{L^2(Y)}^2$ and $\|\langle f, T^n g \rangle_{X|Y}\|_{L^2(Y)}^2$ converge to zero in the Cesàro sense. *Hint*: You will need to show that expressions such as $\langle g, T^n f \rangle_{X|Y} T^n f$ converge in $L^2(X)$ in the Cesàro sense. Apply the van der Corput lemma and use the fact that $\langle g, T^n f \rangle_{X|Y}$ are uniformly bounded in $L^\infty(Y)$ by the conditional Cauchy-Schwarz inequality.

Exercise 2.14.2. Show that the space of conditionally weakly mixing functions in $L^2(X|Y)$ is a module over $L^\infty(Y)$ (i.e., it is closed under addition and multiplication by the "scalars" $L^\infty(Y)$), which is also shift-invariant and topologically closed in the topology of $L^2(X|Y)$ (see Exercise 2.13.2).

Let us now see the first link between conditional weak mixing and conditional almost periodicity (cf. Exercise 2.12.18):

Lemma 2.14.2. *If $f \in L^2(X|Y)$ is conditionally weakly mixing and $g \in L^2(X|Y)$ is conditionally almost periodic, then $\langle f, g \rangle_{X|Y} = 0$ a.e.*

Proof. Since $\langle f, g \rangle_{X|Y} = T^{-n} \langle T^n f, T^n g \rangle_{X|Y}$, it will suffice to show that

$$(2.166) \qquad \text{C-sup}_{n \to \infty} |\langle T^n f, T^n g \rangle_{X|Y}|_{L^2(Y)} = 0.$$

Let $\varepsilon > 0$ be arbitrary. As g is conditionally almost periodic, one can find a finitely generated module zonotope

$$\{c_1 f_1 + \ldots + c_d f_d : \|c_1\|_{L^\infty(Y)}, \ldots, \|c_d\|_{L^\infty(Y)} \leq 1\}$$

with $f_1, \ldots, f_d \in L^2(X|Y)$ such that all the shifts $T^n g$ lie within ε (in $L^2(X|Y)$) of this zonotope. Thus (by the conditional Cauchy-Schwarz inequality) we have
(2.167)
$$\|\langle T^n f, T^n g \rangle_{X|Y}\|_{L^2(Y)} = \|\langle T^n f, c_{1,n} f_1 + \ldots + c_{d,n} f_d \rangle_{X|Y}\|_{L^2(Y)} + O(\varepsilon)$$

for all n and some $c_{1,n}, \ldots, c_{d,n} \in L^\infty(Y)$ with norm at most 1. We can pull these constants out of the conditional inner product and bound the left-hand side of (2.167) by

(2.168) $$\|\langle T^n f, f_1 \rangle\|_{L^2(Y)} + \ldots + \|\langle T^n f, f_1 \rangle\|_{L^2(Y)} + O(\varepsilon).$$

By Exercise 2.14.1, the Cesàro supremum of (2.168) is at most $O(\varepsilon)$. Since ε is arbitrary, the claim (2.166) follows. □

Since all functions in $L^\infty(Y)$ are conditionally almost periodic, we conclude that every conditionally weakly mixing function f is orthogonal to $L^\infty(Y)$, or equivalently that $\mathbf{E}(f|Y) = 0$ a.e. Let us say that f has *relative mean zero* if the latter holds.

Definition 2.14.3. A system X is a *weakly mixing extension* of a factor Y if every $f \in L^2(X|Y)$ with relative mean zero is relatively weakly mixing.

Exercise 2.14.3. Show that a product $X = Y \times Z$ of a system Y with a weakly mixing system Z is always a weakly mixing extension of Y.

Remark 2.14.4. If X is regular, then we can disintegrate the measure μ as an average $\mu = \int_Y \mu_y \, d\nu(y)$; see Theorem 2.9.21. It is then possible to construct a relative product system $X \times_Y X$, which is the product system $X \times X$ but with the measure $\mu \times_\nu \mu := \int_Y \mu_y \times \mu_y \, d\nu(y)$ instead of $\mu \times \mu$. It can then be shown (cf. Exercise 2.12.9) that X is a weakly mixing extension of Y if and only if $X \times_Y X$ is ergodic; see for instance [**Fu1981**] for details. However, in these notes we shall focus instead on the more abstract operator-algebraic approach which avoids the use of disintegrations.

Now we show that the uniform multiple recurrence property (UMR) from Section 2.13 is preserved under weakly mixing extensions (cf. Theorem 2.13.11).

Theorem 2.14.5. *Suppose that $X = (X, \mathcal{X}, \mu, T)$ is a weakly mixing extension of $Y = (Y, \mathcal{Y}, \nu, S)$. If Y obeys UMR, then so does X.*

The proof of this theorem rests on the following analogue of Proposition 2.12.11:

Proposition 2.14.6. *Let $a_1, \ldots, a_k \in \mathbf{Z}$ be distinct integers for some $k \geq 1$. Let $X = (X, \mathcal{X}, \mu, T)$ is a weakly mixing extension of $Y = (Y, \mathcal{Y}, \nu, S)$, and*

let $f_1, \ldots, f_k \in L^\infty(X)$ be such that at least one of $f_1, \ldots, f_k$ has relative mean zero. Then

$$\text{C-lim}_{n\to\infty} T^{a_1 n} f_1 \ldots T^{a_k n} f_k = 0 \tag{2.169}$$

in $L^2(X, \mathcal{X}, \mu)$.

Exercise 2.14.4. Prove Proposition 2.14.6. *Hint*: Modify (or "relativise") the proof of Proposition 2.12.11.

Corollary 2.14.7. *Let $a_1, \ldots, a_k \in \mathbf{Z}$ be distinct integers for some $k \geq 1$. Let $X = (X, \mathcal{X}, \mu, T)$ be a weakly mixing extension of $Y = (Y, \mathcal{Y}, \nu, S)$, and let $f_1, \ldots, f_k \in L^\infty(X)$. Then*

$$
\begin{aligned}
&\text{C-lim}_{n\to\infty} \int_X T^{a_1 n} f_1 \ldots T^{a_k n} f_k \, d\mu \\
&\quad - \int_X T^{a_1 n} \mathbf{E}(f_1|Y) \ldots T^{a_k n} \mathbf{E}(f_k|Y) \, d\mu = 0.
\end{aligned}
\tag{2.170}
$$

Exercise 2.14.5. Prove Corollary 2.14.7. *Hint*: Adapt the proof of Corollary 2.12.13.

Proof of Theorem 2.14.5. Let $f \in L^\infty(X)$ be non-negative with positive mean. Then $\mathbf{E}(f|Y) \in L^\infty(Y)$ is also non-negative with positive mean. Since Y obeys UMR, we have

$$\liminf_{N\to\infty} \frac{1}{N} \sum_{n=0}^{N-1} \mathbf{E}(f|Y) T^n \mathbf{E}(f|Y) \ldots T^{(k-1)n} \mathbf{E}(f|Y) > 0. \tag{2.171}$$

Applying Corollary 2.14.7 we see that the same statement holds with $\mathbf{E}(f|Y)$ replaced by f, and the claim follows. $\qquad\square$

Remark 2.14.8. As the above proof shows, Corollary 2.14.7 lets us replace functions in the weakly mixing extension X by their expectations in Y for the purposes of computing k-fold averages. In the notation of [**FuWe1996**], Corollary 2.14.7 asserts that Y is a *characteristic factor* of X for the average (2.170). The deeper structural theory of such characteristic factors (and in particular, on the minimal characteristic factor for any given average) is an active and difficult area of research, with surprising connections with Lie group actions (and in particular with flows on nilmanifolds), as well as the theory of inverse problems in additive combinatorics (and in particular with inverse theorems for the Gowers norms); see for instance [**Kr2006**] for a survey of recent developments. The concept of a characteristic factor (or more precisely, finitary analogues of this concept) also is fundamental in my work with Ben Green [**GrTa2008**] on primes in arithmetic progression.

2.14.2. The dichotomy between structure and randomness. The remainder of this section is devoted to proving the following "relative" generalisation of Theorem 2.12.14, which is a fundamental ingredient in the proof of the Furstenberg recurrence theorem:

Theorem 2.14.9. *Suppose that $X = (X, \mathcal{X}, \mu, T)$ is an extension of a system $Y = (Y, \mathcal{Y}, \nu, S)$. Then exactly one of the following statements is true:*

(1) *(Structure) X has a factor Z which is a non-trivial compact extension of Y.*

(2) *(Randomness) X is a weakly mixing extension of Y.*

As in Section 2.12, the key to proving this theorem is to prove

Proposition 2.14.10. *Suppose that $X = (X, \mathcal{X}, \mu, T)$ is an extension of a system $Y = (Y, \mathcal{Y}, \nu, S)$. Then a function $f \in L^2(X|Y)$ is relatively weakly mixing if and only if $\langle f, g \rangle_{X|Y} = 0$ a.e. for all relatively almost periodic g.*

The "only if" part of this proposition is Lemma 2.14.2; the harder part is the "if" part, which we will prove shortly. But for now, let us see why Proposition 2.14.10 implies Theorem 2.14.9.

From Lemma 2.14.2, we already know that no non-trivial function can be simultaneously conditionally weakly mixing and conditionally almost periodic, which shows that cases (1) and (2) of Theorem 2.14.9 cannot simultaneously hold. To finish the proof of Theorem 2.14.9, suppose that X is not a weakly mixing extension of Y, thus there exists a function $f \in L^2(X|Y)$ of relative mean zero which is not weakly mixing. By Proposition 2.14.10, there must exist a relatively almost periodic $g \in L^2(X|Y)$ such that $\langle f, g \rangle_{X|Y}$ does not vanish a.e. Since f is orthogonal to all functions in $L^\infty(Y)$, we conclude that g is *not* in $L^\infty(Y)$, thus we have a single relatively almost periodic function. From Exercise 2.13.6, this shows that the maximal compact extension of Y is non-trivial, and the claim follows.

It thus suffices to prove the "if" part of Proposition 2.14.10; we need to show that every non-conditionally-weakly-mixing function correlates with some conditionally almost periodic function. But if $f \in L^2(X|Y)$ is not conditionally weakly mixing, then by definition we have

$$(2.172) \qquad \limsup_{N \to \infty} \frac{1}{N} \sum_{n=0}^{N-1} |\mathbf{E}(T^n f \overline{f} | Y)|^2_{L^2(Y)} > 0.$$

We can rearrange this as

$$(2.173) \qquad \limsup_{N \to \infty} \langle S_{f,N} f, f \rangle_X > 0,$$

where $S_{f,N} : L^2(X|Y) \to L^2(X|Y)$ is the operator

$$(2.174) \qquad\qquad S_{f,N} g := \frac{1}{N} \sum_{n=0}^{N-1} \mathbf{E}(g\overline{T^n f}|Y)T^n f.$$

To prove Proposition 2.14.10, it thus suffices (by weak compactness) to show that

Proposition 2.14.11 (Dual functions are almost periodic). *Suppose that* $X = (X, \mathcal{X}, \mu, T)$ *is an extension of a system* $Y = (Y, \mathcal{Y}, \nu, S)$, *and let* $f \in L^2(X|Y)$. *Let* S_f *be any limit point of* $S_{f,N}$ *in the weak operator topology. Then* $S_f f$ *is relatively almost periodic.*

Remark 2.14.12. By applying the mean ergodic theorem to the dynamical system $X \times_Y X$, one can show that the sequence D_N is in fact convergent in the weak or strong operator topology (at least when X is regular). But to avoid some technicalities we shall present an argument that does not rely on the existence of a strong limit.

As one might expect from the experience with unconditional weak mixing, the proof of Proposition 2.14.11 relies on the theory of conditionally Hilbert-Schmidt operators on $L^2(X|Y)$. We give here a definition of such operators which is suited for our needs.

Definition 2.14.13. Let X, Y be as above. A *suborthonormal set* in $L^2(X|Y)$ is any at most countable sequence $e_\alpha \in L^2(X|Y)$ such that $\langle e_\alpha, e_\beta \rangle_{X|Y} = 0$ a.e. for all $\alpha \neq \beta$ and $\langle e_\alpha, e_\alpha \rangle_{X|Y} \leq 1$ a.e. for all α. A linear operator $A : L^2(X|Y) \to L^2(X|Y)$ is said to be a *conditionally Hilbert-Schmidt operator* if we have the module property

$$(2.175) \qquad\qquad A(cf) = cAf \text{ for all } c \in L^\infty(Y)$$

and the bound

$$(2.176) \qquad\qquad \sum_\alpha \sum_\beta |\langle Ae_\alpha, f_\beta \rangle_{X,Y}|^2 \leq C^2 \text{ a.e.}$$

for all suborthonormal sets $\{e_\alpha\}$, $\{f_\beta\}$ and some constant $C > 0$; the best such C is called the *(uniform) conditional Hilbert-Schmidt norm* $\||\|A\|_{HS(X|Y)}\|_{L^\infty(Y)}$ of A.

Remark 2.14.14. As in Section 2.12, one can also set up the concept of a tensor product of two Hilbert modules, and use that to define conditionally Hilbert-Schmidt operators in a way which does not require suborthonormal sets. But we will not need to do so here. One can also define a pointwise conditional Hilbert-Schmidt norm $\|A\|_{HS(X|Y)}(y)$ for each $y \in Y$, but we will not need this concept.

Example 2.14.15. Suppose Y is just a finite set (with the discrete σ-algebra); then X splits into finitely many fibres $\pi^{-1}(\{y\})$ with conditional measures μ_y, and $L^2(X|Y)$ can be a direct sum (with the l^∞ norm) of the Hilbert spaces $L^2(\mu_y)$. A conditional Hilbert-Schmidt operator is then equivalent to a family of Hilbert-Schmidt operators $A_y : L^2(\mu_y) \to L^2(\mu_y)$ for each y, with the A_y uniformly bounded in the Hilbert-Schmidt norm.

Example 2.14.16. In the skew shift example $X = (\mathbf{R}/\mathbf{Z})^2 = \{(y,z) : y, z \in \mathbf{R}/\mathbf{Z}\}$, $Y = (\mathbf{R}/\mathbf{Z})$, one can show that an operator A is conditionally Hilbert-Schmidt if and only if it takes the form $Af(y,z) = \int_{\mathbf{R}/\mathbf{Z}} K_y(z,z') f(y,z') \, dz'$ a.e. for all $f \in L^2(X|Y)$, with

$$\| \|A\|_{HS(X|Y)} \|_{L^\infty(Y)} = \sup_y \left(\int_{\mathbf{R}/\mathbf{Z}} \int_{\mathbf{R}/\mathbf{Z}} |K_y(z,z')|^2 \, dz \, dz' \right)^{1/2}$$

finite.

Exercise 2.14.6. Let $f_1, f_2 \in L^2(X|Y)$ with $\|f_1\|_{L^2(X|Y)}, \|f_2\|_{L^2(X|Y)} \le 1$ a.e. Show that the rank one operator $g \mapsto \langle g, f_1 \rangle_{X|Y} f_2$ is conditionally Hilbert-Schmidt with norm at most 1.

Observe from (2.174) that the $S_{f,N}$ are averages of rank one operators arising from the functions $T^n f$, and so by Exercise 2.14.6 and the triangle inequality we see that the $S_{f,N}$ are uniformly conditionally Hilbert-Schmidt. Taking weak limits using (2.176) (and Fatou's lemma) we conclude that S_f is also conditionally Hilbert-Schmidt.

Next, we observe from the telescoping identity that for every h, $T^h S_{f,N} - S_{f,N} T^h$ converges to zero in the weak operator topology (and even in the operator norm topology) as $N \to \infty$; taking limits, we see that S_f commutes with T. To show that $S_f f$ is conditionally almost periodic, it thus suffices to show the following analogue of Lemma 2.12.21:

Lemma 2.14.17. *Let $A : L^2(X|Y) \to L^2(X|Y)$ be a conditionally Hilbert-Schmidt operator. Then the image of the unit ball of $L^2(X|Y)$ under A is conditionally precompact.*

Proof. We shall prove this lemma by establishing a sort of conditional *singular value decomposition* for A. We can normalise A to have uniform conditional Hilbert-Schmidt norm 1. We fix $\varepsilon > 0$, and we will also need an integer k and a small quantity $\delta > 0$ depending on ε to be chosen later.

We first consider the quantities $|\langle Ae_1, f_1 \rangle_{X|Y}|^2$ where the pair e_1, f_1 ranges over all suborthonormal sets of cardinality 1. On the one hand, these quantities are bounded pointwise by 1, thanks to (2.176). On the other hand, observe that if $|\langle Ae_1, f_1 \rangle_{X|Y}|^2$ and $|\langle Ae_1', f_1' \rangle_{X|Y}|^2$ are of the above form, then so is the join $\max(|\langle Ae_1, f_1 \rangle_{X|Y}|^2, |\langle Ae_1', f_1' \rangle_{X|Y}|^2)$, as can be seen by taking $\tilde{e}_1 := e_1 1_E + e_1' 1_{E^c}$ and $\tilde{f}_1 := f_1 1_E + f_1' 1_{E^c}$, where E is the set where $|\langle Ae_1, f_1 \rangle_{X|Y}|^2$ exceeds $|\langle Ae_1', f_1' \rangle_{X|Y}|^2$. By using a maximising sequence for the quantity $\int_Y |\langle Ae, f \rangle_{X|Y}|^2 \, d\nu$ and applying joins repeatedly, we can thus (on taking limits) find a pair e_1, f_1 which is near-optimal in the sense that $|\langle Ae_1, f_1 \rangle_{X|Y}|^2 \geq (1-\delta)|\langle Ae_1', f_1' \rangle_{X|Y}|^2$ a.e. for all competitors e_1', f_1'.

Now fix e_1, f_1, and consider the quantity $|\langle Ae_2, f_2 \rangle_{X|Y}|^2$, where $\{e_1, e_2\}$ and $\{f_1, f_2\}$ are suborthonormal sets. By arguing as before we can find a pair e_2, f_2 which is near-optimal in the sense that $|\langle Ae_2, f_2 \rangle_{X|Y}|^2 \geq (1-\delta)|\langle Ae_2', f_2' \rangle_{X|Y}|^2$ a.e. for all competitors e_2', f_2'.

We continue in this fashion k times to obtain suborthonormal sets $\{e_1, \ldots, e_k\}$ and $\{f_1, \ldots, f_k\}$ with the property that $|\langle Ae_i, f_i \rangle_{X|Y}|^2 \geq (1-\delta)|\langle Ae_i', f_i' \rangle_{X|Y}|^2$ whenever $\{e_1, \ldots, e_{i-1}, e_i'\}$ and $\{f_1, \ldots, f_{i-1}, f_i'\}$ are suborthonormal sets. On the other hand, from (2.176) we know that $\sum_i |\langle Ae_i, f_i \rangle_{X|Y}|^2 \leq 1$. From these two facts we soon conclude that $|\langle Ae, f \rangle_{X|Y}|^2 \leq 1/k + O_k(\delta)$ a.e. whenever $\{e_1, \ldots, e_k, e\}$ and $\{f_1, \ldots, f_k, f\}$ are suborthonormal. If k, δ are chosen appropriately, we obtain $|\langle Ae, f \rangle_{X|Y}| \leq \varepsilon$ a.e. Thus (by duality) A maps the unit ball of the orthogonal complement of the span of $\{e_1, \ldots, e_k\}$ to the ε-neighbourhood of the span of $\{f_1, \ldots, f_k\}$ (with notions such as orthogonality, span, and neighbourhood being defined conditionally of course, using the $L^\infty(Y)$-Hilbert module structure of $L^2(X|Y)$). From this it is not hard to establish the desired precompactness. $\qquad\square$

Notes. This lecture first appeared at

terrytao.wordpress.com/2008/03/02.

Thanks to Lior Silberman, Orr, and Liu Xiao Chuan for corrections.

2.15. The Furstenberg-Zimmer structure theorem and the Furstenberg recurrence theorem

In this section—the final lecture on general measure-preserving dynamics—we put together the results from the past few sections to establish the Furstenberg-Zimmer structure theorem for measure-preserving systems, and then use this to finish the proof of the Furstenberg recurrence theorem.

2.15.1. The Furstenberg-Zimmer structure theorem. Suppose $X = (X, \mathcal{X}, \mu, T)$ is a measure-preserving system, and let $Y = (Y, \mathcal{Y}, \nu, S)$ be a factor. In Theorem 2.14.9, we showed that if X was not a weakly mixing extension of Y, then we could find a non-trivial compact extension Z of Y (thus $L^2(Z)$ is a non-trivial superspace of $L^2(Y)$). Combining this with

Zorn's lemma (and starting with the trivial factor $Y = \text{pt}$), one obtains (see [**Fu1977**], [**Zi1976**]) the following theorem.

Theorem 2.15.1 (Furstenberg-Zimmer structure theorem). *Suppose* $(X, \mathcal{X}, \mu, T)$ *is a measure-preserving system. Then there exists an ordinal* α *and a factor* $Y_\beta = (Y_\beta, \mathcal{Y}_\beta, \nu_\beta, S_\beta)$ *for every* $\beta \leq \alpha$ *with the following properties:*

(1) *$Y_\emptyset$ is a point.*

(2) *For every successor ordinal $\beta + 1 \leq \alpha$, $Y_{\beta+1}$ is a compact extension of Y_β.*

(3) *For every limit ordinal $\beta \leq \alpha$, Y_β is the inverse limit of the Y_γ for $\gamma < \beta$, in the sense that $L^2(Y_\beta)$ is the closure of $\bigcup_{\gamma < \beta} L^2(Y_\gamma)$.*

(4) *X is a weakly mixing extension of Y_α.*

Remark 2.15.2. This theorem should be compared with Furstenberg's structure theorem for distal systems in topological dynamics (Theorem 2.7.5). Indeed, in analogy to that theorem, the factors Y_β are known as *distal measure-preserving systems.*

Exercise 2.15.1. Deduce Theorem 2.15.1 from Theorem 2.14.9.

Remark 2.15.3. Since the Hilbert spaces $L^2(Y_\beta)$ are increasing inside the separable Hilbert space $L^2(X)$, it is not hard to see that the ordinal α must be at most countable. Conversely, in [**BeFo1996**] it was shown that every countable ordinal can appear as the minimal length of a Furstenberg tower of a given system. Thus, in some sense, the complexity of a system can be as great as any countable ordinal. This is because the structure theorem roots out every last trace of structure from the system, so much so that every remaining function orthogonal to the final factor $L^2(Y_\alpha)$ is weakly mixing. But in many applications one does not need so much weak mixing; for instance to establish k-fold recurrence for a function f, it would be enough to obtain weak mixing control on just a few combinations of f (such as $T^h f \bar{f}$), as we already saw in the proof of Roth's theorem in Section 2.12. In fact, it is not hard to show that to prove Furstenberg's recurrence theorem for a fixed k, one only needs to analyse the first $k - 2$ steps of the Furstenberg tower. As one consequence of this, it is possible to avoid the use of Zorn's lemma (and the axiom of choice) in the proof of the recurrence theorem.

Remark 2.15.4. Analogues of the structure theorem exist for other actions, such as the action of $\mathbf{Z}^d$ on a measure space (which can equivalently be viewed as the action of d commuting shifts $T_1, \ldots, T_d : X \to X$). There is a new feature in this case, though: instead of having a tower of purely compact extensions, followed by one weakly mixing extension at the end, one has a tower of hybrid extensions (known as primitive extensions), each one of which is compact along one subgroup of $\mathbf{Z}^d$ and weakly mixing along a complementary subgroup. See for instance [**Fu1981**] for details.

2.15.2. The Furstenberg recurrence theorem. The Furstenberg recurrence theorem asserts that every measure-preserving system $(X, \mathcal{X}, \mu, T)$ has the uniform multiple recurrence (UMR) property, thus

$$(2.177) \qquad \liminf_{N \to \infty} \frac{1}{N} \sum_{n=0}^{N-1} \int_X f T^n f \dots T^{(k-1)n} f \, d\mu > 0$$

whenever $k \geq 1$ and $f \in L^\infty(X)$ is non-negative with positive mean. The UMR property is trivially true for a point, and we have already shown that UMR is preserved by compact extensions (Theorem 2.13.11) and by weakly mixing extensions (Theorem 2.14.5). The former result lets us climb the successor ordinal steps of the tower in Theorem 2.15.1, while the latter lets us jump from the final distal system Y_α to X. But to clinch the proof of the recurrence theorem, we also need to deal with the limit ordinals. More precisely, we need to prove

Theorem 2.15.5 (Limits of chains). *Let $(Y_\beta)_{\beta \in B}$ be a totally ordered family of factors of a measure-preserving system X (thus $L^2(Y_\beta)$ is increasing with β), and let Y be the inverse limit of the Y_β. If each of the Y_β obeys the UMR, then Y does also.*

With this theorem, the Furstenberg recurrence theorem (Theorem 2.11.4) follows from the previous theorems and transfinite induction.

The main difficulty in establishing Theorem 2.15.5 is that while each Y_β obeys the UMR separately, we do not know that this property holds uniformly in β. The main new observation needed to establish the theorem is that there is another way to leverage the UMR from a factor to an extension... if the support of the function f is sufficiently "dense". We motivate this by first considering the unconditional case.

Proposition 2.15.6 (UMR for densely supported functions). *Let $(X, \mathcal{X}, \mu, T)$ be a measure-preserving system, let $k \geq 1$ be an integer, and let $f \in L^\infty(X)$ be a non-negative function whose support $\{x : f(x) > 0\}$ has measure greater than $1 - 1/k$. Then (2.177) holds.*

Proof. By monotone convergence, we can find $\varepsilon > 0$ such that $f(x) > \varepsilon$ for all x outside of a set E of measure at most $1/k - \varepsilon$. For any n, this implies that $f(x) T^n f(x) \dots T^{(k-1)n} f(x) > \varepsilon^k$ for all x outside of the set $E \cup T^n E \cup \dots \cup T^{(k-1)n} E$, which has measure at most $1 - k\varepsilon$. In particular we see that

$$(2.178) \qquad \int_X fT^n f \dots T^{(k-1)n} f \, d\mu > k\varepsilon^{k+1}$$

for all n, and the claim follows. □

As with the other components of the proof of the recurrence theorem, we will need to upgrade the above proposition to a "relative" version:

Proposition 2.15.7 (UMR for relatively densely supported functions). *Let $(X, \mathcal{X}, \mu, T)$ be an extension of a factor $(Y, \mathcal{Y}, \nu, S)$ with the UMR, let $k \geq 1$ be an integer, and let $f \in L^\infty(X)$ be a non-negative function whose support $\Omega := \{x : f(x) > 0\}$ is such that the set $\{y \in Y : \mathbf{E}(1_\Omega|Y) > 1 - 1/k\}$ has positive measure in Y. Then (2.177) holds.*

Proof. By monotone convergence again, we can find $\varepsilon > 0$ such that the set $E := \{x : f(x) > \varepsilon\}$ is such that the set $F := \{y \in Y : \mathbf{E}(1_E|Y) > 1 - 1/k + \varepsilon\}$ has positive measure. Since Y has the UMR, this implies that (2.177) holds for 1_F. In other words, there exists $c > 0$ such that

$$(2.179) \qquad \nu(F \cap T^n F \cap \dots \cap T^{(k-1)n} F) > c$$

for all n in a set of positive lower density.

Now we turn to f. We have the pointwise lower bound $f(x) \geq \varepsilon 1_E(x)$, and so

$$(2.180) \qquad fT^n f \dots T^{(k-1)n} f(x) \geq \varepsilon^k 1_{E \cap T^n E \cap \dots \cap T^{(k-1)n} E}(x).$$

We have the crude lower bound

$$(2.181) \qquad 1_{E \cap T^n E \cap \dots \cap T^{(k-1)n} E}(x) \geq 1 - \sum_{j=0}^{k-1} 1_{T^{jn} E^c}(x);$$

inserting this into (2.180) and taking conditional expectations, we conclude that

$$(2.182) \qquad \mathbf{E}(fT^n f \dots T^{(k-1)n} f|Y)(y) \geq \varepsilon^k \left(1 - \sum_{j=0}^{k-1} \mathbf{E}(1_{T^{jn} E^c}|Y)(y)\right)$$

a.e. On the other hand, we have

$$(2.183) \qquad \mathbf{E}(1_{T^{jn} E^c}|Y) = 1 - \mathbf{E}(1_{T^{jn} E}|Y) = 1 - T^{jn}\mathbf{E}(1_E|Y).$$

By definition of F, we thus see that if y lies in $F \cap T^n F \cap \dots \cap T^{(k-1)n} F$, then

$$(2.184) \qquad \mathbf{E}(fT^n f \dots T^{(k-1)n} f|Y)(y) \geq \varepsilon^k \times k\varepsilon.$$

Integrating this and using (2.179), we obtain

$$(2.185) \qquad \int_X fT^n f \dots T^{(k-1)n} f \, d\mu \geq c\varepsilon^k \times k\varepsilon$$

for all n in a set of positive lower density, and (2.177) follows. $\square$

Proof of Theorem 2.15.5. Let $f \in L^\infty(Y)$ be non-negative with positive mean $\int_X f \, d\mu = c > 0$; we may normalise f to be bounded by 1. Since Y is the inverse limit of the Y_β, we see that the orthogonal projections $\mathbf{E}(f|Y_\beta)$ converge in the $L^2(X)$ norm to $\mathbf{E}(f|Y) = f$. Thus, for any ε, we can find β such that

$$(2.186) \qquad \|f - \mathbf{E}(f|Y_\beta)\|_{L^2(X)} \leq \varepsilon.$$

Now $\mathbf{E}(f|Y_\beta)$ has the same mean c as f, and is also bounded by 1. Thus the set $E := \{y : \mathbf{E}(f|Y_\beta)(y) \geq c/2\}$ must have measure at least $c/2$ in Y_β. Now if $\Omega := \{x : f(x) > 0\}$, then we have the pointwise bound

$$(2.187) \qquad |f - \mathbf{E}(f|Y_\beta)| \geq \frac{c}{2} 1_{\Omega^c} 1_E;$$

squaring this and taking conditional expectations we obtain

$$(2.188) \qquad \mathbf{E}(|f - \mathbf{E}(f|Y_\beta)|^2|Y_\beta)(y) \geq \frac{c^2}{4}(1 - \mathbf{E}(1_\Omega|Y_\beta)(y))1_E(y),$$

and so by (2.186) and Markov's inequality we see that $1 - \mathbf{E}(1_\Omega|Y_\beta)(y)1_E(y)$ $< 1/k$ on a set of measure $O_c(\varepsilon^2)$. Choosing ε sufficiently small depending on c, we conclude (from the lower bound $\mu(E) \geq c/2$) that $\mathbf{E}(1_\Omega|Y_\beta)(y) > 1 - 1/k$ on a set of positive measure. The claim now follows from Proposition 2.15.7. $\square$

The proof of the Furstenberg recurrence theorem (and thus Szemerédi's theorem) is finally complete.

Remark 2.15.8. The same type of argument yields many further recurrence theorems, and thus (by the correspondence principle) many combinatorial results also. For instance, in [**Fu1977**] it was noted that the above arguments allow one to strengthen (2.177) to

$$(2.189) \qquad \liminf_{N \to \infty} \inf_M \frac{1}{N} \sum_{n=M}^{M+N-1} \int_X f T^n f \ldots T^{(k-1)n} f \, d\mu > 0,$$

which allows one to conclude that in a set A of positive upper density, the set of n for which $A \cap (A + n) \cap \ldots \cap (A + (k-1)n)$ has positive upper density is syndetic for every k. One can also extend the argument to higher dimensions, and to polynomial recurrence without too many changes in the structure of the proof. But some more serious modifications to the argument are needed for other recurrence results involving IP systems or Hales-Jewett type results; see Section 2.10 for more discussion.

Notes. This lecture first appeared at

terrytao.wordpress.com/2008/03/05.

Thanks to Lior Silberman, Nate Chandler, and Liu Xiao Chuan for corrections.

2.16. A Ratner-type theorem for nilmanifolds

This section and the next will be on *Ratner's theorems* on equidistribution of orbits on homogeneous spaces (see also Section 1.11 of *Structure and Randomness* for an introduction to this family of results). Here, I will discuss two special cases of Ratner-type theorems. In this section, I will talk about Ratner-type theorems for discrete actions (of the integers **Z**) on nilmanifolds; this case is much simpler than the general case, because there is a simple criterion in the nilmanifold case to test whether any given orbit is equidistributed or not. Ben Green and I had need recently [**GrTa2009c**] to develop quantitative versions of such theorems for a number-theoretic application. In Section 2.17, I will discuss Ratner-type theorems for actions of $SL_2(\mathbf{R})$, which is simpler in a different way (due to the semisimplicity of $SL_2(\mathbf{R})$, and lack of compact factors).

2.16.1. Nilpotent groups. Before we can get to Ratner-type theorems for nilmanifolds, we will need to set up some basic theory for these nilmanifolds. We begin with a quick review of the concept of a nilpotent group, a generalisation of that of an abelian group. Our discussion here will be purely algebraic (no manifolds, topology, or dynamics will appear at this stage).

Definition 2.16.1 (Commutators). Let G be a (multiplicative) group. For any two elements g, h in G, we define the commutator $[g, h]$ to be $[g, h] := g^{-1}h^{-1}gh$ (thus g and h commute if and only if the commutator is trivial). If H and K are subgroups of G, we define the *commutator* $[H, K]$ to be the group generated by all the commutators $\{[h, k] : h \in H, k \in K\}$.

For future reference we record some trivial identities regarding commutators:

(2.190)
$$gh = hg[g, h] = [g^{-1}, h^{-1}]hg,$$

(2.191)
$$h^{-1}gh = g[g, h] = [h, g^{-1}]g,$$

(2.192)
$$[g, h]^{-1} = [h, g].$$

Exercise 2.16.1. Let H, K be subgroups of a group G.

    (1) Show that $[H, K] = [K, H]$.

    (2) Show that H is abelian if and only if $[H, H]$ is trivial.

(3) Show that H is central if and only if $[H, G]$ is trivial.

(4) Show that H is normal if and only if $[H, G] \subset H$.

(5) Show that $[H, G]$ is always normal.

(6) If $L \triangleleft H, K$ is a normal subgroup of both H and K, show that $[H, K]/([H, K] \cap L) \equiv [H/L, K/L]$.

(7) Let HK be the group generated by $H \cup K$. Show that $[H, K]$ is a normal subgroup of HK, and when one *quotients* by this subgroup, $H/[H, K]$ and $K/[H, K]$ become abelian.

Exercise 2.16.2. Let G be a group. Show that the group $G/[G, G]$ is abelian, and is the *universal* abelianisation of G in the sense that every *homomorphism* $\phi : G \to H$ from G to an abelian group H can be uniquely factored as $\phi = \tilde{\phi} \circ \pi$, where $\pi : G \to G/[G, G]$ is the quotient map and $\tilde{\phi} : G/[G, G] \to H$ is a homomorphism.

Definition 2.16.2 (Nilpotency). Given any group G, define the *lower central series*

$$(2.193) \qquad G = G_0 = G_1 \triangleright G_2 \triangleright G_3 \triangleright \ldots$$

by setting $G_0, G_1 := G$ and $G_{i+1} := [G_i, G]$ for $i \geq 1$. We say that G is *nilpotent* of step s if G_{s+1} is trivial (and G_s is non-trivial).

Examples 2.16.3. A group is nilpotent of step 0 if and only if it is trivial. It is nilpotent of step 1 if and only if it is non-trivial and abelian. Any subgroup or homomorphic image of a nilpotent group of step s is nilpotent of step at most s. The direct product of two nilpotent groups is again nilpotent, but the semidirect product of nilpotent groups is merely solvable in general. If G is any group, then G/G_{s+1} is nilpotent of step at most s.

Example 2.16.4. Let $n \geq 1$ be an integer, and let

$$(2.194) \qquad U_n(\mathbf{R}) = \begin{pmatrix} 1 & \mathbf{R} & \ldots & \mathbf{R} \\ 0 & 1 & \ldots & \mathbf{R} \\ \vdots & \vdots & \ddots & \vdots \\ 0 & 0 & \ldots & 1 \end{pmatrix}$$

be the group of all upper-triangular $n \times n$ real matrices with 1s on the diagonal (i.e., the group of unipotent upper-triangular matrices). Then $U_n(\mathbf{R})$ is nilpotent of step n. Similarly if $\mathbf{R}$ is replaced by other fields.

Exercise 2.16.3. Let G be an arbitrary group.

(1) Show that each element G_i of the lower central series is a *characteristic subgroup* of G, i.e., $\phi(G_i) = G_i$ for all automorphisms[51] $\phi : G \to G$.

---

[51]Specialising to *inner automorphisms*, we see in particular that the G_i are all normal subgroups of G.

(2) Show the *filtration property* $[G_i, G_j] \subset G_{i+j}$ for all $i, j \geq 0$. *Hint*: Induct on $i + j$; then, holding $i + j$ fixed, quotient by G_{i+j}, and induct on (say) i. Note that once one quotients by G_{i+j}, all elements of $[G_{i-1}, G_j]$ are central (by the first induction hypothesis), while G_{i-1} commutes with $[G, G_j]$ (by the second induction hypothesis). Use these facts to show that all the generators of $[G, G_{i-1}]$ commute with G_j.

Exercise 2.16.4. Let G be a nilpotent group of step 2. Establish the identity

(2.195) $$g^n h^n = (gh)^n [g, h]^{\binom{n}{2}}$$

for any integer n and any $g, h \in G$, where $\binom{n}{2} := \frac{n(n-1)}{2}$. (This can be viewed as a discrete version of the first two terms of the *Baker-Campbell-Hausdorff formula*.) Conclude in particular that the space of *Hall-Petresco sequences* $n \mapsto g_0 g_1^n g_2^{\binom{n}{2}}$, where $g_i \in G_i$ for $i = 0, 1, 2$, is a group under pointwise multiplication (this group is known as the *Hall-Petresco group* of G). There is an analogous identity (and an analogous group) for nilpotent groups of higher step; see for instance [**Le1998**] for details. The Hall-Petresco group is rather useful for understanding multiple recurrence and polynomial behaviour in nilmanifolds; we will not discuss this in detail, but see Exercise 2.16.5 below for a hint as to the connection.

Exercise 2.16.5 (Arithmetic progressions in nilspaces are constrained). Let G be a nilpotent group of step $s \leq 2$, and consider two arithmetic progressions $x, gx, \ldots, g^{s+1}x$ and $y, hy, \ldots, h^{s+1}y$ of length $s+2$ in G, where $x, y \in X$ and $g, h \in G$. Show that if these progressions agree in the first $s + 1$ places (thus $g^i x = h^i y$ for all $i = 0, \ldots, s$), then they also agree in the last place. *Hint*: The only tricky case is $s = 2$. For this, either use direct algebraic computation, or experiment with the group of Hall-Petresco sequences from the previous exercise. The claim is in fact true for general s; see e.g. [**GrTa2009d**].

Remark 2.16.5. By Exercise 2.16.3.2, the lower central series is a filtration with respect to the commutator operation $g, h \mapsto [g, h]$. Conversely, if G admits a filtration $G = G_{(0)} = G_{(1)} \geq \ldots$ with $[G_{(i)}, G_{(j)}] \subset G_{(i+j)}$ and $G_{(j)}$ trivial for $j > s$, then it is nilpotent of step at most s. It is sometimes convenient for inductive purposes to work with filtrations rather than the lower central series (which is the "minimal" filtration available to a group G); see for instance [**GrTa2009c**].

Remark 2.16.6. Let G be a nilpotent group of step s. Then $[G, G_s] = G_{s+1}$ is trivial and so G_s is central (by Exercise 2.16.1), thus abelian and normal. By another application of Exercise 2.16.1, we see that G/G_s is nilpotent of step $s - 1$. Thus we see that any nilpotent group G of step s is an *extension* of a nilpotent group G/G_s of step $s - 1$, in the sense that we have a short exact sequence

$$(2.196) \qquad\qquad 0 \to G_s \to G \to G/G_s \to 0,$$

where the kernel G_s is abelian. Conversely, every abelian extension of an $s - 1$-step nilpotent group is nilpotent of step at most s. In principle, this gives a recursive description of s-step nilpotent groups as an s-fold iterated tower of abelian extensions of the trivial group. Unfortunately, while abelian groups are of course very well understood, abelian *extensions* are a little inconvenient to work with algebraically; the sequence (2.196) is not quite enough, for instance, to assert that G is a semidirect product of G_s and G/G_s (this would require some means of embedding G/G_s back into G, which is not available in general). One can identify G (using the axiom of choice) with a product set $G/G_s \times G_s$ with a group law $(g, n) \cdot (h, m) = (gh, nm\phi(g, h))$, where $\phi : G/G_s \times G/G_s \to G_s$ is a map obeying various cocycle-type identities, but the algebraic structure of ϕ is not particularly easy to exploit. Nevertheless, this recursive tower of extensions seems to be well suited for understanding the *dynamical* structure of nilpotent groups and their quotients, as opposed to their *algebraic* structure (cf. our use of recursive towers of extensions in our previous lectures in dynamical systems and ergodic theory).

In our applications we will not be working with nilpotent groups G directly, but rather with their *homogeneous spaces* X, i.e., spaces with a transitive left action of G. (Later we will also add some topological structure to these objects, but let us work in a purely algebraic setting for now.) Such spaces can be identified with group quotients $X \equiv G/\Gamma$, where $\Gamma \leq G$ is the *stabiliser* $\Gamma = \{g \in G : gx = x\}$ of some point x in X. (By the transitivity of the action, all stabilisers are conjugate to each other.) It is important to note that in general, Γ is not normal, and so X is not a group; it has a left action of G but not right action of G. Note though that any central subgroup of G acts on either the left or the right.

Now let G be s-step nilpotent, and let us temporarily refer to $X = G/\Gamma$ as an *s-step nilspace*. Then G_s acts on the right in a manner that commutes with the left action of G. If we set $\Gamma_s := G_s \cap \Gamma \lhd G_s$, we see that the right action of Γ_s on G/Γ is trivial; thus we in fact have a right action of the abelian group $T_s := G_s/\Gamma_s$. (In our applications, T_s will be a torus.) This action can be easily verified to be free. If we let $\overline{X} := X/T_s$ be the quotient space, then we can view X as a principal T_s-bundle over $\overline{X}$. It is not hard to see that $\overline{X} \equiv \pi(G)/\pi(\Gamma)$, where $\pi : G \to G/G_s$ is the quotient map. Observe that $\pi(G)$ is nilpotent of step $s - 1$, and $\pi(\Gamma)$ is a subgroup. Thus we have expressed an arbitrary s-step nilspace as a principal bundle (by some abelian group) over an $s-1$-step nilspace, and so s-step nilspaces can be viewed as towers of abelian principal bundles, just as s-step nilpotent groups can be viewed as towers of abelian extensions.

2.16.2. Nilmanifolds. It is now time to put some topological structure (and in particular, Lie structure) on our nilpotent groups and nilspaces.

Definition 2.16.7 (Nilmanifolds). An *s-step nilmanifold* is a nilspace G/Γ, where G is a finite-dimensional Lie group which is nilpotent of step s, and Γ is a discrete subgroup which is *cocompact* or *uniform* in the sense that the quotient G/Γ is compact.

Remark 2.16.8. In the literature, it is sometimes assumed that the nilmanifold G/Γ is connected, and that the group G is connected, or at least that its group $\pi_0(G) := G/G^\circ$ of connected components ($G^\circ \lhd G$ being the identity component of G) is finitely generated (one can often easily reduce to this case in applications). It is also convenient to assume that G° is simply connected (again, one can usually reduce to this case in applications, by passing to the universal cover of G° if necessary), as this implies (by the *Baker-Campbell-Hausdorff formula*) that the nilpotent Lie group G° is *exponential*, i.e., the exponential map $\exp : \mathfrak{g} \to G^\circ$ is a homeomorphism.

Example 2.16.9 (Skew torus). If we define

$$(2.197) \qquad G := \begin{pmatrix} 1 & \mathbf{R} & \mathbf{R} \\ 0 & 1 & \mathbf{Z} \\ 0 & 0 & 1 \end{pmatrix}; \quad \Gamma := \begin{pmatrix} 1 & \mathbf{Z} & \mathbf{Z} \\ 0 & 1 & \mathbf{Z} \\ 0 & 0 & 1 \end{pmatrix}$$

(thus G consists of the upper-triangular unipotent matrices whose middle right entry is an integer, and Γ is the subgroup in which all entries are integers), then G/Γ is a 2-step nilmanifold. If we write

$$(2.198) \qquad [x,y] := \begin{pmatrix} 1 & x & y \\ 0 & 1 & 0 \\ 0 & 0 & 1 \end{pmatrix} \Gamma,$$

then we see that G/Γ is isomorphic to the square $\{[x,y] : 0 \le x, y \le 1\}$ with the identifications $[x,1] \equiv [x,0]$ and $[0,y] := [1,y+x \bmod 1]$. (Topologically, this is homeomorphic to the ordinary 2-torus $(\mathbf{R}/\mathbf{Z})^2$, but the skewness will manifest itself when we do dynamics.)

Example 2.16.10 (Heisenberg nilmanifold). If we set

$$(2.199) \qquad G := \begin{pmatrix} 1 & \mathbf{R} & \mathbf{R} \\ 0 & 1 & \mathbf{R} \\ 0 & 0 & 1 \end{pmatrix}, \quad \Gamma := \begin{pmatrix} 1 & \mathbf{Z} & \mathbf{Z} \\ 0 & 1 & \mathbf{Z} \\ 0 & 0 & 1 \end{pmatrix},$$

then G/Γ is a 2-step nilmanifold. It can be viewed as a three-dimensional cube with the faces identified in a somewhat skew fashion, similarly to the skew torus in Example 2.16.9.

Let $\mathfrak{g}$ be the Lie algebra of G. Every element g of G acts linearly on $\mathfrak{g}$ by conjugation. Since G is nilpotent, it is not hard to see (by considering the iterated commutators of g with an infinitesimal perturbation of the identity) that this linear action is unipotent, and in particular has determinant 1. Thus, any constant volume form on this Lie algebra will be preserved by conjugation, which by basic differential geometry allows us to create a volume form (and hence a measure) on G which is invariant under both left and right translation; this Haar measure is clearly unique up to scalar multiplication. (In other words, nilpotent Lie groups are unimodular.) Restricting this measure to a fundamental domain of G/Γ and then descending to the nilmanifold we obtain a left-invariant Haar measure, which (by compactness) we can normalise to be a Borel probability measure. (Because of the existence of a left-invariant probability measure μ on G/Γ, we refer to the discrete subgroup Γ of G as a lattice.) One can show that this left-invariant Borel probability measure is unique.

Definition 2.16.11 (Nilsystem). An *s-step nilsystem (or nilflow)* is a topological measure-preserving system (i.e., both a topological dynamical system and a measure-preserving system) with underlying space G/Γ an s-step nilmanifold (with the Borel σ-algebra and left-invariant probability measure), with a shift T of the form $T : x \mapsto gx$ for some $g \in G$.

Example 2.16.12. The Kronecker systems $x \mapsto x + \alpha$ on compact abelian Lie groups are 1-step nilsystems.

Example 2.16.13. The skew shift system $(x, y) \mapsto (x + \alpha, y + x)$ on the torus $(\mathbf{R}/\mathbf{Z})^2$ can be identified with a nilflow on the skew torus (Example 2.16.9), after identifying (x, y) with $[x, y]$ and using the group element

$$(2.200) \qquad\qquad g := \begin{pmatrix} 1 & \alpha & 0 \\ 0 & 1 & 1 \\ 0 & 0 & 1 \end{pmatrix}$$

to create the flow.

Example 2.16.14. Consider the Heisenberg nilmanifold (Example 2.16.10) with a flow generated by a group element

$$(2.201) \qquad\qquad g := \begin{pmatrix} 1 & \gamma & \beta \\ 0 & 1 & \alpha \\ 0 & 0 & 1 \end{pmatrix}$$

for some real numbers α, β, γ. If we identify

(2.202)
$$[x, y, z] := \begin{pmatrix} 1 & z & y \\ 0 & 1 & x \\ 0 & 0 & 1 \end{pmatrix} \Gamma$$

then one can verify that
(2.203)
$$T^n : [x, y, z] \mapsto \Big[\{x + n\alpha\},$$

$$y + n\beta + \frac{n(n+1)}{2}\alpha\gamma - \lfloor x + n\alpha \rfloor (z + n\gamma) \bmod 1,$$

$$z + n\gamma \bmod 1\Big],$$

where $\lfloor \ \rfloor$ and $\{ \ \}$ are the integer part and fractional part functions respectively. Thus we see that orbits in this nilsystem are vaguely quadratic in n, but for the presence of the not-quite-linear operators $\lfloor \ \rfloor$ and $\{ \ \}$. (These expressions are known as *bracket polynomials*, and are intimately related to the theory of nilsystems.)

Given that we have already seen that nilspaces of step s are principal abelian bundles of nilspaces of step $s - 1$, it should be unsurprising that nilsystems of step s are abelian extensions of nilsystems of step $s - 1$. But in order to ensure that topological structure is preserved correctly, we do need to verify one point:

Lemma 2.16.15. *Let G/Γ be an s-step nilmanifold, with G connected and simply connected. Then $\Gamma_s := G_s \cap \Gamma$ is a discrete cocompact subgroup of G_s. In particular, $T_s := G_s/\Gamma_s$ is a compact connected abelian Lie group (in other words, it is a torus).*

Proof. Recall that G is exponential and thus identifiable with its Lie algebra $\mathfrak{g}$. The commutators G_i can be similarly identified with the Lie algebra commutators $\mathfrak{g}_i$; in particular, the G_i are all connected, simply connected Lie groups.

The key point is to verify the cocompact nature of Γ_s in G_s; all other claims are straightforward. We first work in the abelianisation G/G_2, which is identifiable with its Lie algebra and thus isomorphic to a vector space. The image of Γ under the quotient map $G \to G/G_2$ is a cocompact subgroup of this vector space; in particular, it contains a basis of this space. This implies that Γ contains an "abelianised" basis $e_1, \ldots, e_d$ of G in the sense that every element of G can be expressed in the form $e_1^{t_1} \ldots e_d^{t_d}$ modulo an element of the normal subgroup G_2 for some real numbers $t_1, \ldots, t_d$, where we take advantage of the exponential nature of G to define real exponentiation $g^t := \exp(t \log(g))$. Taking commutators s times (which eliminates all the "modulo G_2" errors), we then see that G_s is generated by expressions of the form $[e_{i_1}, [e_{i_2}, [\ldots, e_{i_s}]\ldots]^t$ for $i_1, \ldots, i_s \in \{1, \ldots, d\}$ and real t. Observe that these expressions lie in Γ_s if t is an integer. As G_s is abelian, we conclude that each element in G_s can be expressed as an element of Γ_s, times a bounded number of elements of the form $[e_{i_1}, [e_{i_2}, [\ldots, e_{i_s}]\ldots]^t$ with $0 \leq t < 1$. From this we conclude that the quotient map $G_s \mapsto G_s/\Gamma_s$ is already surjective on some bounded set, which we can take to be compact, and so G_s/Γ_s is compact as required. $\square$

As a consequence of this lemma, we see that if $X = G/\Gamma$ is an s-step nilmanifold with G connected and simply connected, then X/T_s is an $s-1$-step nilmanifold (with G still connected and simply connected), and that X is a principal T_s-bundle over X/T_s in the topological sense as well as in the purely algebraic sense. One consequence of this is that every s-step nilsystem (with G connected and simply connected) can be viewed as a *toral extension* (i.e., a group extension by a torus) of an $s-1$-step nilsystem (again with G connected and simply connected). Thus for instance the skew shift system (Example 2.16.13) is a circle extension of a circle shift, while the Heisenberg nilsystem (Example 2.16.14) is a circle extension of an abelian 2-torus shift.

Remark 2.16.16. One should caution though that the converse of the above statement is not necessarily true; an extension $X \times_\phi T$ of an $s-1$-step nilsystem X by a torus T using a cocycle $\phi : X \to T$ need not be isomorphic to an s-step nilsystem (the cocycle ϕ has to obey an additional equation (or more precisely, a system of equations when $s > 2$), known as the *Conze-Lesigne equation*, before this is the case. See for instance [**Zi2007**] for further discussion.

Exercise 2.16.6. Show that Lemma 2.16.15 continues to hold if we relax the condition that G is connected and simply connected, to instead require that G/Γ is connected, that G/G° is finitely generated, and that G° is simply connected.

Exercise 2.16.7. Show that Lemma 2.16.15 continues to hold if G_s and Γ_s are replaced by G_i and $\Gamma_i = G_i \cap \Gamma$ for any $0 \leq i \leq s$. In particular, setting $i = 2$, we obtain a projection map $\pi : X \to X_2$ from X to the Kronecker nilmanifold $X_2 = (G/G_2)/(\Gamma G_2/G_2)$.

Remark 2.16.17. One can take the structural theory of nilmanifolds much further, in particular developing the theory of *Mal'cev bases* (of which the elements $e_1, \ldots, e_d$ used to prove Lemma 2.16.15 were a very crude prototype). See the foundational paper [**Ma1951**] of Mal'cev for details, as well as the later paper [**Le2005**], which addresses the case in which G is not necessarily connected.

2.16.3. A criterion for ergodicity. We now give a useful criterion to determine when a given nilsystem is ergodic.

Theorem 2.16.18. *Let $(X, T) = (G/\Gamma, x \mapsto gx)$ be an s-step nilsystem with G connected and simply connected, and let (X_2, T_2) be the underlying Kronecker factor, as defined in Exercise 2.16.7. Then X is ergodic if and only if X_2 is ergodic.*

This result was first proven in [**Gr1961**], using spectral theory methods. We will apply an argument of Parry [**Pa1969**] (adapted in [**Le2005**]), relying on "vertical" Fourier analysis and topological arguments, which we have already used for the skew shift in Proposition 2.9.11. An alternate proof also appears in Section 1.3.

Proof. If X is ergodic, then the factor X_2 is certainly ergodic. To prove the converse implication, we induct on s. The case $s \leq 1$ is trivial, so suppose $s > 1$ and the claim has already been proven for $s-1$. Then if X_2 is ergodic, we already know from induction hypothesis that X/T_s is ergodic. Suppose for contradiction that X is not ergodic; then we can find a non-constant shift-invariant function on X. Using Fourier analysis (or representation theory) of the vertical torus T_s as in Proposition 2.9.11, we may thus find a non-constant shift-invariant function f which has a *single vertical frequency* χ in the sense that $f(g_s x) = \chi(g_s) f(x)$ for all $x \in X$, $g_s \in G_s$, and some character $\chi : G_s \to S^1$. If the character χ is trivial, then f descends to a non-constant shift-invariant function on X/T_s, contradicting the ergodicity there, so we may assume that χ is non-trivial. Also, $|f|$ descends to a shift-invariant function on X/T_s and is thus constant by ergodicity; by normalising we may assume $|f| = 1$.

Now let $g_{s-1} \in G_{s-1}$, and consider the function
$$F_{g_{s-1}}(x) := f(g_{s-1}x)\overline{f(x)}.$$
As G_s is central, we see that $F_{g_{s-1}}$ is G_s-invariant and thus descends to X/T_s. Furthermore, as f is shift-invariant (so $f(gx) = f(x)$), and $[g_{s-1}, g] \in G_s$, some computation reveals that $F_{g_{s-1}}$ is an eigenfunction:

(2.204) $$F_{g_{s-1}}(gx) = \chi([g_{s-1}, g])F_{g_{s-1}}(x).$$

In particular, if $\chi([g_{s-1}, g]) \neq 1$, then $F_{g_{s-1}}$ must have mean zero. On the other hand, by continuity (and the fact that $|f| = 1$) we know that $F_{g_{s-1}}$ has non-zero mean for g_{s-1} close enough to the identity. We conclude that $\chi([g_{s-1}, g]) = 1$ for all g_{s-1} close to the identity; as the map $g_{s-1} \mapsto \chi([g_{s-1}, g])$ is a homomorphism, we conclude in fact that $\chi([g_{s-1}, g]) = 1$ for all g_{s-1}. In particular, from (2.204) and ergodicity we see that $F_{g_{s-1}}$ is constant, and so $f(g_{s-1}x) = c(g_{s-1})f(x)$ for some $c(g_{s-1}) \in S^1$.

Now let $h \in G$ be arbitrary. Observe that

(2.205)
$$
\begin{aligned}
\int_G f(hg_{s-1}x)\overline{f(x)}\,d\mu &= \int_G f(hy)\overline{f(g_{s-1}^{-1}y)}\,d\mu \\
&= c(g_{s-1})\int_G f(hy)\overline{f(y)}\,d\mu \\
&= \int_G f(g_{s-1}hy)\overline{f(y)}\,d\mu \\
&= \chi([g_{s-1},h])\int_G f(hg_{s-1}y)\overline{f(y)}\,d\mu.
\end{aligned}
$$

For h and g_{s-1} close enough to the identity, the integral is non-zero, and we conclude that $\chi([g_{s-1},h]) = 1$ in this case. The map $(g_{s-1},h) \mapsto \chi([g_{s-1},h])$ is a homomorphism in each variable and so is constant. Since $G_s = [G_{s-1},G]$, we conclude that χ is trivial, a contradiction. $\qquad\square$

Remark 2.16.19. The hypothesis that G is connected and simply connected can be dropped; see [**Le2005**] for details.

One pleasant fact about nilsystems, as compared with arbitrary dynamical systems, is that ergodicity can automatically be upgraded to unique ergodicity:

Theorem 2.16.20. *Let (X,T) be an ergodic nilsystem. Then (X,T) is also uniquely ergodic. Equivalently, for every $x \in X$, the orbit $(T^n x)_{n \in \mathbf{Z}}$ is equidistributed.*

Exercise 2.16.8. By inducting on step and adapting the proof of Proposition 2.9.14, prove Theorem 2.16.20.

2.16.4. A Ratner-type theorem. A *subnilsystem* of a nilsystem $(X,T) = (G/\Gamma, T)$ is a compact subsystem (Y,S) which is of the form $Y = Hx$ for some $x \in X$ and some closed subgroup $H \leq G$. One easily verifies that a subnilsystem is indeed a nilsystem.

From the above theorems we quickly obtain

Corollary 2.16.21 (Dichotomy between structure and randomness). *Let (X,T) be a nilsystem with the group G connected and simply connected, and* *let $x \in X$. Then exactly one of the following statements is true:*

(1) *(Structure) The orbit $(T^n x)_{n \in \mathbf{Z}}$ is contained in a proper subnilsystem (Y, S) with the group H connected and simply connected, and with the dimension strictly smaller than that of G.*

(2) *(Randomness) The orbit $(T^n x)_{n \in \mathbf{Z}}$ is equidistributed.*

Proof. It is clear that (1) and (2) cannot both be true. Now suppose that (1) is false. By Theorem 2.16.20, this means that (X, T) is not ergodic; by Theorem 2.16.18, this implies that the Kronecker system (X_2, T_2) is not ergodic. Expanding functions on $X_2 \equiv G/G_2$ into characters and using Fourier analysis, we conclude that there is a non-trivial character $\chi : G/G_2 \to S^1$ which is T_2-invariant. If we let $\pi : G \to G/G_2$ be the canonical projection, then $\chi : G \to S^1$ is a continuous homomorphism, and the kernel H is a closed connected subgroup of G of strictly lower dimension. Furthermore, Hx is equal to a level set of χ and is thus compact. Since χ is T_2 invariant, we see that $T^n x \in Hx$ for all n, and the claim follows. $\square$

Iterating this corollary, we obtain

Corollary 2.16.22 (Ratner-type theorem for nilmanifolds). *Let (X, T) be a nilsystem with group G connected and simply connected, and let $x \in X$. Then the orbit $(T^n x)_{n \in \mathbf{Z}}$ is equidistributed in some subnilmanifold (Y, S) of (X, T). (In particular, this orbit is dense in Y.) Furthermore, $Y = Hx$ for some closed connected subgroup H of G.*

Remark 2.16.23. Analogous claims also hold when G is not assumed to be connected or simply connected, and when the orbit $(T^n x)_{n \in \mathbf{Z}}$ is replaced with a polynomial orbit $(T^{p(n)} x)_{n \in \mathbf{Z}}$; see [**Le2005**], [**Le2005b**]. In a different direction, such discrete Ratner-type theorems have been extended to other unipotent actions on finite volume homogeneous spaces by Shah [**Sh1996**]. Quantitative versions of this theorem have also been obtained by Ben Green and myself [**GrTa2009c**].

Notes. This lecture first appeared at

`terrytao.wordpress.com/2008/03/09`.

Thanks to Jordi-Lluis Figueras Romero for corrections.

2.17. A Ratner-type theorem for $SL_2(R)$ orbits

In this final section of this chapter, we establish a Ratner-type theorem for actions of the special linear group $SL_2(\mathbf{R})$ on homogeneous spaces. More precisely, we prove the following.

Theorem 2.17.1. *Let G be a Lie group, let $\Gamma < G$ be a discrete subgroup, and let $H \leq G$ be a subgroup isomorphic to $SL_2(\mathbf{R})$. Let μ be an H-invariant probability measure on G/Γ which is ergodic with respect to H (i.e., all H-invariant sets either have full measure or zero measure). Then μ is* homogeneous *in the sense that there exists a closed connected subgroup $H \leq L \leq G$ and a closed orbit $Lx \subset G/\Gamma$ such that μ is L-invariant and supported on Lx.*

This result is a special case of a more general theorem of Ratner, which addresses the case when H is generated by elements which act unipotently on the Lie algebra $\mathfrak{g}$ by conjugation, and when G/Γ has finite volume. To prove this theorem we shall follow an argument of Einsiedler [**Ei2006**], which uses many of the same ingredients used in Ratner's arguments but in a simplified setting (in particular, taking advantage of the fact that H is semisimple with no non-trivial compact factors). These arguments have since been extended and made quantitative in [**EiMaVe2007**].

2.17.1. Representation theory of $SL_2(\mathbf{R})$. Theorem 2.17.1 concerns the action of $H \equiv SL_2(\mathbf{R})$ on a homogeneous space G/Γ. Before we are able to tackle this result, we must first understand the linear actions of $H \equiv SL_2(\mathbf{R})$ on real or complex vector spaces—in other words, we need to understand the representation theory of the Lie group $SL_2(\mathbf{R})$ (and its associated Lie algebra $\mathfrak{sl}_2(\mathbf{R})$).

Of course, this theory is very well understood, and by using the machinery of weight spaces, raising and lowering operators, etc., one can completely classify all the finite-dimensional representations of $SL_2(\mathbf{R})$; in fact, all such representations are isomorphic to direct sums of *symmetric powers* of the standard representation of $SL_2(\mathbf{R})$ on $\mathbf{R}^2$. This classification quickly yields all the necessary facts we will need here. However, we will use only a minimal amount of this machinery here, to obtain as direct and elementary a proof of the results we need as possible.

The first fact we will need is that finite-dimensional representations of $SL_2(\mathbf{R})$ are completely reducible:

Lemma 2.17.2 (Complete reducibility). *Let $SL_2(\mathbf{R})$ act linearly (and smoothly) on a finite-dimensional real vector space V, and let W be an $SL_2(\mathbf{R})$-invariant subspace of V. Then there exists a complementary subspace W' to W which is also $SL_2(\mathbf{R})$-invariant (thus V is isomorphic to the direct sum of W and W').*

Proof. We will use *Weyl's unitary trick* to create the complement W', but in order to invoke this trick, we first need to pass from the non-compact group $SL_2(\mathbf{R})$ to a compact counterpart. This is done in several stages.

First, we linearise the action of the Lie group $SL_2(\mathbf{R})$ by differentiating to create a corresponding linear action of the Lie algebra $\mathfrak{sl}_2(\mathbf{R})$ in the usual manner.

Next, we complexify the action. Let $V^{\mathbf{C}} := V \otimes \mathbf{C}$ and $W^{\mathbf{C}} := W \otimes \mathbf{C}$ be the complexifications of V and W respectively. Then the complexified Lie algebra $\mathfrak{sl}_2(\mathbf{C})$ acts on both $V^{\mathbf{C}}$ and $W^{\mathbf{C}}$, and in particular the *special unitary* Lie algebra $\mathfrak{su}_2(\mathbf{C})$ does also.

Since the *special unitary group*

$$(2.206) \qquad SU_2(\mathbf{C}) = \left\{ \begin{pmatrix} \alpha & \beta \\ -\overline{\beta} & \overline{\alpha} \end{pmatrix} : \alpha, \beta \in \mathbf{C}, \ |\alpha|^2 + |\beta|^2 = 1 \right\}$$

is topologically equivalent to the 3-sphere S^3 and is thus simply connected, a standard homotopy argument allows one to exponentiate the $\mathfrak{su}_2(\mathbf{C})$ action to create an $SU_2(\mathbf{C})$ action, thus creating the desired compact action.[52]

Now we can apply the unitary trick. Take any Hermitian form $\langle , \rangle$ on $V^{\mathbf{C}}$. This form need not be preserved by the $SU_2(\mathbf{C})$ action, but if one defines the averaged form

$$(2.207) \qquad \langle u, v \rangle_{SU_2} := \int_{SU_2(\mathbf{C})} \langle gu, gv \rangle \, dg,$$

where dg is Haar measure on the compact Lie group $SU_2(\mathbf{C})$, then we see that $\langle , \rangle_{SU_2}$ is a Hermitian form which is $SU_2(\mathbf{C})$-invariant; thus this form endows $V^{\mathbf{C}}$ with a Hilbert space structure with respect to which the $SU_2(\mathbf{C})$-action is unitary. If we then define $(W')^{\mathbf{C}}$ to be the orthogonal complement of $W^{\mathbf{C}}$ in this Hilbert space, then this vector space is invariant under the $SU_2(\mathbf{C})$ action, and thus (by differentiation) by the $\mathfrak{su}_2(\mathbf{C})$ action. But observe that $\mathfrak{su}_2(\mathbf{C})$ and $\mathfrak{sl}_2(\mathbf{R})$ have the same complex span (namely, $\mathfrak{sl}_2(\mathbf{C})$); thus the complex vector space $(W')^{\mathbf{C}}$ is also $\mathfrak{sl}_2(\mathbf{R})$-invariant.

The last thing to do is to undo the complexification. If we let W' be the space of real parts of vectors in $(W')^{\mathbf{C}}$ which are real modulo $W^{\mathbf{C}}$, then one easily verifies that W' is $\mathfrak{sl}_2(\mathbf{R})$-invariant (hence $SL_2(\mathbf{R})$-invariant, by exponentiation) and is a complementary subspace to W, as required. $\qquad\square$

Remark 2.17.3. We can of course iterate the above lemma and conclude that every finite-dimensional representation of $SL_2(\mathbf{R})$ is the direct sum of irreducible representations, which explains the term "complete reducibility". Complete reducibility of finite-dimensional representations of a Lie algebra (over a field of characteristic zero) is equivalent to that Lie algebra being *semisimple*. The situation is slightly more complicated for Lie groups, though, if such groups are not simply connected.

---

[52]This trick is not restricted to $\mathfrak{sl}_2(\mathbf{R})$, but can be generalised to other semisimple Lie algebras using the *Cartan decomposition*.

An important role in our analysis will be played by the one-parameter unipotent subgroup $U := \{u^t : t \in \mathbf{R}\}$ of $SL_2(\mathbf{R})$, where

$$(2.208) \qquad\qquad u^t := \begin{pmatrix} 1 & t \\ 0 & 1 \end{pmatrix}.$$

Clearly, the elements of U are unipotent when acting on $\mathbf{R}^2$. It turns out that they are unipotent when acting on all other finite-dimensional representations also:

Lemma 2.17.4. *Suppose that $SL_2(\mathbf{R})$ acts on a finite-dimensional real or complex vector space V. Then the action of any element of U on V is unipotent.*

Proof. By complexifying V if necessary we may assume that V is complex. The action of the Lie group $SL_2(\mathbf{R})$ induces a Lie algebra homomorphism $\rho : \mathfrak{sl}_2(\mathbf{R}) \to \text{End}(V)$. To show that the action of U is unipotent, it suffices to verify that $\rho(\log u)$ is nilpotent, where

$$(2.209) \qquad\qquad \log u = \begin{pmatrix} 0 & 1 \\ 0 & 0 \end{pmatrix}$$

is the infinitesimal generator of U. To this end, we will exploit the fact that $\log u$ induces a *raising operator*. We introduce the diagonal subgroup $D := \{d^t : t \in \mathbf{R}\}$ of $SL_2(\mathbf{R})$, where

$$(2.210) \qquad\qquad d^t := \begin{pmatrix} e^t & 0 \\ 0 & e^{-t} \end{pmatrix}.$$

This group has infinitesimal generator

$$(2.211) \qquad\qquad \log d = \begin{pmatrix} 1 & 0 \\ 0 & -1 \end{pmatrix}.$$

Observe that $[\log d, \log u] = 2 \log u$, and thus (since ρ is a Lie algebra homomorphism)

$$(2.212) \qquad\qquad [\rho(\log d), \rho(\log u)] = 2\rho(\log u).$$

We can rewrite this as

$$(2.213) \qquad (\rho(\log d) - \lambda - 2)\rho(\log u) = \rho(\log u)(\rho(\log d) - \lambda)$$

for any $\lambda \in \mathbf{C}$, which on iteration implies that

$$(2.214) \qquad (\rho(\log d) - \lambda - 2r)^m \rho(\log u)^r = \rho(\log u)^r (\rho(\log d) - \lambda)^m$$

for any non-negative integers m, r. But this implies that $\rho(\log u)^r$ raises generalised eigenvectors of $\rho(\log d)$ with eigenvalue λ to generalised eigenvectors of $\rho(\log d)$ with eigenvalue $\lambda + 2m$. But as V is finite-dimensional, there are only finitely many eigenvalues of $\rho(\log d)$, and so $\rho(\log u)$ is nilpotent on each of the generalised eigenvectors of $\rho(\log d)$. By the Jordan normal form (see

Section 1.13 of *Structure and Randomness*), these generalised eigenvectors span V, and we are done. □

Exercise 2.17.1. By carrying the above analysis further (and also working with the adjoint of U to create lowering operators) show (for complex V) that $\rho(\log d)$ is diagonalisable, and the eigenvalues are all integers. For an additional challenge: deduce from this that the representation is isomorphic to a direct sum of the representations of $SL_2(\mathbf{R})$ on the symmetric tensor powers $\operatorname{Sym}^k(\mathbf{R}^2)$ of $\mathbf{R}^2$ (or, if you wish, on the space of homogeneous polynomials of degree k on two variables).

The group U is merely a subgroup of the group $SL_2(\mathbf{R})$, so it is not a priori evident that any vector (in a space that $SL_2(\mathbf{R})$ acts on) which is U-invariant, is also $SL_2(\mathbf{R})$-invariant. But, thanks to the highly non-commutative nature of $SL_2(\mathbf{R})$, this turns out to be the case, even in infinite dimensions, once one restricts attention to *continuous unitary* actions:

Lemma 2.17.5 (Mautner phenomenon). *Let $\rho : SL_2(\mathbf{R}) \to U(V)$ be a continuous unitary action on a Hilbert space V (possibly infinite-dimensional). Then any vector $v \in V$ which is fixed by U, is also fixed by $SL_2(\mathbf{R})$.*

Proof. We use an argument of Margulis. We may of course take v to be non-zero. Let $\varepsilon > 0$ be a small number. Then even though the matrix $w^\varepsilon := \left(\begin{smallmatrix} 1 & 0 \\ \varepsilon & 1 \end{smallmatrix}\right)$ is very close to the identity, the double orbit $Uw^\varepsilon U$ can stray very far away from U. Indeed, from the algebraic identity

$$(2.215) \qquad \begin{pmatrix} e^t & 0 \\ \varepsilon & e^{-t} \end{pmatrix} = u^{(e^t-1)/\varepsilon} w^\varepsilon u^{(e^{-t}-1)/\varepsilon},$$

which is valid for any $t \in \mathbf{R}$, we see that this double orbit in fact comes very close to the diagonal group D. Applying (2.215) to the U-invariant vector v and taking inner products with v, we conclude from unitarity that

$$(2.216) \qquad \left\langle \rho\left(\begin{pmatrix} e^t & 0 \\ \varepsilon & e^{-t} \end{pmatrix}\right) v, v \right\rangle = \langle \rho(w^\varepsilon)v, v\rangle.$$

Taking limits as $\varepsilon \to 0$ (taking advantage of the continuity of ρ) we conclude that $\langle \rho(d^t)v, v\rangle = \langle v, v\rangle$. Since $\rho(d^t)v$ has the same length as v, we conclude from the converse Cauchy-Schwarz inequality that $\rho(d^t)v = v$, i.e., that v is D-invariant. Since U and D generate $SL_2(\mathbf{R})$, the claim follows. □

Remark 2.17.6. The key fact about U being used here is that its Lie algebra is not trapped inside any proper ideal of $\mathfrak{sl}_2(\mathbf{R})$, which, in turn, follows from the fact that this Lie algebra is simple. One can do the same thing for semisimple Lie algebras provided that the unipotent group U is non-degenerate in the sense that it has non-trivial projection onto each simple factor.

This phenomenon has an immediate dynamical corollary:

Corollary 2.17.7 (Moore ergodic theorem). *Suppose that $SL_2(\mathbf{R})$ acts in a measure-preserving fashion on a probability space $(X, \mathcal{X}, \mu)$. If this action is ergodic with respect to $SL_2(\mathbf{R})$, then it is also ergodic with respect to U.*

Proof. Apply Lemma 2.17.5 to $L^2(X, \mathcal{X}, \mu)$. $\square$

2.17.2. Proof of Theorem 2.17.1. Having completed our representation-theoretic preliminaries, we are now ready to begin the proof of Theorem 2.17.1. The key is to prove the following dichotomy:

Proposition 2.17.8 (Lack of concentration implies additional symmetry). *Let G, H, μ, Γ be as in Theorem 2.17.1. Suppose there exists a closed connected subgroup $H \leq L \leq G$ such that μ is L-invariant. Then exactly one of the following statements hold:*

(1) *(Concentration) μ is supported on a closed orbit Lx of L.*

(2) *(Additional symmetry) There exists a closed connected subgroup $L < L' \leq G$ such that μ is L'-invariant.*

Iterating this proposition (noting that the dimension of L' is strictly greater than that of L) we will obtain Theorem 2.17.1. So it suffices to establish the proposition.

We first observe that the ergodicity allows us to obtain the concentration conclusion (2.206) as soon as μ assigns any non-zero mass to an orbit of L:

Lemma 2.17.9. *Let the notation and assumptions be as in Proposition 2.17.8. Suppose that $\mu(Lx_0) > 0$ for some x_0. Then Lx_0 is closed and μ is supported on Lx_0.*

Proof. Since Lx_0 is H-invariant and μ is H-ergodic, the set Lx_0 must either have full measure or zero measure. It cannot have zero measure by hypothesis, thus $\mu(Lx_0) = 1$. Therefore, if we show that Lx_0 is closed, we automatically have that μ is supported on Lx_0. As G/Γ is a homogeneous space, we may assume without loss of generality (conjugating L if necessary) that x_0 is at the origin; then $Lx_0 \equiv L/(\Gamma \cap L)$. The measure μ on this set can then be pulled back to a measure m on L by the formula

$$(2.217) \qquad \int_L f(g)\, dm(g) = \int_{L/(\Gamma \cap L)} \sum_{g \in x(\Gamma \cap L)} f(g)\, d\mu(x).$$

By construction, m is left-L-invariant (i.e., a left Haar measure) and is right-$(\Gamma \cap L)$-invariant. From uniqueness of left Haar measure up to constants, we see that for any g in L there is a constant $c(g) > 0$ such that $m(Eg) = c(g)m(E)$ for all measurable E. It is not hard to see that $c : L \to \mathbf{R}^+$ is

a character, i.e., it is continuous and multiplicative, thus $c(gh) = c(g)c(h)$ for all g, h in L. Also, it is the identity on $(\Gamma \cap L)$ and thus descends to a continuous function on $L/(\Gamma \cap L)$. Since μ is L-invariant, we have
(2.218)
$$\int_{L/(\Gamma \cap L)} c(g)\, d\mu(g) = \int_{L/(\Gamma \cap L)} c(hg)\, d\mu(g) = \int_{L/(\Gamma \cap L)} c(h)c(g)\, d\mu(g)$$
for all h in L, and thus c is identically 1 (i.e., L is unimodular). Thus m is right-invariant, which implies that μ obeys the right-invariance property $\mu(Kx_0) = \mu(Kgx_0)$ for any g in L and any sufficiently small compact set $K \subset L$ (small enough to fit inside a single fundamental domain of $L/(\Gamma \cap L)$).

Recall that $\mu(Lx_0) = 1$. By partitioning L into countably many small sets as above, we can thus find a small compact set $K \subset L$ such that $\mu(Kx_0) > 0$. Now consider a maximal set of disjoint translates Kg_1x_0, $Kg_2x_0, \ldots, Kg_kx_0$ of Kx_0; since all of these sets have the same positive measure, such a maximal set exists and is finite. Then for any g in L, Kgx_0 must intersect one of the sets Kg_ix_0, which implies that $Lx_0 = \bigcup_{i=1}^{k} K^{-1}Kg_ix_0$. But the right-hand side is compact, and so Lx_0 is closed as desired. $\square$

We return to the proof of Proposition 2.17.8. In view of Lemma 2.17.9, we may assume that μ is totally non-concentrated on L-orbits in the sense that

(2.219) $\mu(Lx) = 0$ for all $x \in G/\Gamma$.

In particular, for μ-almost every x and y, y does not lie in the orbit Lx of x and vice versa; informally, the group elements in G that are used to move from x to y should be somehow "transverse" to L. On the other hand, we are given that μ is ergodic with respect to H, and thus (by Corollary 2.17.7) ergodic with respect to U. This implies (cf. Proposition 2.9.13) that μ-almost every point x in G/Γ is *generic* (with respect to U) in the sense that

(2.220) $\int_{G/\Gamma} f\, d\mu = \lim_{T \to +\infty} \frac{1}{T} \int_0^T f(u^t x)\, dt$

for all continuous compactly supported $f : G/\Gamma \to \mathbf{R}$.

Exercise 2.17.2. Prove this claim. *Hint*: Obtain continuous analogues of the theory from Sections 2.8, 2.9.

Equation (2.220) (and the *Riesz representation theorem*) lets us describe the measure μ in terms of the U-orbit of a generic point. On the other hand, from (2.219) and the ensuing discussion we see that any two generic points are likely to be separated from each other by some group element "transverse" to L. It is the interaction between these two facts which is going to generate the additional symmetry needed for Proposition 2.17.8. We illustrate this with a model case, in which the group element centralises U:

Proposition 2.17.10 (Central case). *Let the notation and assumptions be as in Proposition 2.17.8. Suppose that x, y are generic points such that $y = gx$ for some $g \in G$ that centralises U (i.e., it commutes with every element of u). Then μ is invariant under the action of g.*

Proof. Let $f : G/\Gamma \to \mathbf{R}$ be continuous and compactly supported. Applying (2.220) with x replaced by $y = gx$ we obtain

$$(2.221) \qquad \int_{G/\Gamma} f \, d\mu = \lim_{T \to +\infty} \frac{1}{T} \int_0^T f(u^t gx) \, dt.$$

Commuting g with u^t and using (2.220) again, we conclude that

$$(2.222) \qquad \int_{G/\Gamma} f \, d\mu = \int_{G/\Gamma} f(gy) \, d\mu(y),$$

and the claim follows from the Riesz representation theorem. $\square$

Of course, we do not just want invariance under one group element g; we want a whole group L' of symmetries for which one has invariance. But it is not hard to leverage the former to the latter, provided one has enough group elements:

Lemma 2.17.11. *Let the notation and assumptions be as in Proposition 2.17.8. Suppose one has a sequence g_n of group elements tending to the identity, such that the action of each of the g_n preserves μ, and such that none of the g_n lies in L. Then there exists a closed connected subgroup $L < L' \leq G$ such that μ is L-invariant.*

Proof. Let S be the stabiliser of μ, i.e., the set of all group elements g whose action preserves μ. This is clearly a closed subgroup of G which contains L. If we let L' be the identity connected component of S, then L' is a closed connected subgroup containing L which will contain g_n for all sufficiently large n, and in particular is not equal to L. The claim follows. $\square$

From Proposition 2.17.10 and Lemma 2.17.11 we see that we are done if we can find pairs $x_n, y_n = g_n x_n$ of nearby generic points with g_n going to the identity such that $g_n \notin L$ and that g_n centralises U. Now we need to consider the non-central case; thus suppose for instance that we have two generic points x, $y = gx$ in which g is close to the identity but does not centralise U. The key observation here is that we can use the U-invariance of the situation to pull x and y slowly apart from each other. More precisely, since x and y are generic, we observe that $u^t x$ and $u^t y$ are also generic for any t, and that these two points differ by the conjugated group element

$g^t := u^t g u^{-t}$. Taking logarithms (which are well defined as long as g^t stays close to the identity), we can write

$$(2.223) \qquad \log(g^t) = u^t \log(g) u^{-t} = \exp(t \operatorname{ad}(\log u)) \log(g),$$

where ad is the *adjoint representation*. From Lemma 2.17.4, we know that $\operatorname{ad}(\log u) : \mathfrak{g} \to \mathfrak{g}$ is nilpotent, and so (by the Taylor expansion of the exponential) $\log(g^t)$ depends polynomially on t. In particular, if g does not centralise U, then $\log(g^t)$ is non-constant and thus must diverge to infinity as $t \to +\infty$. In particular, given a small ball B around the origin in $\mathfrak{g}$ (with respect to an arbitrary norm), we see that whenever $\log g$ lies inside B around the origin and is not central, there must be a first time $t = t_g$ such that $\log g^{t_g}$ reaches the boundary ∂B of this ball. We write $g^* := g^{t_g} \in \partial B$ for the location of g when it escapes. We now have the following variant of Proposition 2.17.10:

Proposition 2.17.12 (Non-central case). *Let the notation and assumptions be as in Proposition 2.17.8. Suppose that $x_n, y_n \in G$ are generic points such that $y_n = g_n x_n$ for some $g_n \in G$, which do not centralise u but are such that g_n converge to the identity (in particular, $g_n \in B$ for all sufficiently large n). Suppose furthermore that x_n, y_n are uniformly generic in the sense that for any continuous compactly supported $f : G/\Gamma \to \mathbf{R}$, the convergence of (2.220) (with x replaced by x_n or y_n) is uniform in n. Then μ is invariant under the action of any limit point $g^* \in \partial B$ of the g_n^*.*

Proof. By passing to a subsequence if necessary we may assume that g_n^* converges to g^*. For each sufficiently large n, we write $T_n := t_{g_n}$, thus $g_n^t \in B$ for all $0 \le t \le T_n$, and $g_n^{T_n} = g_n^*$. We rescale this by defining the functions $h_n : [0, 1] \to B$ by $h_n(s) := g_n^{sT_n}$. From the unipotent nature of U, these functions are polynomial (with bounded degree), and also bounded (as they live in B), and are thus equicontinuous (since all norms are equivalent on finite-dimensional spaces). Thus, by the Arzelà-Ascoli theorem, we can assume (after passing to another subsequence) that h_n is uniformly convergent to some limit f, which is another polynomial. Since we already have $h_n(2.206) = g_n^*$ converging to g_*, this implies that for any $\varepsilon > 0$ there exists $\delta > 0$ such that $h_n(s) = g_* + O(\varepsilon)$ for all $1 - \delta \le s \le 1$ and all sufficiently large n. In other words, we have

$$(2.224) \qquad u^t g_n u^{-t} = g_* + O(\varepsilon)$$

for sufficiently large n, whenever $(1 - \delta)T_n \le t \le T_n$.

This is good enough to apply a variant of the Proposition 2.17.10 argument. Namely, if $f : G/\Gamma \to \mathbf{R}$ is continuous and compactly supported, then by the uniform genericity we have for T sufficiently large that

$$(2.225) \qquad \int_{G/\Gamma} f \, d\mu = \frac{1}{\delta T} \int_{(1-\delta)T}^{T} f(u^t y_n) \, dt + O(\varepsilon)$$

for all n. Applying (2.224) we can write $u^t y_n = g_* u^t x_n + O(\varepsilon)$ on the support of f, and so by the uniform continuity of f,

$$(2.226) \qquad \int_{G/\Gamma} f \, d\mu = \frac{1}{\delta T} \int_{(1-\delta)T}^{T} f(g_* u^t x_n) \, dt + o(1),$$

where $o(1)$ goes to zero as $\varepsilon \to 0$, uniformly in n. Using (2.220) again and then letting $\varepsilon \to 0$, we obtain the g_*-invariance of μ as desired. $\qquad \square$

Now we have all the ingredients to prove Proposition 2.17.8, and thus Theorem 2.17.1.

Proof of Proposition 2.17.8. We know that μ-almost every point is generic. Applying *Egorov's theorem*, we can find sets $E \subset G/\Gamma$ of measure arbitrarily close to 1 (e.g., $\mu(E) \geq 0.9$) on which the points are uniformly generic.

Now let V be a small neighbourhood of the origin in L. Observe from the Fubini-Tonelli theorem that

$$(2.227) \qquad \int_X \frac{1}{m(V)} \int_V 1_E(x) 1_E(gx) \, dm(g) \, d\mu(x) \geq 2\mu(E) - 1 \geq 0.8,$$

where m is the Haar measure on the unimodular group L, from which one can find a set $E' \subset E$ of positive measure such that $m(\{g \in V : gx \in E\}) = 0.7m(V)$ for all $x \in E'$; one can view E' as "points of density" of E in some approximate sense (and with regard to the L action).

Since E' has positive measure, it is not hard to find, using (2.219), sequences $x_n, y_n \in E'$ with $y_n \notin Lx_n$ for any n and with $\text{dist}(x_n, y_n) \to 0$ (using some reasonable metric on G/Γ).

Exercise 2.17.3. Verify this. *Hint*: G/Γ can be covered by countably many balls of a fixed radius.

Next, recall that $H \equiv SL_2(\mathbf{R})$ acts by conjugation on the Lie algebra $\mathfrak{g}$ of G, and also leaves the Lie algebra $\mathfrak{l} \subset \mathfrak{g}$ of L invariant. By Lemma 2.17.2, this implies that there is a complementary subspace W of $\mathfrak{l}$ in $\mathfrak{g}$ which is also H-invariant (and in particular, U-invariant). From the inverse function theorem, we conclude that for any group element g in G sufficiently close to the identity, we can factor $g = \exp(w)l$, where $l \in L$ is also close to the identity, and $w \in W$ is small (in fact this factorisation is unique). We let $\pi_L : g \mapsto l$ be the map from g to l; this is well defined and smooth near the identity.

Let n be sufficiently large, and write $y_n = g_n x_n$, where g_n goes to the identity as n goes to infinity. Pick $l_n \in V$ at random (using the measure m conditioned to V). Using the inverse function theorem and continuity, we see that the random variable $\pi_L(l_n g_n)$ is supported in a small neighbourhood of V, and that its distribution converges to the uniform distribution of V (in, say, total variation norm) as $n \to \infty$. In particular, we see that $y'_n := l_n y_n \in E$ with probability at least 0.7 and $x'_n := \pi_L(l_n g_n) x_n \in E$ with probability at least 0.6 (say) if n is large enough. In particular we can find an $l_n \in V$ such that y'_n, x'_n both lie in E. Also by construction we see that $y'_n = \exp(w_n) x'_n$ for some $w_n \in W$; since $y_n \notin L x_n$, we see that w_n is non-zero. On the other hand, since W is transverse to l and the distance between x_n, y_n goes to zero, we see that w_n goes to zero.

There are now two cases. If $\exp(w_n)$ centralises U for infinitely many n, then from Proposition 2.17.10 followed by Lemma 2.17.11 we obtain conclusion (2) of Proposition 2.17.8 as required. Otherwise, we may pass to a subsequence and assume that none of the $\exp(w_n)$ centralises U. Since W is preserved by U, we see that the group elements $\exp(w_n)^*$ also lie in $\exp(K)$ for some compact set K in W, and also on the boundary of B. This space is compact, and so by Proposition 2.17.12 we see that μ is invariant under some group element $g \in \exp(K) \cap \partial B$, which cannot lie in L. Since the ball B can be chosen arbitrarily small, we can thus apply Lemma 2.17.11 to again obtain conclusion 2 of Proposition 2.17.8 as required. $\square$

Notes. This lecture first appeared at

terrytao.wordpress.com/2008/03/15.

Chapter 3

Lectures in Additive Prime Number Theory

3.1. Structure and randomness in the prime numbers

This talk concerns the subject of *additive prime number theory*—which, roughly speaking, is the theory of additive patterns contained inside the prime numbers $\mathcal{P} = \{2, 3, 5, 7, 11, \ldots\}$. This is a very old subject in mathematics; for instance, the *twin prime conjecture*, which asserts that there are infinitely many patterns of the form $n, n+2$ in the primes, may have been considered in one form or another by Euclid (although the modern version of the conjecture probably dates to [**Br1915**], which showed the first non-trivial progress towards the problem). It remains open today, although there are some important partial results. Another well-known conjecture in the subject is the *odd Goldbach conjecture* (dating from 1742), which asserts that every odd number n greater than 5 is the sum of three primes. A famous theorem of Vinogradov [**Vi1937**] asserts that this conjecture is true for all sufficiently large n; Vinogradov's original argument did not explicitly say how large is "sufficiently large", but later authors did quantify the argument; currently, it is known [**LiWa2002**] that the odd Goldbach conjecture is true for all odd $n > 10^{1346}$. The conjecture is also known [**Sa1998**] for all odd $5 < n < 10^{20}$, by a completely different method.

In this lecture, I will present the following result of myself and Ben Green [**GrTa2008**] in this subject:

Theorem 3.1.1 (Green-Tao Theorem). *The prime numbers $\mathcal{P} = \{2, 3, 5, 7, \ldots\}$ contain arbitrarily long arithmetic progressions.*

More specifically, I want to talk about three basic ingredients in the proof, and how they come together to prove the theorem:

(1) Random models for the primes.

(2) Sieve theory and almost primes.

(3) Szemerédi's theorem on arithmetic progressions.

3.1.1. Random models for the primes. One of the most fundamental results in this field is the *prime number theorem*, which asserts that the number of primes less than any large integer N is asymptotically equal to $(1 + o(1))\frac{N}{\log N}$ as N goes to infinity. This theorem is proven by using the *Euler product formula*[1]

$$\zeta(s) = \sum_{n=1}^{\infty} \frac{1}{n^s} = \prod_{p} \left(1 - \frac{1}{p^s}\right)^{-1},$$

which relates the primes to the *Riemann zeta function* $\zeta(s)$. By combining this formula with some non-trivial facts about the Riemann zeta function (and in particular, in the zeroes of that function), one can eventually obtain the prime number theorem.

One way to view the prime number theorem is as follows: if one picks an integer n from 1 to N at random, then that number n has a probability of $\frac{1+o(1)}{\log N}$ of being prime.

With this in mind, one can propose the following heuristic "proof" of the twin prime conjecture:

(1) Take a large number N, and let n be a randomly chosen integer from 1 to N. By the prime number theorem, the event that n is prime has probability $\mathbf{P}(n \text{ is prime}) = \frac{1+o(1)}{\log N}$.

(2) By another application of the prime number theorem, the event that $n+2$ is prime also has probability $\mathbf{P}(n+2 \text{ is prime}) = \frac{1+o(1)}{\log N}$. (The shift by 2 causes some additional correction terms, but these can be easily absorbed into the $o(1)$ term.)

---

[1]Incidentally, this formula, if rewritten using the *geometric series formula* as

$$\sum_{n=1}^{\infty} \frac{1}{n^s} = \prod_{p} \left(1 + \frac{1}{p^s} + \frac{1}{p^{2s}} + \cdots\right),$$

is a restatement (via *generating functions*) of the *fundamental theorem of arithmetic*; if instead one rewrites it as

$$\left(1 - \frac{1}{2^s}\right)\left(1 - \frac{1}{3^s}\right)\left(1 - \frac{1}{5^s}\right) \cdots \times \sum_{n=1}^{\infty} \frac{1}{n^s} = 1,$$

one can view this as a restatement of (a variant of) the *sieve of Eratosthenes*.

(3) Assuming that these two events are independent, we conclude that $\mathbf{P}(n, n+2 \text{ are both prime}) = (\frac{1+o(1)}{\log N})^2$. In other words, the number of twin primes less than N is $(1 + o(1))\frac{N}{\log^2 N}$.

(4) Since $(1 + o(1))\frac{N}{\log^2 N}$ goes to infinity as N goes to infinity, there are infinitely many twin primes.

Unfortunately, this argument does not work. One way to see this is to observe that the same argument could be trivially modified to imply that there are infinitely many pairs of *adjacent* primes n, $n + 1$, which is clearly absurd.

OK, so the above argument is broken; can we fix it? Well, we can try to accomodate the above objection. Why is it absurd to have infinitely many pairs of adjacent primes? Ultimately, it is because of the obvious fact that the prime numbers are all odd (with one exception). In contrast, the above argument was implicitly using a *random model* for the primes (first proposed by Cramér [**Cr1936**]) in which every integer from 1 to N—odd or even— had an equal chance of $\frac{1+o(1)}{\log N}$ of being prime; this model is clearly at odds with the parity structure of the primes. But we can repair this by replacing the above random model with a more sophisticated model in which parity is taken into account. More precisely, we observe that a randomly chosen odd number from 1 to N has a probability of $\frac{2+o(1)}{\log N}$ of being prime, while a randomly chosen even number has a probability of $\frac{0+o(1)}{\log N}$ (one can be more precise than this, of course). In the language of probability theory, we have the *conditional probabilities*

$$\mathbf{P}(n \text{ is prime}|n \text{ is odd}) = \frac{2 + o(1)}{\log N},$$

$$\mathbf{P}(n \text{ is prime}|n \text{ is even}) = \frac{0 + o(1)}{\log N},$$

and similarly

$$\mathbf{P}(n + 2 \text{ is prime}|n \text{ is odd}) = \frac{2 + o(1)}{\log N},$$

$$\mathbf{P}(n + 2 \text{ is prime}|n \text{ is even}) = \frac{0 + o(1)}{\log N}.$$

Now, instead of assuming that the events "n is prime" and "$n + 2$ is prime" are absolutely independent, let us assume that they are *conditionally* independent, relative to the parity of n. Then we conclude that

$$\mathbf{P}(n, n + 2 \text{ are both prime}|n \text{ is odd}) = \frac{4 + o(1)}{\log^2 N},$$

$$\mathbf{P}(n, n + 2 \text{ are both prime}|n \text{ is even}) = \frac{0 + o(1)}{\log^2 N},$$

and a little computation (using the *law of total probability*) then shows that the number of twin primes less than N is now $(2 + o(1))\frac{N}{\log^2 N}$, which still goes to infinity, and so we recover the twin prime conjecture again. Or do we?

Well, the above random model is still flawed. It now correctly asserts that there are extremely few pairs n, $n + 1$ of adjacent primes, but it also erroneously predicts that there are infinitely many triplets of primes of the form n, $n + 2$, $n + 4$, when in fact there is only one, 3, 5, 7, since exactly one of n, $n + 2$, $n + 4$ must be divisible by 3. But we can refine the random model further by taking mod 3 structures into account as well as mod 2 structures. Indeed, if we partition the integers from 1 to N using both the mod 2 partition and the mod 3 partition, we obtain the six residue classes $\{1 \leq n \leq N : n = i \bmod 6\}$ for $i = 0, 1, 2, 3, 4, 5$. From the prime number theorem in arithmetic progressions (a common generalisation of the prime number theorem and *Dirichlet's theorem*) one can show that

$$\mathbf{P}(n \text{ is prime}|n = i \bmod 6) = \frac{3 + o(1)}{\log N}$$

for $i = 1, 5$, and

$$\mathbf{P}(n \text{ is prime}|n = i \bmod 6) = \frac{0 + o(1)}{\log N}$$

for $i = 0, 2, 3, 4$. By repeating the previous analysis, the predicted count of twin primes less than N now drops from $(2 + o(1))\frac{N}{\log^2 N}$ to $(1.5 + o(1))\frac{N}{\log^2 N}$.

Now, it turns out that this model is still not correct—it fails to account for the mod 5 structure of the primes. But it is not hard to incorporate that structure into this model also, which revises the twin prime count downward a bit to $(1.40625 + o(1))\frac{N}{\log^2 N}$. And then the mod 7 structure also changes the predicted number of twin primes a little bit more, and so on and so forth. But one notices that as one continues to input in all this structural information about the primes, the predicted count of twin primes begins to converge to a limit, namely $(2\Pi_2 + o(1))\frac{N}{\log^2 N} \approx 1.32 \frac{N}{\log^2 N}$, where

$$\Pi_2 := \prod_{p \geq 3, \text{ prime}} \left(1 - \frac{1}{(p-1)^2}\right) = 0.66016\ldots$$

is known as the *twin prime constant*. More generally, Hardy and Littlewood proposed a general conjecture [**HaLi1923**], now known as the *Hardy-Littlewood prime k-tuples conjecture*, that predicted asymptotic counts for a general class of additive patterns in the primes; this conjecture (and further refinements) would imply the twin prime conjecture, Vinogradov's theorem, my theorem with Ben Green, and many other results and conjectures in the subject also.

Roughly speaking, these conjectures assert that apart from the "obvious" structure in the primes, arising from the prime number theorem and from the local behaviour of the primes mod 2, mod 3, etc., there are no other structural patterns in the primes, and so the primes behave "pseudorandomly" once all the obvious structures are taken into account. The conjectures are plausible, and backed up by a significant amount of numerical evidence; unfortunately, nobody knows how to enforce enough pseudorandomness in the primes to make the conjectures rigorously proven. (One cannot simply take limits of the above random models as one inputs more and more mod p information, because the $o(1)$ error terms grow rapidly and soon overwhelm the main term that one is trying to understand.) The problem is that the primes may well contain "exotic structures" or "conspiracies", beyond the obvious structures listed above, which could further distort things like the twin prime count, perhaps so much so that only finitely many twin primes remain. This seems extremely unlikely, but we cannot rule it out completely yet; how can one disprove a conspiracy?

Some numerics may help illustrate what I mean by the primes becoming random after the mod 2, mod 3, etc. structures are "taken into account" (though I should caution against reading *too* much into such small-scale computations, as there are many opportunities in small data sets for random coincidences or transient phenomena to create misleading artefacts; cf. the "law of small numbers" [**Gu1988**]). Here are the first few natural numbers, with the primes underlined, the odd numbers in the starred columns, and the even numbers in dotted columns:

*	.	*	.	*	.	*	.	*	.	*	.	*	.
1	2	**3**	4	**5**	6	**7**	8	**9**	10	**11**	12	**13**	14
15	16	**17**	18	**19**	20	**21**	22	**23**	24	**25**	26	**27**	28
29	30	**31**	32	**33**	34	**35**	36	**37**	38	**39**	40	**41**	42
43	44	**45**	46	**47**	48	**49**	50	**51**	52	**53**	54	**55**	56

It is then clear that the primes have mod 2 structure; they cluster in the odd numbers (the starred columns) rather than the even numbers (the dotted columns), and are thus distributed quite non-randomly. But suppose we "zoom in" on the odd numbers, discarding the even numbers:

```
*   .   *   .   *   .   *   .   *   .   *   .   *   .   *   .
1   3   5   7   9  11  13  15  17  19  21  23  25  27  29  31
33  35  37  39  41  43  45  47  49  51  53  55  57  59  61  63
65  67  69  71  73  75  77  79  81  83  85  87  89  91  93  95
```

Then it seems that there is no further parity structure; the starred columns (which have numbers that are 1 mod 4) and the dotted columns (which have numbers that are 3 mod 4), seem equally likely to contain primes. (Indeed, this is a proven fact, being a special case of the prime number theorem in arithmetic progressions.) But look what happens if we highlight the mod 3 structure instead:

```
    *       .       *       .       *       .       *       .
1   3   5   7   9  11  13  15  17  19  21  23
25  27  29  31  33  35  37  39  41  43  45  47
49  51  53  55  57  59  61  63  65  67  69  71
73  75  77  79  81  83  85  87  89  91  93  95
```

Then the dotted columns (whose entries are 0 mod 3) are devoid of primes other than 3 itself. But if we instead zoom into (say) the starred columns (whose entries are 1 mod 3), we eliminate the mod 3 structure, making the remaining primes more randomly distributed:

```
*                       .       *                       .
1    7   13   19   25   31   37   43   49   55
61   67   73   79   85   91   97  103  109  115
121  127  133  139  145  151  157  163  169  175
```

Now the primes exhibit obvious mod 5 structure (the dotted columns, whose entries are 0 mod 5, have no primes) but look fairly randomly distributed otherwise. Indeed, if we zoom in to (say) the 1 mod 5 residue class, which corresponds to the starred columns above, there seems to be very little structure at all:

```
                .
1    31   61   91  121  151  181
211  241  271  301  331  361  391
421  451  481  511  541  571  601
631  661  691  721  751  781  811
```

Apart from the dotted column, which has all entries divisible by 7 and thus not prime, the primes seem fairly randomly distributed. In the above examples I always zoomed into the residue class 1 mod p for $p = 2, 3, 5, \ldots,$

but if one picks other residue classes (other than 0 mod p, of course), one also sees the primes become increasingly randomly distributed, with no obvious pattern within each class (and no obvious relation between pairs of classes, triplets of classes, etc.). One can view the prime k-tuples conjecture as a precise formalisation of this assertion of increasingly random distribution.[2]

3.1.2. Sieve theory and almost primes. We have talked about random models for the primes, which seem to be very accurate, but difficult to rigorously justify. However, there is a concept closely related to a prime, namely an *almost prime*, for which we *can* show the corresponding random models to be accurate, using the elementary yet surprisingly useful technique of *sieve theory*.

The most elementary sieve of all is the *sieve of Eratosthenes*. This sieve uncovers (or "sifts out") all the prime numbers in a given range, say between $N/2$ and N for some large number N, by starting with all the integers in this range, and then discarding all the multiples of 2, then the multiples of 3, the multiples of 5, and so forth. After all multiples of prime numbers less than $\sqrt{N}$ are discarded, the remaining set is precisely the set of primes between $N/2$ and N.

It is tempting to use this sieve to count patterns in primes, such as twin primes. After all, it is easy to count how many twins there are in the integers from $N/2$ to N; if one then throws out all the multiples of 2, it is still straightforward to count the number of twins remaining. Similarly, if one then throws out multiples of 3, of 5, and so forth; not surprisingly, the computations here bear some resemblance to those used to predict twin prime counts from the random models just mentioned. However, as with the random models, the error terms begin to accumulate rapidly, and one loses control of these counts long before one reaches the end of the sieve. More advanced sieves (in which one does not totally exclude the multiples of small numbers, but instead adjusts the "weight" or "score" of a number upward or downward depending on its factors) can improve matters significantly, but even the best sieves still only work if one stops sieving well before the $\sqrt{N}$ mark (sieve levels such as $N^{1/4}$ are typical). (The reason for this has to do with the *parity problem*, which I will not discuss further here, but see Section 3.10 of *Structure and Randomness*.)

I like to think of the sieving process as being analogous to carving out a sculpture (analogous to the primes) from a block of granite (analogous to the integers). To begin with, one uses crude tools (such as a mallet) to remove large, simple pieces from the block; but after a while, one has to make finer and finer adjustments, replacing the mallet by a chisel, and

---

[2]In [**GrTa2008**], we rely quite crucially on this zooming-in trick to improve the pseudorandomness of the almost primes, referring to it as the "W-trick".

then by a pick, removing lots and lots of very small pieces, until the final sculpture is complete. Initially, the structure is simple enough that one can easily pick out patterns (such as twins, or arithmetic progressions); but there is the (highly unlikely) possibility that the many small sets removed at the end are just distributed perversely enough to knock out all the patterns one discerns in the initial stage of the process.

Because we cannot exclude this possibility, sieve theory alone does not seem to be able to count patterns in primes. However, we can stop the sieve at an earlier level. If we do so, we obtain good counts of patterns, not in primes, but in the larger set of almost primes—which, for the purposes of this talk, one can think of as being defined as those numbers with no small factors (e.g. no factors less than N^ε for some $\varepsilon > 0$). (This is an oversimplification—one needs to use the weights mentioned above—but it will suffice for this discussion.) Then it turns out that (to oversimplify some more) everything we want to show about the primes, we can show about the almost primes. For instance, it was shown by Chen that there are infinitely many twins $n, n+2$, one of which is a prime and the other is the product of at most two primes. Similarly, given any k, one can show using sieve theory that there are infinitely many arithmetic progressions of length k, each element of which has at most $O_k(1)$ prime factors. More generally, the almost primes behave the way we expect the primes to, distributed pseudorandomly, after taking into account the obvious structures (for instance, almost primes, like the primes, tend to be almost all coprime to 2, coprime to 3, etc.)

Unfortunately, there is still a gap between finding patterns in the almost primes and finding patterns in the primes themselves, because the primes are only a subset of the almost primes. For instance, while the number of primes less than N is roughly $N/\log N$, the number of numbers less than N with no factors less than (say) $N^{1/100}$ is (very) roughly $100N/\log N$. Thus the primes only form a small fraction of the almost primes.

3.1.3. Szemerédi's theorem on arithmetic progressions. Thus far, we have discussed the general problem of how to find patterns in sets such as the primes or almost primes. This problem in general seems to be very difficult, because we do not know how structured or pseudorandom the primes are. There is however one type of pattern which is special—it necessarily shows up in just about any kind of set—structured or random. This type of set is an *arithmetic progression*. In fact, we have the following important theorem:

Theorem 3.1.2 (Szemerédi's theorem [**Sz1975**]). *Let $A \subset \mathbf{Z}$ be a set of integers of positive upper density (i.e., $\limsup_{N\to\infty} \frac{1}{2N+1}|A \cap \{-N, \ldots, N\}| > 0$). Then A contains arbitrarily long arithmetic progressions.*

This is a remarkable theorem: it says that if one picks any set of integers at all, so long as it is large enough to occupy a positive fraction of all integers, that is enough to guarantee the existence of arithmetic progressions of any length inside that set. This is in contrast with just about any other pattern, such as twins: for instance, the multiples of three have positive density, but contain no twins.

There are several proofs of this difficult theorem known. It would be too technical to discuss any of them here in detail, but *very* roughly speaking, all the proofs proceed by dividing the set A into "structured" components (such as sets which are periodic) and "pseudorandom" components (which, roughly, are those components for which the random model gives accurate predictions). One can show that the structured components always generate a lot of arithmetic progressions, and that the pseudorandom components do not significantly disrupt this number of progressions.

Unfortunately, Szemerédi's theorem does not apply to the primes, because they have zero density. (There are some quantitative versions of that theorem that apply to some sets of zero density, but they are not yet strong enough to directly deal with sets as sparse as the primes.)

3.1.4. Putting it all together. To summarise: random models predict arbitrarily long progressions in the primes, but we cannot verify these models. Sieve theory does let us establish long progressions in the almost primes, but the primes are only a fraction of the almost primes. Szemerédi's theorem gives progressions, but only for sets of positive density within the integers.

To proceed, we exploit the fact that the primes have positive *relative* density inside the almost primes, by the following argument (inspired, incidentally, by Furstenberg's ergodic-theoretic proof [**Fu1977**] of Szemerédi's theorem; see Sections 2.10–2.15). We conjecture that the primes obey a certain random model, in which the only structure present being mod p structure for small p. If this is the case, then we are done. If not, it means that there is some specific obstruction to pseudorandomness in the primes, much as the mod 2 or mod 3 obstructions we discussed earlier prevented the most naive random models of the primes from being accurate. We do not know exactly what that obstruction is, but it turns out that it is possible nevertheless to use that obstruction to modify the random model for the primes to be more accurate, much as we used the mod 2 and mod 3 information previously. We then repeat this process, locating obstructions to pseudorandomness and incorporating them into our random model, until no major obstructions remain. (This can be done after only a bounded number of steps, because one can show (with some effort) that each addition to the random model increases its "energy" by a certain amount, and one can show that the total amount of energy in the model must stay bounded.)

As a consequence we know that we can model the primes accurately (at least for the purposes of counting arithmetic progressions) by *some* random model.[3]

The above procedure does not give a very tractable formula for what this model is. However, because the primes are a dense subset of the almost primes, which behave like a random subset of the integers, one can show (by a "comparison principle", and oversimplifying somewhat) that the primes must then behave like a random subset of a dense subset B of the integers. But then Szemerédi's theorem applies, and it shows that this set B contains plenty of progressions; a random subset of B will then also contain many progressions, and thus the primes will also.

Remark 3.1.3. A simplified version of the above argument, using game theory instead of the above ergodic theory motivated approach, was recently presemted in [**ReTrTuVa2008**], [**Go2008**].

Notes. This talk first appeared at `terrytao.wordpress.com/2008/01/07`, and was given in the annual joint meeting of the American Mathematical Society and Mathematical Association of America in San Diego in January of 2008. A similar talk also formed the first in my series of four Marker Lectures in Penn State University in November of 2008.

3.2. Linear equations in primes

In the previous lecture, I focused on finding a specific type of pattern inside the prime numbers, namely that of an arithmetic progression $n, n + r, \ldots,$ $n + (k - 1)r$. The main reason why the analysis there is specific to progressions is because of its reliance on *Szemerédi's theorem*, which shows that arithmetic progressions are necessarily abundant in sufficiently "large" sets of integers. There are several other variants and generalisations of this theorem known to a few other types of patterns (e.g. polynomial progressions $n + P_1(r), \ldots, n + P_k(r)$, where $P_1, \ldots, P_k$ are polynomials from the integers to the integers with $P_1(0) = \ldots = P_k(0) = 0$ and $r \neq 0$), and in some cases the analogous results about primes are known (e.g. in [**TaZi2008**] we showed that for any given $P_1, \ldots, P_k$ as above, there are infinitely many polynomial progressions of primes).

However, for most patterns, there is no analogue of Szemerédi's theorem, and the strategy used in the previous lecture cannot be directly applied. For instance, it is certainly not true that any subset of integers with positive

---

[3]I have not precisely defined what I mean here by a "random model", but very roughly speaking, any such model consists of a partition (or σ-*algebra*) of the integers from 1 to N into a bounded number of sets, a specified density for the primes on each such set, and an assumption that the primes behave on each set like a random set with the specified density. Readers familiar with the *Szemerédi regularity lemma* may see a parallel here.

upper density contains any twins $n, n+2$; the multiples of three, for instance, form a counterexample, among many others.[4]

Furthermore, even in the cases when these methods do work, for instance in demonstrating for each k that there are infinitely many progressions of length k inside the primes, they do not settle the more quantitative problem of how many progressions of length k there are asymptotically in any given finite range of primes, e.g. the primes less than a number N in the asymptotic limit $N \to \infty$. This is because Szemerédi's theorem provides a lower bound for the number of progressions in a large finite set, but not a matching upper bound. (For instance, given a subset A of $\{1, \ldots, N\}$ of density $1/2$, the number of progressions of length 3 in A can be as large as $(1/4 + o(1))N^2$ (if A consists of the even integers from 1 to N, for instance) and as small as $(1/8+o(1))N^2$ (if A is a randomly chosen set of density $1/2$), and can even be a little bit smaller by perturbing this example slightly.)

On the other hand, as discussed in the previous lecture, one can use standard random models for the primes to predict what the correct asymptotic for these questions should be. For instance, the number of arithmetic progressions $n, n+r, \ldots, n+(k-1)r$ of a fixed length k consisting of primes less than N should be asymptotically

$$(3.1) \qquad \left(\frac{1}{2(k-1)} \left(\prod_p \beta_p \right) + o(1) \right) \frac{N^2}{\log^k N},$$

where the product is over all primes p, and the quantity β_p is defined as

$$\beta_p := \frac{1}{p} \left(\frac{p}{p-1} \right)^{k-1}$$

for $p \leq k$, and

$$\beta_p := \left(1 - \frac{k-1}{p} \right) \left(\frac{p}{p-1} \right)^{k-1}$$

for $p \geq k$.

The various terms in this complicated-looking formula can be explained as follows. The "Archimedean" factors $\frac{1}{2(k-1)}$ and N^2 come from the fact that the number of arithmetic progressions $n, n+r, \ldots, n+(k-1)r$ of *natural numbers* less than N is $(\frac{1}{2(k-1)} + o(1))N^2$. The "density" factor $\frac{1}{\log^k N}$ comes from the *prime number theorem*, which roughly speaking asserts that each of the k elements $n, n+r, \ldots, n+(k-1)r$ in a typical arithmetic progression has a $(1+o(1))\frac{1}{\log N}$ "probability" of being prime. The "local" factors β_p measure how much bias arithmetic progressions with respect to being coprime to a fixed prime p, which is relevant for progressions of primes, since primes

---

[4]In fact, there are so many counterexamples here, that it looks unlikely that the twin prime conjecture can be attacked by this method without a significant new idea; see Section 2.1 of *Structure of Randomness* for further discussion.

of course tend to be coprime to p. More precisely, β_p can be defined as the probability that a random arithmetic progression of length k has all entries coprime to p, divided by the probability that a random collection of k independent numbers are all coprime to p. It is not difficult to show that the product $\prod_p \beta_p$ converges to some finite non-zero number for each k.

Similar heuristic asymptotic formulae exist for the number of many other patterns of primes; for instance, the number of representations $N = p_1 + p_2$ of a large integer N as the sum of two primes should be equal to $(\prod_p \beta_{p,N} + o(1))N$, where $\beta_{p,N}$ is[5] the probability that two randomly chosen numbers *conditioned* to be coprime to p sum to N modulo p, divided by the probability that two randomly chosen numbers sum to N modulo p. A more general prediction for counting linear patterns inside primes exists, and is essentially the *Hardy-Littlewood prime tuples conjecture* [**HaLi1923**]. This conjecture, which is widely believed to be true, would imply many other conjectures in the subject, such as the twin primes conjecture and the *Goldbach conjecture* (for sufficiently large even numbers). Unfortunately, the cases of the prime tuples conjecture which would have these consequences remain out of reach of current technology.

Using some elementary linear algebra, one can recast the prime tuples conjecture not as a question of finding linear patterns inside primes, but rather that of solving linear equations in which all the unknowns are required to be prime, subject to some additional linear inequalities. For instance, finding progressions of length k consisting entirely of primes less than N is essentially the same as asking for solutions to the system of equations

$$p_2 - p_1 = p_3 - p_2 = \ldots = p_k - p_{k-1}$$

and inequalities

$$0 \leq p_1, \ldots, p_k \leq N$$

where the unknowns $p_1, \ldots, p_k$ are required to be primes. More generally, one could imagine the question of asking the number of k-tuples $(p_1, \ldots, p_k)$ that consist entirely of primes and are contained in a convex set B in $\mathbf{R}^k$ of some intermediate dimension[6] $1 \leq d \leq k$, which lies in a ball of radius $O(N)$ around the origin. For instance, in the above example B is the 2-dimensional set

$$\{(x_1, \ldots, x_k) \in \mathbf{R}^k : x_2 - x_1 = \ldots = x_k - x_{k-1}; 0 \leq x_1, \ldots, x_k \leq N\}.$$

---

[5]In particular, $\beta_{2,N} = 0$ for odd N, reflecting the fact that it is very difficult for an odd number to be representable as the sum of two primes. More generally, one can compute that $\beta_{p,N} = 1 + \frac{1}{p-1}$ when p divides N, and $\beta_{p,N} = 1 - \frac{1}{(p-1)^2}$ otherwise.

[6]We make the technical assumption that the linear coefficients of the equations defining the d-dimensional subspace that B lives in are independent of N; k and d are of course also assumed to be independent of N. The constant coefficients, however, are allowed to vary with N; this is the situation that comes up for instance in the Goldbach conjectures.

One can think of the problem of finding points in B as that of solving $k - d$ equations in k unknowns. One can also generalise this problem slightly by enforcing some residue constraints $x_j = a_j \mod q_j$ on the unknowns, but we will ignore this minor extension to simplify the discussion.

The prime tuples conjecture for this problem can roughly speaking be phrased as follows. Suppose that the number of k-tuples $(n_1, \ldots, n_k)$ in B consisting of *natural numbers* is known to be

$$(\beta_\infty + o(1))N^d$$

for some constant β_∞ independent of N (one can think of this constant as the normalised volume of B). Suppose also that for any fixed prime p, the number of k-tuples in B consisting of natural numbers coprime to p is known to be[7]

$$(\beta_\infty \beta_p + o(1))\left(1 - \frac{1}{p}\right)^k N^d$$

for some constant β_p independent of n. Then the prime tuples conjecture asserts that the number of k-tuples in B consisting of primes should be

(3.2)
$$\left(\beta_\infty \prod_p \beta_p + o(1)\right) \frac{N^d}{\log^k N}.$$

In particular, this can be shown to imply that if $\beta_\infty > 0$ (thus there are no obstructions to solving the system of equations at infinity) and if $\beta_p > 0$ for all p (thus there are no obstructions to solvability mod p for any p), then there will exist many solutions to the system of equations in primes when N is large enough.

As mentioned earlier, this conjecture remains open in several important cases, most particularly in the one-dimensional case $d = 1$. For instance, the twin prime conjecture would follow from the case

$$B := \{(x_1, x_2) : x_2 - x_1 = 2, 0 \le x_1, x_2 \le N\},$$

but this case remains open. However, there has now been significant progress in the higher-dimensional cases $d \ge 2$, especially when the *codimension* $k - d$ (representing the number of equations in the system) is low. Firstly, the prime number theorem settles the zero codimension case $d = k$ (and is pretty much the only situation in which we can handle a $d = 1$ case). The Hardy-Littlewood circle method, based on Fourier analysis, settles all "non-degenerate" cases when $d \ge \max(k - 1, 2)$, where "non-degenerate" roughly speaking means that the problem does not secretly contain a $d = 1$ problem inside it as a lower-dimensional projection (e.g. $B := \{(x_1, x_2, x_3) : x_2 - x_1 = 2, 0 \le x_1, x_2, x_3 \le N\}$ would be degenerate; more generally, B is

---

[7]The factor $(1 - \frac{1}{p})^k$ is natural, as it represents the proportion of tuples of k natural numbers in which all the entries are coprime to p.

non-degenerate if it is not contained in any hyperplane that can be defined using at most two of the unknowns). It can also handle some cases in which the codimension $k - d$ exceeds 1 (e.g. one could take the Cartesian product of some codimension 1 examples); the precise description of what problems are within reach of this method is a little technical to state and will not be given here.

Ben Green and I were able to establish the following partial result towards this conjecture (see [**GrTa2009**], [**GrTa2009c**], [**GrTa2009d**], [**GrTa2009e**], [**GrTa2009f**]):

Theorem 3.2.1. *The prime tuples conjecture is true in all non-degenerate situations in which* $d \geq \max(k - 2, 2)$. *If the* inverse conjecture for the Gowers norms over the integers *is true, then the prime tuples conjecture is true in all non-degenerate situations in which* $d \geq 2$.

I will say a little bit more about what the inverse conjecture for the Gowers norms is later. This theorem unfortunately does not touch the most interesting case $d = 1$ (when the patterns one is seeking only have one degree of freedom), but it does largely settle all the other cases. For instance, this theorem implies the asymptotic (3.1) for prime progressions of length k for $k \leq 4$ (the cases $k \leq 3$ were established earlier by van der Corput using the circle method), and the case of higher k would also follow from the theorem once the inverse conjecture is proven. As with the circle method, we can also unconditionally handle some cases in which d is less than $\max(k - 2, 2)$, but the precise statement here is technical and will be omitted. (Details and further examples can be found in [**GrTa2009**].)

Now I would like to turn to the proof of this theorem. At first glance, the result looks like it is going to be quite complicated, due to the presence of all the different factors in the asymptotic (3.2) that one is trying to prove. However, most of the factors can be dealt with by various standard tricks. The Archimedean factor β_∞ can be eliminated from the problem by working locally (with respect to the infinite *place*), covering B by cubes of sidelength $o(N)$. For similar reasons, the local factors β_p can be eliminated by working locally mod p (i.e., restricting to a single residue class mod p for various small p). The factors $\frac{1}{\log^k N}$, which come from the density $\frac{1}{\log N}$ of primes in the region of interest (i.e., from the *prime number theorem*), can largely be compensated for (with some effort) from the transference principle technology developed in our earlier paper on long progressions in the primes, which was discussed in the previous lecture. After all this, the problem basically boils down to the following. We have a certain subset A of the integers $\{1, \ldots, N\}$ with some density $0 < \delta < 1$ (one should think of A as being a "model" for the primes, after all the distorting structure coming from local obstructions has been stripped out; in actuality, one has to replace the set A by a weight function $f : \{1, \ldots, N\} \to [0, 1]$ of mean value δ, but let us ignore this technicality to simplify the discussion). We pick a random instance of some linear pattern inside $\{1, \ldots, N\}$ (for the sake of concreteness, let us pick a random arithmetic progression $n, n+r, \ldots, n+(k-1)r$ and ask what is the probability that all the elements of this pattern lie in A. Since A has density δ, we expect each element $n+jr$ of our random progression to have a probability[8] $\delta + o(1)$ to lie in A. Since our pattern consists of k elements, we thus expect

$$(3.3) \qquad \mathbf{P}(n, n+r, \ldots, n+(k-1)r \in A) = \delta^k + o(1).$$

Roughly speaking, the key issue in proving the theorem is to work out some "easily checkable" conditions on A that would guarantee that the heuristic (3.3) is in fact valid. One then verifies that these "easily checkable" conditions do indeed hold for the set A of interest (which is a proxy for the set of primes).

As stated before, we expect $\mathbf{P}(n + jr \in A) = \delta + o(1)$ for each $0 \le j < k$. Thus (3.3) is asserting in some sense that the events $n + jr \in A$ are "approximately independent". This would be a reasonable assertion if A was *pseudorandom* (i.e., it behaved like a random subset of $\{1, \ldots, N\}$ of the given density δ), and is consistent with the general heuristic from number theory that we expect the prime numbers to behave randomly once all the "obvious" irregularities in distribution (in particular, irregularity modulo p for small p) has been dealt with. But if A exhibits certain types of *structure* (or at least some bias towards structure), then (3.3) can fail. For instance, suppose that A consists entirely of odd numbers. Then, if the first two elements n, $n + r$ of an arithmetic progression lie in A, they are necessarily odd, which then forces the rest of the elements of this progression to be odd. As A is concentrated entirely in these odd numbers, these elements of the progression are thus expected to have an elevated probability of lying in A, and so the left-hand side of (3.3) would be expected to significantly exceed the former once $k \ge 3$. (The asymptotic (3.3) becomes trivially true for $k < 3$.) A similar distorting effect occurs if A is not entirely contained in the odd numbers, but is merely *biased* towards them, in that odd numbers are more likely to lie in A than even numbers. In this example, the bias in A caused the number of progressions to go up from the expected number predicted by (3.3); it is also possible (but more tricky) to concoct examples in which bias in A forces the number of progressions to go down somewhat, though Szemerédi's theorem does prevent one from extinguishing these progressions completely when N is large.

---

[8]This is not quite the case if A is biased to lie on one side of $\{1, \ldots, N\}$ than on the other, but it turns out that one can ignore this possibility.

Bias towards odd or even numbers is equivalent to a correlation between A and the *linear character* $\chi(n) := (-1)^n$; the algebraic constraints between the $\chi(n + jr)$, and in particular the relationship

$$(3.4) \qquad\qquad \chi(n + 2r)\chi(n + r)^{-2}\chi(n) = 1$$

can be viewed as the underlying source of the distorting effect that can prevent (3.3) from holding for $k \geq 3$. The same algebraic constraint holds for any other linear character, e.g. the Fourier character $\chi(n) := e(\xi n)$ (where $e(x) := e^{2\pi i x}$) for some fixed frequency $\xi \in \mathbf{R}$, for much the same reason that two points on a line determine the rest of the line (it is also closely related to the fact that the second derivative of a linear function vanishes). Because of this, we expect (3.3) to be distorted when A *correlates* with such a character (which means that the Fourier coefficient $\sum_{n \in A} \chi(n)$ is unexpectedly large in magnitude).

It turns out that in the case $k = 3$ of progressions of length three, correlation with a linear character is the *only* source of distortion in the count (3.3). A sign of this can be seen from the identity

$$\#\{(n, r) : n, n + r, n + 2r \in A\}$$
$$= \int_0^1 \Big(\sum_{n_1 \in A} e(\xi n_1) \Big) \Big(\sum_{n_2 \in A} e(-2\xi n_2) \Big) \Big(\sum_{n_3 \in A} e(\xi n_3) \Big) \, d\xi,$$

which can be viewed as a "Fourier transform" of the algebraic identity (3.4). One can formalise this using (a slight variant of) the above identity and some other Fourier-analytic tools (in particular, the *Plancherel identity*) to conclude

Theorem 3.2.2 (Inverse theorem for length 3 progressions; informal). *Let $k = 3$. Suppose that A is a subset of $\{1, \ldots, N\}$ of density δ for which (3.3) fails. Then A correlates with a non-trivial linear character $\chi(n) = e(\xi n)$. ("Non-trivial" basically means that χ oscillates at least once on the interval $\{1, \ldots, N\}$.)*

Applying this theorem in the contrapositive, we conclude that we can justify the asymptotic (3.3) in the $k = 3$ case as long as we can show that A does not correlate with a linear character. In the case when A is a proxy for the primes, this task essentially boils down to that of establishing non-trivial estimates for exponential sums over primes, such as

$$\sum_{p < N} e(\xi p);$$

for technical reasons it is more convenient to deal with slight modifications[9] of this sum such as

$$(3.5) \qquad \sum_{n=1}^{N} \Lambda(n)e(\xi n),$$

where Λ is the *von Mangoldt function*, or

$$(3.6) \qquad \sum_{n=1}^{N} \mu(n)e(\xi n),$$

where μ is the *Möbius function*. The reason for using these functions instead is that they enjoy a number of very useful identities, such as

$$(3.7) \qquad \sum_{n=1}^{\infty} \frac{\Lambda(n)\chi(n)}{n^s} = -\frac{L'(s,\chi)}{L(s,\chi)}$$

and

$$(3.8) \qquad \sum_{n=1}^{\infty} \frac{\mu(n)\chi(n)}{n^s} = \frac{1}{L(s,\chi)}$$

for any *Dirichlet character* χ (where $L(s,\chi)$ is the *Dirichlet L-function*), and also multiplicative identities such as

$$(3.9) \qquad \Lambda(n) = \sum_{d|n} \mu(d) \log \frac{n}{d}$$

and

$$(3.10) \qquad \mu(n) = \sum_{n=abc} \mu(a)\mu(b).$$

To cut a long story short, identities such as (3.7), (3.8) are useful for estimating (3.5), (3.6), respectively, in the *major arc* case when ξ is rational or close to rational (with small denominator), while (variants of) identities such as (3.9) or (3.10) (in particular, certain truncated versions of (3.9) and (3.10) such as *Vaughan's identity*) are useful for estimating (3.5), (3.6), respectively, in the *minor arc* case when ξ is far from a rational with small denominator (or close to a rational with large denominator). This theory was pioneered by Vinogradov (and also Hardy and Littlewood), and refined and simplified over the years with many contributions by Vaughan, Davenport, Heath-Brown, and others, with the upshot being that we now have a fairly good understanding of sums such as (3.5) and (3.6), and in particular that the sum (3.6) exhibits a strong cancellation (by a factor of $O_A(\log^{-A} N)$ for

---

[9]There are various elementary identities, such as summation by parts, that allow us to express one of these sums in terms of the others. There is plenty of flexibility in here as long as one retains a factor in the sum, such as $\Lambda(n)$ or $\mu(n)$, which is somehow sensitive to the prime factorisation of n.

any fixed A) uniformly in ξ (i.e., we can handle both major and minor arcs with a uniform estimate). Combining this with the inverse theorem and the previous reductions, one can eventually establish the asymptotic (3.1) in the $k = 3$ case.[10]

Now we turn to progressions of longer length, such as the case $k = 4$. Here, linear characters $\chi(n) = e(\xi n)$ continue to cause bias that distorts the expected asymptotic (3.3), and so it is still necessary to control sums such as (3.5) or (3.6) to prevent such bias from occurring. However, a major new difficulty arises that we are faced with new sources of bias. For instance, if one takes a quadratic character $\chi(n) := e(\xi n^2)$ for some ξ, then one easily verifies the identity

$$(3.11) \qquad \chi(n)\chi(n+r)^{-3}\chi(n+2r)^3\chi(n+3r)^{-1} = 1,$$

which reflects the fact that the third derivative of a quadratic function (such as $n \mapsto \xi n^2$) is zero (it also reflects the fact that three points on the graph of a quadratic function (i.e., a parabola) determine the entire parabola). One consequence of this is that if $\chi(n), \chi(n+r), \chi(n+2r)$ are all close to 1 (say), then $\chi(n+3r)$ will be also. This constraint between the four values of χ along an arithmetic progression suggests that if A exhibits significant correlation with χ, then the event that $n + 3r$ lies in A will be influenced in some non-trivial manner by whether n, $n + r$, and $n + 2r$ already lie in A, which will lead to some distortion in (3.3). Thus one will need to update the inverse theorem by taking quadratic characters into account.[11] The most optimistic conjecture in this regard would be

Theorem 3.2.3 (Proposed inverse theorem for length 4 progressions; informal). *Let $k = 4$. Suppose that A is a subset of $\{1, \ldots, N\}$ of density δ for which (3.3) fails. Then A correlates with a non-trivial quadratic character $\chi(n) = e(\xi_2 n^2 + \xi_1 n)$.*

Unfortunately, this conjecture fails. The easiest way to see this is to consider a *bracket quadratic character*, such as $\chi(n) = e(\lfloor \sqrt{2}n \rfloor \sqrt{3}n)$, where $\lfloor \ \rfloor$ is the *greatest integer function*. This is not quite a quadratic character, because $\lfloor \ \rfloor$ is not quite a linear function. However, this function is linear "a positive fraction of the time"; if one picks x and y to be some generic real numbers, one expects $\lfloor x + y \rfloor$ to equal $\lfloor x \rfloor + \lfloor y \rfloor$ about half of the time. Because of this, we see that while the identity (3.11) certainly does not hold all the time for $\chi(n)$, it does hold a positive fraction of the time, and this

---

[10]This is not exactly how the original proof of (3.1) by van der Corput proceeded in this case, but both proofs use the same general ingredients and method, i.e., the Hardy-Littlewood circle method and the Vinogradov method for estimating exponential sums.

[11]Easy examples show that it is possible for a set to correlate with a quadratic character without exhibiting any correlation with linear characters, by choosing a quadratic character with irrational coefficients.

is enough to still cause significant bias to disrupt (3.3) if A correlates with this object. It is furthermore possible to concoct examples of sets A that correlate with bracket quadratic characters such as $e(\lfloor\sqrt{2}n\rfloor\sqrt{3}n)$ but not with any linear or quadratic characters.[12] Once one throws in these bracket quadratics, it turns out that these do in fact constitute all the possible obstructions to (3.3) holding in the $k = 4$ case, as shown in [**GrTa2009d**].

Theorem 3.2.4 (Inverse theorem for length 4 progressions; informal). *Let $k = 4$. Suppose that A is a subset of $\{1, \dots, N\}$ of density δ for which (3.3) fails. Then A correlates with a non-trivial bracket quadratic character $\chi(n) = e(\sum_{j=1}^{J}\lfloor\alpha_j n\rfloor\beta_n j + \xi_2 n^2 + \xi_1 n)$ for some real numbers $\alpha_j, \beta_j, \xi_1, \xi_2$ and bounded J.*

The proof of this result involves both Fourier analysis and additive combinatorics, relying heavily on ideas from a paper of Gowers [**Go1998**] on Szemerédi's theorem for progressions of length 4. It will not be discussed here.

In view of the inverse theorem, the problem of establishing the asymptotic (3.1) for length 4 progressions then reduces (by suitable generalisations of the various methods discussed previously) to that of estimating exponential sums, of which

$$\sum_{n=1}^{N}\mu(n)e(\lfloor\sqrt{2}n\rfloor\sqrt{3}n)$$

is a typical example. One can begin to apply the methods of Vinogradov and Vaughan to control this type of expression. But one is soon faced with the problem of understanding the distribution of quadratic phases such as $\lfloor\sqrt{2}n\rfloor\sqrt{3}n$, and in particular to estimate exponential sums such as

$$\sum_{n=1}^{N}e(\lfloor\sqrt{2}n\rfloor\sqrt{3}n).$$

This turns out to be somewhat unpleasant; the standard technology of Weyl differencing and the van der Corput lemma (see Section 1.3) eventually works [**GrTa2009e**], but does not scale well to bracket polynomials of higher degree such as $e(\lfloor\lfloor\sqrt{2}n\rfloor\sqrt{3}n\rfloor\sqrt{5}n)$, which would be necessary if one were to extend (3.1) to k beyond 4.

To resolve this, and inspired by the work in ergodic theory done by [**FuWe1996**], [**HoKr2005**], [**BeLe2007**], and others, we re-interpreted

---

[12]The same phenomenon is not visible at the linear level; a bracket linear phase such as $e(\lfloor\sqrt{2}n\rfloor\sqrt{3})$ can be rewritten as $e(\sqrt{6}n)e(-\sqrt{3}\{\sqrt{2}n\})$, which by Fourier series can be expressed as a linear combination of linear characters $e((\sqrt{6} + k\sqrt{2})n)$ for integer k. Note that the same trick does not work for $e(\lfloor\sqrt{2}n\rfloor\sqrt{3}n)$.

bracket polynomials from a more dynamical systems perspective. To motivate this, observe that the linear character $\chi(n) = e(\alpha n)$ is closely tied to the circle rotation $T : x \mapsto x + \alpha$ on the unit circle $\mathbf{R}/\mathbf{Z}$, in that the character χ can be described as a function $\chi(n) = F(T^n x_0)$ of an orbit $(T^n x_0)_{n \in \mathbf{Z}}$ on this system, where $x_0 = 0$ is the origin and $F(x) := e(x)$ is the exponential function. In a similar spirit, a quadratic character such as $\chi(n) = e(\alpha \frac{n(n-1)}{2})$ can be expressed in terms of the *skew shift system* $(x, y) \mapsto (x + \alpha, y + x)$ on the torus $(\mathbf{R}/\mathbf{Z})^2$, being of the form $\chi(n) = F(T^n x_0)$ where $x_0 := (0, 0)$ and $F(x, y) := e(y)$. More generally, one can also express bracket quadratic polynomials such as $e(\lfloor \sqrt{2}n \rfloor \sqrt{3}n)$ in the form $\chi(n) = F(T^n x_0)$, where T is now the action of a group element $x \mapsto gx$ on a 2-step nilmanifold G/Γ, and F is some reasonable (e.g. piecewise smooth) function on this nilmanifold. (See Section 2.16 or [**BeLe2007**], [**GrTa2009c**] for details.) The relevance of 2-step nilpotent groups and nilmanifolds to length 4 progressions can be glimpsed in the identity

$$(g^n x)(g^{n+r} x)^{-3}(g^{n+2r} x)^3 (g^{n+3r} x)^{-1} = 1,$$

which is valid for all g, x in a 2-step nilpotent group G (compare this with (3.11)); it is an instructive exercise to prove this identity and to see how the 2-step nilpotency is used.[13] Indeed, one can reformulate the inverse theorem for length 4 progressions in an equivalent form:

Theorem 3.2.5 (Inverse theorem for length 4 progressions, again informal). *Let $k = 4$. Suppose that A is a subset of $\{1, \ldots, N\}$ of density δ for which (3.3) fails. Then A correlates with a non-trivial 2-step nilsequence $\chi(n) = F(T^n x_0)$ for some 2-step nilmanifold G/Γ (of bounded "complexity"), some group rotation $T : x \mapsto gx$, some starting point $x \in G/\Gamma$, and some function $F : G/\Gamma \to \mathbf{C}$ (also of "bounded complexity"; e.g. bounded Lipschitz norm will do).*

The precise formulation of the theorem is a little technical; for details, see [**GrTa2009d**]. Using this theorem and all the standard machinery, the task of establishing asymptotics such as (3.1) in the $k = 4$ case now reduces to that of understanding sums such as

$$\sum_{n=1}^{N} F(T^n x_0).$$

At this point, one can start using the existing theory of equidistribution of orbits on homogeneous spaces G/Γ (of which nilmanifolds are an important example). It turns out that the existing theory is not quite quantitative enough for our purposes, and we had to develop a quantitative analogue

---

[13]To make the connection more precise, one needs a variant of this constraint in which x lies in G/Γ rather than G, and which is harder to state; see [**Zi2007**], [**GrTa2009d**] for details.

of this theory; see [**GrTa2009c**], [**GrTa2009f**] for more discussion. Anyway, it all works, and gives asymptotics for progressions of length 4 in the primes, as well as other linear patterns of similar "complexity" (e.g. any non-degenerate system of two equations in four prime unknowns is OK). To handle higher patterns, what we need is

Conjecture 3.2.6 (Inverse conjecture for arithmetic progressions; informal). *Let $k \geq 3$. Suppose that A is a subset of $\{1, \ldots, N\}$ of density δ for which (3.3) fails. Then A correlates with a non-trivial $k-2$-step nilsequence $\chi(n) = F(T^n x_0)$ for some $(k-2)$-step nilmanifold G/Γ (of bounded "complexity"), some group rotation $T : x \mapsto gx$, some starting point $x \in G/\Gamma$, and some function $F : G/\Gamma \to \mathbf{C}$ (also of "bounded complexity").*

This is a consequence of (and very closely related to) the *inverse conjecture for the Gowers norm*. It is already known for $k = 3$ and $k = 4$, and hopefully the higher k cases will be resolved in the near future, presumably using a mix of techniques from Fourier analysis, additive combinatorics, and ergodic theory. At the time of writing, this is the only remaining obstacle before we can understand the asymptotics of linear patterns in primes which genuinely involve two or more free parameters (as mentioned earlier, one-parameter problems such as the twin prime conjecture seem well out of reach of these methods for a number of reasons, one of which is that there is definitely no analogue of the inverse conjecture for such one-parameter patterns).

Notes. This talk first appeared at `terrytao.wordpress.com/2008/11/18`, and was given as the second talk in my series of four Marker Lectures in Penn State University in November of 2008.

3.3. Small gaps between primes

In this lecture, I would like to discuss the recent progress, particularly by Goldston, Pintz, and Yıldırım, on finding small gaps $p_{n+1} - p_n$ between consecutive primes. (See also the surveys [**GoPoYi2005**], [**Gr2006**], [**So2006**]; the material here is based to some extent on these prior surveys.)

The *twin prime conjecture* can be rephrased as the assertion that $p_{n+1} - p_n$ attains the value of 2 infinitely often, where $p_1 = 2, p_2 = 3, p_3 = 5, \ldots$ are the primes. As discussed in previous lectures, this conjecture remains out of reach at present, at least with the techniques centred around counting solutions to linear equations in primes. However, there is another direction to pursue towards the twin prime conjecture which has shown significant progress recently (though, again, there appears to be a significant difficulty in pushing it all the way to the full conjecture). This is to try to show that $p_{n+1} - p_n$ is unexpectedly small for many n. Let us make this a bit more quantitative by posing the following question:

Question 3.3.1. *Let N be a large number. What is the smallest value of $p_{n+1} - p_n$, where p_n is a prime between N and $2N$?*

The twin prime conjecture (or more precisely, the quantitative form of this conjecture coming from the prime tuples conjecture) would assert that the answer to this question is 2 for all sufficiently large N.

There are various ways to get upper bounds on this question. For instance, from *Bertrand's postulate* (which can be proven by elementary means) we know that $p_{n+1} - p_n = O(N)$ for all $N \le p_n \le 2N$. The *prime number theorem* asserts that $p_n = (1 + o(1))n \log n$, which gives $p_{n+1} - p_n = o(N)$; using various non-trivial facts known about the zeroes of the zeta function, one can improve this to $O(N^c)$ for various c (the best value of c known unconditionally is 0.525; see [**BaHaPi2001**]). The *Riemann hypothesis* gives a significantly more precise asymptotic formula for p_n, which ultimately leads to the bound $p_{n+1} - p_n = O(\sqrt{N} \log N)$. These bounds hold for all n in the given range, and so in fact bound the *largest* value of $p_{n+1} - p_n$, not just the smallest. As far as I know, the $O(\sqrt{N} \log N)$ bound for the largest prime gap has not been improved even after one assumes the Riemann hypothesis, though this gap is generally expected to be much smaller than this.[14] In the converse direction, the best result is due to Rankin [**Ra1962**], who proved the somewhat unusual bound

$$p_{n+1} - p_n \ge c \log N \frac{(\log \log N)(\log \log \log \log N)}{(\log \log \log N)^2}$$

for some n and some absolute constant $c > 0$ (in fact one can take c arbitrarily close to $2e^\gamma$). Remarkably, this type of right-hand side appears to be a genuine limit of what current methods can achieve (Paul Erdős in fact offered \$10,000 to anyone who could improve the rate of growth of the right-hand side in N).

But for the smallest value of $p_{n+1} - p_n$, much more is known. The prime number theorem already tells us that there are $(1 + o(1))N/\log N$ primes between N and $2N$, so from the *pigeonhole principle* we have

$$(3.12) \qquad\qquad p_{n+1} - p_n \le (1 + o(1)) \log N$$

for some n.

This bound should not be sharp, since this would imply that the primes are almost equally spaced by $\log N$, which is suspiciously regular behaviour

---

[14]In particular, the old conjecture that there always exists at least one prime between two consecutive square numbers remains open, even assuming RH. Cramér conjectured [**Cr1936**] a bound of $(1 + o(1)) \log^2 N$, though it is possible that the constant 1 here may need to be revised upwards to $2e^\gamma \approx 1.1229$, where γ is the Euler-Mascheroni constant; see [**Gr1995**] for details.

for a sequence as irregular as the primes. To get some intuition as to what to expect, we turn to random models of the primes. In particular, we begin with *Cramér's random model* [**Cr1936**] for the primes, which asserts that the primes between N and $2N$ behave as if each integer in this range had an independent chance of about $1/\log N$ of being prime. Standard probability theory then shows that the primes are distributed like a *Poisson process* of intensity $1/\log N$. In particular, if one takes a random interval I_λ in $[N, 2N]$ of length $\lambda \log N$ for some $\lambda > 0$, the number of primes $|I_\lambda \cap \mathcal{P}|$ that I_λ captures is expected to behave like a Poisson random variable of mean λ; in other words, we expect

$$(3.13) \qquad \mathbf{P}(|I_\lambda \cap \mathcal{P}| = k) \approx \frac{e^{-\lambda} \lambda^k}{k!}$$

for $k = 0, 1, 2, \dots$. (In contrast, the prime number theorem only gives the much weaker statement $\mathbf{E}|I_\lambda \cap \mathcal{P}| = \lambda + o(1)$.)

Now, as discussed in previous lectures, Cramér's random model is not a completely accurate model for the primes, because it does not reflect the fact that the primes very strongly favour the odd numbers, the numbers coprime to 3, and so forth. However, it turns out that even after one corrects for these local irregularities, the predicted Poisson random variable behaviour (3.13) does not change significantly for any fixed λ (e.g. a Poisson process of intensity $2/\log N$ on the odd numbers looks much the same as a Poisson process of intensity $1/\log N$ on the natural numbers, when viewed at scales comparable to $\log N$). This computation was worked out fully by Gallagher [**Ga1976**] as a rigorous consequence of the Hardy-Littlewood prime tuples conjecture.[15]

Applying (3.13) for small values of λ (but still independent of N), we see that intervals of length $\lambda \log N$ are still expected to contain two or more primes with non-zero probability, which in particular would imply that $p_{n+1} - p_n \leq \lambda \log N$ for at least one value of n. So one path to creating small gaps between primes is to show that $|I_\lambda \cap \mathcal{P}|$ can exceed 1 for as small a value of λ as one can manage.

One approach to this proceeds by controlling the *second moment*

$$(3.14) \qquad \mathbf{E}|I_\lambda \cap \mathcal{P}|^2;$$

the heuristic (3.13) predicts that this second moment should be $\lambda^2 + \lambda + o(1)$ (reflecting the fact that the Poisson distribution has both mean and variance equal to λ). On the other hand, the prime number theorem gives the first moment estimate $\mathbf{E}|I_\lambda \cap \mathcal{P}| = \lambda + o(1)$. Also, if $|I_\lambda \cap \mathcal{P}|$ never exceeds 1, then the first and second moments are equal. Thus if one could get

---

[15]On the other hand, these corrections to the Cramér model *do* disrupt (3.13) for very large values of λ; see [**So2007**] for more discussion.

the right bound for the second moment, one would be able to show that $p_{n+1} - p_n \le \lambda \log N$ is possible for arbitrarily small λ.

Second moments such as (3.14) are very amenable to tools from Fourier analysis or complex analysis; applying such tools, we soon see that (3.14) can be re-expressed easily in terms of zeroes of the Riemann zeta function, and one can use various standard facts (or hypotheses) about these zeroes to gain enough control on (3.14) to obtain non-trivial improvements to (3.12). This approach was pursued by many authors [**Ra1937**], [**Er1940**], [**BoDa1966**], leading to non-trivial unconditional results for any $\lambda \ge 1/2$, but it seems difficult to push the method much beyond this. It was later shown in [**GoMo1987**] that the correct asymptotic for (3.14) is essentially equivalent to the Riemann hypothesis combined with a certain statement on pair correlations between zeroes, and thus well out of reach of current technology.

Another method, introduced by Maier [**Ma1985**], is based on finding some (rare) intervals I_λ of numbers of length $\lambda \log N$ for some moderately large λ which contain significantly more primes than average value of λ; if for instance one can find such an interval with over $(k+1)\lambda$ primes in it, then from the pigeonhole principle one must be able to find a prime gap of size at most $\frac{1}{k} \log N$. The ability to do this stems from the remarkable and unintuitive fact that the regularly distributed nature of primes in long arithmetic progressions, together with the tendency of primes to avoid certain residue classes, forces the primes to be *irregularly* distributed in *short* intervals. This phenomenon (which has now been systematically studied as an "uncertainty principle" for equidistribution [**GrSo2007**]), is related to the following curious failure of naive probabilistic heuristics to correctly predict the prime number theorem. Indeed, consider the question of asking how likely it is that a randomly chosen number n between N and $2N$ is to be prime. Well, n will have about a $1 - \frac{1}{2}$ chance of being coprime to 2, a $1 - \frac{1}{3}$ chance of being coprime to 3, and so forth; the *Chinese remainder theorem* also suggests that these events behave independently. Thus one might expect that the probability would be something like

$$\prod_{p<N} \left(1 - \frac{1}{p}\right).$$

We then invoke *Mertens' theorem*, which provides the asymptotic

(3.15)
$$\prod_{p<N} \left(1 - \frac{1}{p}\right) = (e^{-\gamma} + o(1)) \frac{1}{\log N}.$$

But this is off by a factor of e^γ from what the prime number theorem says the true probability of being prime is, which is $(1 + o(1))\frac{1}{\log N}$. This discrepancy reflects the difficulty in cutting off the product in primes (3.15) at the right place (for instance, the *sieve of Eratosthenes* suggests that one might want to cut off at $\sqrt{N}$ instead). At any rate, this e^γ discrepancy can be exploited to find intervals with an above-average number of primes by a "first moment" argument (known as the *Maier matrix method*) that we sketch as follows. Let w be a moderately large number, and let W be the product of all the primes less than w. If we pick a random number n between N and $2N$, then as mentioned before, the prime number theorem says that this number will be prime with probability about $\frac{1}{\log N}$. But if in addition we know that the number is coprime to W, then by the prime number theorem in arithmetic progressions this information boosts the probability of being prime to about $\prod_{p<w}(1-\frac{1}{p})^{-1}\frac{1}{\log N}$, which by (3.15) is about $e^\gamma \frac{\log w}{\log N}$.

Now we restrict attention to numbers n which are equal to $a \bmod W$ for some $1 \le a \le w$. By the prime number theorem, about $\frac{w}{\log w}$ of the a are prime and thus coprime to W. Combining this with the previous discussion, we see that the total probability that such a number n is prime is about $\frac{1}{\log w} \times e^\gamma \frac{\log w}{\log N} = e^\gamma \frac{1}{\log N}$.

On the other hand, the set of numbers n which are equal to $a \bmod W$ for some $1 \le a \le w$ is given by a sequence of intervals of length w. By the pigeonhole principle, we must therefore have an interval of length w on which the density of primes is at least $e^\gamma \frac{1}{\log N}$, which is greater than the expected density by a factor of e^γ.

One can make the above arguments rigorous for certain ranges of interval length, and by combining this with the pigeonhole principle one can eventually improve (3.12) by a factor of e^γ:

$$p_{n+1} - p_n \le (e^{-\gamma} + o(1)) \log N.$$

This is better than the bound of $(\frac{1}{2} + o(1)) \log N$ obtained by the second moment method, but on the other hand the latter method establishes that a positive proportion of primes have small gaps; Maier's method, by its very nature, is restricted to a very sparse set of primes (note that w is much smaller than W).

In a series of papers, Goldston and Yıldırım improved the numerical constants in these results by a variety of methods including those mentioned above, as well as replacing some of the reliance on information on zeroes of the zeta function with tools from sieve theory instead. To oversimplify substantially, the latter idea is to try to control the set of primes $\mathcal{P}$ in terms of a larger set $\mathcal{AP}$ of *almost primes*—numbers with few prime factors.[16] For instance, to control the second moment $\mathbf{E}|I_\lambda \cap \mathcal{P}|^2$, one can take advantage

---

[16]Strictly speaking, one does not actually work with a set of almost primes, but rather with a weight function or sieve which is large on almost primes and small for non-almost primes; however, we ignore this important technical detail to simplify the exposition.

of the Cauchy-Schwarz inequality

$$\mathbf{E}|I_\lambda \cap \mathcal{P}|^2 \geq \frac{\mathbf{E}|I_\lambda \cap \mathcal{P}||I_\lambda \cap \mathcal{AP}|}{|I_\lambda \cap \mathcal{AP}|^2}.$$

The denominator on the right-hand side involves only almost primes and can be computed easily by sieve theory methods. The numerator involves primes, but only one prime at a time; note that this quantity is roughly counting the set of pairs p, q where p is prime, q is almost prime, and p and q differ by at most $\lambda \log N$. This is in contrast to the left-hand side, which is counting pairs p, q that are both prime and differ by at most $\lambda \log N$. We do not know how to use sieve theory to count the latter type of pattern (involving more than one prime); but sieve theory is perfectly capable of counting the former type of pattern, so long as we understand the distribution of primes in relatively sparse arithmetic progressions. The standard tool for this is the *Bombieri-Vinogradov theorem* [**Bo1987**], which roughly speaking asserts that the primes from N to $2N$ are well distributed in "most" residue classes q, as long as q stays significantly smaller than $\sqrt{N}$; it can be viewed as an averaged version of the *generalised Riemann hypothesis* that can be proven unconditionally.[17] Using such tools, various improvements to (3.12) were established. Finally, in [**GoPoYi2005a**] it was shown that

$$(3.16) \qquad\qquad p_{n+1} - p_n \leq \lambda \log N$$

held for some n and any $\lambda > 0$ (provided N was large enough depending on λ), or equivalently that $p_{n+1} - p_n = o(\log N)$; this was later improved in [**GoPoYi2007**] to $p_{n+1} - p_n = O(\sqrt{\log N}(\log \log N)^2)$. Assuming a strong version of the Bombieri-Vinogradov theorem (in which q is allowed now to get close to N rather than to $\sqrt{N}$), known as the *Elliott-Halberstam conjecture*, this was improved further to the striking result

$$p_{n+1} - p_n \leq 16,$$

thus there are infinitely many pairs of primes which differ by at most 16. This is a remarkable "near miss" to the twin prime conjecture, though it seems clear that substantial new ideas would be needed to reduce 16 all the way to 2.

Let us now discuss some of the ideas involved. As with the previous arguments, the key idea is to find groups of integers which tend to contain more primes than average. Suppose for instance one could find a certain random

---

[17]The key point here is that, while it is possible for the primes to be irregular with respect to a few such small moduli, the "orthogonality" of these moduli with respect to each other makes it impossible for the primes to be simultaneously irregular with respect to many of these moduli at once.

distribution of integers n where[18] n, $n+2$, and $n+6$ each had a probability strictly greater than $1/3$ of being prime. By linearity of expectation, we thus see that the expected number of primes in the set $\{n, n+2, n+6\}$ exceeds 1; thus, with positive probability, there will be at least two primes in this set, which then necessarily differ by at most 6.

Now, of course, the prime number theorem tells us that for n chosen uniformly at random from N to $2N$, the probability that n, $n+2$, or $n+6$ are prime is only about $1/\log N$. So for this type of strategy to work, one would have to pick a highly non-uniform distribution for n, in which n, $n+2$, and $n+6$ are already close to being prime. The extreme choice would be to pick n uniformly among all choices for which n, $n+2$, and $n+6$ are simultaneously prime, but we of course do not even know that any such primes exist (this is strictly harder than the twin prime conjecture!). But what we can do instead is pick n uniformly among all choices such that n, $n+2$, and $n+6$ are *almost prime* (where we shall be vague for now about what "almost prime" means). Thanks to sieve theory, we *can* assert the existence of many numbers n of this form, and get a good count as to how many there are. Also, since the primes have positive density inside the almost primes, it is quite reasonable that the conditional probabilities

$$\mathbf{P}(n \text{ prime}|n, n+2, n+6 \text{ almost prime}),$$

$$\mathbf{P}(n+2 \text{ prime}|n, n+2, n+6 \text{ almost prime}),$$

$$\mathbf{P}(n+6 \text{ prime}|n, n+2, n+6 \text{ almost prime})$$

are large. Indeed, using sieve theory techniques (and the Bombieri-Vinogradov theorem), we can bound each of these probabilities from below by a positive constant (plus a $o(1)$ error). Unfortunately, even if we optimise the sieve that produces the almost primes, this constant is too small (typically one gets numbers of the order of $1/20$ or so, rather than $1/3$). Assuming the Elliot-Halberstam conjecture (which allows us to raise the level of sieving substantially) yields a significant improvement, but one that still falls short of the desired goal. One can of course also add more numbers to the mix than just $n, n+2, n+6$, e.g. looking at those n for which $n+h_1, \ldots, n+h_k$ are simultaneously almost prime, for some suitably chosen $h_1, \ldots, h_k$; on the one hand, this lowers the threshold of probability (currently at $1/3$) that one needs to obtain, but unfortunately this is more than canceled out by the multidimensional sieving one needs to do when restricting all of these numbers to be almost prime.

To get around this, a new idea was introduced: instead of requiring numbers such as n, $n+2$, and $n+6$ to separately be almost prime, ask for

---

[18]We choose these separations for our discussion because it is not possible to make three large prime numbers bunch up any closer than this; for instance, n, $n+2$, $n+4$ cannot be all be prime for $n > 3$, since at least one of these numbers must be divisible by 3.

the *product* $n(n+2)(n+6)$ to be almost prime (for a somewhat more relaxed notion of "almost prime"). This turns out to be more efficient, as it lowers the number of summations involved in the sieve. One has to carefully select how one defines almost prime here (it is roughly like asking for the product $(n + h_1) \ldots (n + h_k)$ to have at most $2k + o(k)$ prime factors, with the $o(k)$ factor being remarkably crucial to the delicate analysis); but to cut a long story short, one can establish probability bounds of the form[19]

$$(3.17) \quad \mathbf{P}(n + h_j \text{ prime}|(n + h_1) \ldots (n + h_k) \text{ almost prime}) \geq \frac{c - o(1)}{k}$$

for all $1 \leq j \leq k$ and some absolute constant $c > 0$.

As soon as one assumes any non-trivial portion of the Elliott-Halberstam conjecture, the quantity c in the above inequality can be made to exceed 1 (for k large enough), leading to the conclusion that there exist infinitely many bounded prime gaps, $p_{n+1} - p_n = O(1)$; pushing the machinery to the limit (taking $\{h_1, \ldots, h_k\} = \{7, 11, 13, 17, 19, 23\}$ to be the first six primes larger than 6), one obtains the bound of 16. But without this conjecture, and just by using the Bombieri-Vinogradov theorem, one can, after optimising everything in sight, get c arbitrarily close to 1, but not quite exceeding 1. To compensate for this, the authors also started looking at the nearby numbers $n + h$ where h was not equal to $h_1, \ldots, h_k$. Here, of course, there is no particularly good reason for $n + h$ to be prime, since it is not involved as a factor in the almost prime quantity $(n + h_1) \ldots (n + h_k)$; but one can show that, for generic values of h, one has

$$\mathbf{P}(n + h \text{ prime}|(n + h_1) \ldots (n + h_k) \text{ almost prime}) \geq \frac{c' + o(1)}{\log N}$$

for some $c' > 0$ (there is an additional singular series factor involving the prime factors of $h - h_j$ that can be easily dealt with, which I am suppressing here). Thus, for any $\lambda > 0$, we expect (from linearity of expectation) that for $(n+h_1) \ldots (n+h_k)$ almost prime, the expected number of $h = O(\lambda \log N)$ (including the k values $h_1, \ldots, h_k$) for which $n + h$ is prime is at least

$$k\left(\frac{c}{k} + o(1)\right) + \lambda \log N \frac{c' + o(1)}{\log N} = c + c'\lambda + o(1).$$

---

[19]More precisely, one needs to compute sums such as

$$\sum_{n=1}^{N} \Lambda(n + h_j)\Lambda_R((n + h_1) \ldots (n + h_k))^2,$$

where Λ is the *von Mangoldt function* and Λ_R is a *Selberg sieve*-type approximation to that function.

Since we can make c arbitrarily close to 1, the extra term $c'\lambda$ can push this expectation to exceed 1 for any choice of λ, and this leads to the desired bound[20] (3.16) for any $\lambda > 0$.

It is tempting to continue to optimise these methods to improve the various constants, and I would imagine that the bound of 16, in particular, can be lowered somewhat (still assuming the Elliott-Halberstam conjecture). But there seems to be a significant obstacle to pushing things all the way to 2. Indeed, the parity problem (see Section 3.10 of *Structure and Randomness*) tells us that for any reasonable definition of "almost prime" which is amenable to sieve theory, the primes themselves can have density at most $1/2$ in these almost primes. Since we need the density to exceed $1/k$ in order for the above argument to work, it is necessary to play with at least three numbers (e.g. n, $n+2$, $n+6$), which forces the bound on the prime gaps to be at least[21] 6. But it may be that a combination of these techniques with some substantially new ideas may push things even further.

Notes. This talk first appeared at `terrytao.wordpress.com/2008/11/19`, and was given as the third talk in my series of four Marker Lectures in Penn State University in November of 2008.

Emmanuel Kowalski pointed out that Gallagher's conditional argument [**Ga1976**] can in fact be extended to give Poisson-type statistics for (say) twin primes in intervals of size $O(\log^2 N)$ in $[N, 2N]$, and also mentioned his Bourbaki exposé [**Kow2006**] on the above work.

3.4. Sieving for almost primes and expanders

In this final lecture, I discuss the recent work of Bourgain, Gamburd, and Sarnak on how *arithmetic combinatorics* and *expander graphs* were used to sieve for *almost primes* in various *algebraic sets*.

In previous lectures, we considered the problem of detecting tuples of primes in various linear or convex sets; in particular, we considered the size of sets of the form $V \cap \mathcal{P}^k$, where $\mathcal{P} = \{2, 3, 5, \ldots\}$ is the set of primes, and V is some *affine subspace* of $\mathbf{R}^k$. For instance, the twin prime conjecture would correspond to the case when $k = 2$ and $V = \{(x, x+2) : x \in \mathbf{R}\}$, while Theorem 3.1.1 would correspond to the case $V = \{(x, x+r, \ldots, x+(k-1)r) : x, r \in \mathbf{R}\}$.

---

[20]The subsequent improvement to (3.16) proceeds by enlarging k substantially, and by a preliminary sieving of the small primes, but we will not discuss these technical details here.

[21]Indeed, this bound has been obtained for *semiprimes* (products of two primes), which are subject to the same parity problem restriction as primes, but are slightly better distributed; see [**GrRoSp1980**].

We refer to elements of $\mathcal{P}^k$ as *prime points*. The *prime tuples conjecture* [**HaLi1923**] implies the following qualitative criterion for when such a set of prime points should be "large":

Conjecture 3.4.1 (Qualitative prime tuples conjecture). *Let V be an affine subspace of $\mathbf{R}^k$. Suppose that*

(1) *(No obstructions at infinity) For any N, $V \cap \mathbf{Z}_{>N}^k$ affinely spans all of V, where $\mathbf{Z}_{>N} := \{n \in \mathbf{Z} : n > N\}$. (In particular, $V \cap \mathbf{Z}_{>N}^k$ is non-empty.)*

(2) *(No obstructions at q) For any $q > 1$, $V \cap (\mathbf{Z}_q^*)^k$ affinely spans all of V, where $\mathbf{Z}_q^* := \{n \in \mathbf{Z} : (n, q) = 1\}$. (In particular, $V \cap (\mathbf{Z}_q^*)^k$ is non-empty.)*

Then $V \cap \mathcal{P}^k$ affinely spans all of V. (In particular, V contains at least one prime point.)

Both of the hypotheses in this conjecture are easily verified for any given V, the first by (integer) linear programming and the second by modular arithmetic. This conjecture would imply several other results and conjectures in number theory, including the twin prime conjecture and Theorem 3.1.1. Needless to say, it remains open in general (though the results mentioned in the previous lecture give partial results in the case when V is at least two-dimensional and non-degenerate).

Now we attempt to generalise the above conjecture to the setting in which V is an *algebraic variety* rather than an affine subspace. (This would cover some famous open problems in number theory, for instance the *Landau problem*, which asks whether there are infinitely many primes of the form $n^2 + 1$.) The notion of a set affinely spanning V is then naturally replaced by the notion of a set being *Zariski dense* in V, which means that the set is not contained in any strictly smaller subvariety of V. One could then formulate a naive generalisation of the above conjecture by replacing "affine space" and "affinely spans all of" with "algebraic variety" and "is Zariski dense in" respectively. However, the hypotheses are no longer easy to verify; indeed, just the problem of determining whether V contains an integer point from $\mathbf{Z}^k$ is essentially *Hilbert's tenth problem*, which by *Matiyasevich's theorem* [**Ma1970**] is known to be undecidable[22] for general V. Indeed, since one can encode any computable set in terms of the integer points of a variety V, it is not too difficult to see that this conjecture fails in general.

Since arbitrary algebraic varieties are far too general to have any hope of a reasonable theory, one should look for prime points in much more special

---

[22]An amusing historical connection here: one of the first demonstrations [**DaPuRo1961**] of the undecidability of Hilbert's tenth problem was conditional on the existence of arbitrarily long progressions of primes (i.e., Theorem 3.1.1), although subsequent proofs did not need this fact.

sets. An important class[23] here is that of an orbit Λb in $\mathbf{Z}^k$, where b is some vector in $\mathbf{Z}^k$ and Λ is some finitely generated subgroup of $SL_k(\mathbf{Z})$. Of course one should take b to be primitive (not a multiple of any smaller vector), since one clearly will have a difficult time finding prime points in Λb otherwise.

The orbit Λb will be Zariski dense in some algebraic variety V, and is clearly a collection of integer points (though it may not cover all of $V \cap \mathbf{Z}^k$). Assuming no local obstructions at infinity or at q (which means that $\Lambda b \cap \mathbf{Z}^k_{>N}$ and $\Lambda b \cap (\mathbf{Z}^*_q)^k$ are Zariski dense in V), one could then conjecture that $\Lambda b \cap \mathbf{P}^k$ is also Zariski dense in V (which, if V is infinite, would in particular imply that the orbit Λb contains infinitely many prime points).

For simplicity let us restrict attention to the two-dimensional case $k = 2$, which is already highly non-trivial; Bourgain, Gamburd and Sarnak have recently begun to get some preliminary results in $k = 3$, but I will not discuss them here. Thus Λ is now a finitely generated subgroup of $SL_2(\mathbf{Z})$. If this subgroup is *elementary*—e.g. if it is cyclic—then the orbit Λb can be exponentially sparse (a ball of radius R may only contain $O(\log R)$ points), and it becomes extremely difficult to do any sieving or primality detection.[24] It thus makes sense to restrict attention to non-elementary subgroups of $SL_2(\mathbf{Z})$, that is, groups which contain a copy of the free non-abelian group on two generators, or equivalently any group whose Zariski closure is all of SL_2 (or equivalently yet again, a group whose limit set consists of more than one point). In this situation, Bourgain, Gamburd, and Sarnak [**BoGaSa2006**] conjectured:

Conjecture 3.4.2. *Let Λ be a non-elementary subgroup of $SL_2(\mathbf{Z})$, and let b be a primitive element of $\mathbf{Z}^2$. Suppose that there are no local obstructions at infinity or at finite places q. Then $\Lambda b \cap \mathcal{P}^2$ is Zariski dense in the plane (in particular, $\Lambda b \cap \mathcal{P}^2$ is infinite).*

This conjecture remains open. However, as in the linear situation, one can make progress[25] if one replaces primes with *almost primes*—products of at most r primes for some bounded r. In particular, Bourgain, Gamburd, and Sarnak [**BoGaSa2006**] were able to prove the following.

Theorem 3.4.3. *Let Λ, b be as in Conjecture 3.4.2. Then there exists an r such that $\Lambda b \cap \mathcal{P}_r^2$ is Zariski dense in the plane, where $\mathcal{P}_r$ is the set of numbers that are the product of at most r primes.*

---

[23]One can also consider the slightly more general set of images $F(\Lambda b)$ under a polynomial map, but for simplicity let us stick to just orbits.

[24]In this case, the problem becomes comparable to such notoriously difficult questions as whether there are infinitely many Mersenne primes.

[25]There are certainly prior results for non-linear patterns in the almost primes; for instance, it is a famous result of Iwaniec [**Iw1978**] that there are infinitely many numbers of the form $n^2 + 1$ that are the product of at most two primes.

Several further generalisations and extensions of this result, with a similar flavour, are known, but will not be discussed here. There are a number of amusing special cases of these results, for instance one can show that there exist infinitely many *Apollonian circle packings* of the unit circle by four other mutually tangent circles, all of whose radii are the reciprocals of an almost prime, or infinitely many *Pythagorean triples* whose area is an almost prime (for a sufficiently large r in the definition of "almost prime").

Let me now discuss some of the key ideas in the proof of this theorem. One begins by rephrasing the question in a more quantitative (or finitary) manner. In the linear case, this would be done by counting the number of points in $\Lambda b \cap \mathcal{P}_r^2$ that lie inside some large Euclidean ball, thus using the Euclidean (or Archimedean) notion of distance to localise the problem. This can also be done here, but it turns out to be more convenient to instead use the *word metric* induced by the finite generating set S of Λ (which we can take to be symmetric for convenience, thus $S = S^{-1}$). One thus looks at sets of the form $B_R b \cap \mathcal{P}_r^2$, where $B_R \subset \Lambda$ consists of all words formed by products of at most R elements of S. A major new difficulty here compared to the linear theory is the exponential growth[26] of B_R (a consequence of the non-elementary nature of Λ).

The next step is to use sieve theory. Recall the *sieve of Eratosthenes*, which expresses the set of all (large) primes as the integers, minus the multiples of two, minus the multiples of three, and so forth. Using the *inclusion-exclusion principle*, we can thus view the indicator function $1_{\mathcal{P}}$ of the primes, when restricted to an interval such as $[N, 2N]$, as equal to 1, minus the indicator function $1_{2\mathbf{Z}}$ of the even numbers, minus the indicator function $1_{3\mathbf{Z}}$ of the multiples of three, plus the indicator function $1_{6\mathbf{Z}}$ of the multiples of six, and so forth. This leads to the *Legendre sieve*

$$(3.18) \qquad\qquad 1_{\mathcal{P}} = \sum_d \mu(d) 1_{d\mathbf{Z}},$$

valid in an interval $[N, 2N]$ as long as one restricts d to those integers which are products of primes less than N. Here $\mu(d)$ is the *Möbius function*.

The basic idea of sieve theory is to replace the indicator function of the primes (or almost primes) by a more general divisor sum

$$\sum_d c_d 1_{d\mathbf{Z}},$$

where the sieve weights c_d are chosen in order to optimise the final bounds in the sieve (they typically resemble "smoothed out" versions of the Möbius function in order that these sieves be large on the almost primes and small

---

[26]Though it is not immediately apparent, the same problem also arises if one uses Euclidean balls instead of word metric balls, due to the multiplicative rather than additive nature of the group Λ.

elsewhere). For the sieve to be practical, one wants to restrict d in this sum to be relatively small, for instance $d \leq N^\theta$ for some absolute constant $0 < \theta < 1$ (values such as $\theta = 1/4$ are fairly typical). The selection of the sieve weights c_d is now a well-developed science (see Section 3.10 of *Structure and Randomness* for further discussion), and Bourgain, Gamburd and Sarnak basically use off-the-shelf sieves (in particular, combinatorial sieves and the Selberg sieve) in their work. Inserting these standard sieves into the problem at hand, the task of counting almost primes in the finite set $B_R b$ then quickly reduces to the question of getting good estimates on sets such as $B_R b \cap (q\mathbf{Z})^2$ for various q. This amounts to much the same thing as asking for good equidistribution bounds for B_R modulo q, thus we project the generating set S, and the ball B_R it produces, from $SL_2(\mathbf{Z})$ to $SL_2(\mathbf{Z}_q)$. For sieving purposes it turns out to be necessary to consider all square-free moduli q, but for simplicity we shall only discuss the (massively easier) case when q is prime.

The reduction to an equidistribution problem converts the original sieving problem to a more combinatorial one, involving the *Cayley graph* G on $SL_2(\mathbf{Z}_q)$ induced by the set S, thus two vertices $x, y \in SL_2(\mathbf{Z}_q)$ are connected by an edge in G if yx^{-1} lie in S (modulo q). The image of the ball B_R in $SL_2(\mathbf{Z}_q)$ is then the set of points one can reach in the graph G from the origin by walking on a path of length at most R. The desired equidistribution result one needs can then be viewed as a mixing result for the random walk along the graph G.

Standard graph theory tells us that the task reduces to showing that the graphs G form a family of *expander graphs* as $q \to \infty$ (recall that we are restricting q to be prime for simplicity). There are many equivalent definitions of what an expander graph is, but let us give a spectral definition that is specialised to Cayley graphs. The symmetric generating set S induces a natural measure

$$\mu := \frac{1}{|S|} \sum_{s \in S} \delta_s$$

that is the uniform distribution on S, which controls the random walk along G; note that if $f : SL_2(\mathbf{Z}_q) \to \mathbf{C}$ is a function, then the convolution $f * \mu : SL_2(\mathbf{Z}_q) \to \mathbf{C}$ is another function, whose value at any vertex is the average value of f at all the neighbours of x. The operation $f \mapsto f * \mu$ is then a self-adjoint contraction on $l^2(SL_2(\mathbf{Z}_q))$ which leaves the function 1 invariant, so its largest eigenvalue λ_1 is equal to 1. The expander graph condition is then equivalent to the existence of a *spectral gap* $\lambda_2 \leq 1 - c$ for the second largest eigenvalue, where $c > 0$ is a constant independent of q.

Of course, to have a spectral gap, one necessary condition is that λ_2 be strictly less than 1. This can be easily seen to be equivalent to the statement that G is connected, which in turn is equivalent to the statement that the projection of S to $SL_2(\mathbf{Z}_q)$ generates all of $SL_2(\mathbf{Z}_q)$. This statement can be verified to be true, either by direct consideration of all possible subgroups of $SL_2(\mathbf{Z}_q)$, or by the *strong approximation property*. However, mere connectedness is not enough to ensure that a Cayley graph is an expander family (which can be viewed as a sort of "robust" version of connectedness, which can survive the deletion of large numbers of edges). For instance, the Cayley graph of the generating set $\{-1, +1\}$ in $\mathbf{Z}/N\mathbf{Z}$ is connected, but does not form an expander family as $N \to \infty$; the second largest eigenvalue[27] is about $1 - O(1/N^2)$ only.

Obtaining the spectral gap property requires more work. When the original subgroup Λ of $SL_2(\mathbf{Z})$ is as large as a finite index subgroup (in particular, if it is a *congruence subgroup*), this gap follows from a celebrated theorem of Selberg [**Se1965**] providing a similar spectral gap for arithmetic quotients of the upper half-plane. Smaller examples (in which the index is now infinite) were first constructed in [**Sh1997**], [**Ga2002**], with the latter following the method of [**SaXu1991**]. Then in [**BoGa2008**], this method was extended using additional tools from additive combinatorics to handle all non-elementary subgroups in the case of prime q.

Let us now describe the method of proof. As mentioned briefly earlier, the existence of a spectral gap implies a strong mixing property: the iterated convolutions $\mu^{(n)} := \mu * \ldots * \mu$ of the probability measure μ (which can be interpreted as the probability distribution of a random walk on n steps) converges exponentially fast to the constant distribution on $SL_2(\mathbf{Z}_q)$. Since the latter distribution has an l^2 norm of $O(q^{-3/2})$, we see in particular that for any fixed $\varepsilon > 0$, we will have

$$(3.19) \qquad \|\mu^{(n)}\|_{l^2(SL_2(\mathbf{Z}_q))} = O(q^{-3/2+\varepsilon})$$

once n is a sufficiently large multiple of $\log q$. This can also be seen explicitly from the trace formula

$$(3.20) \qquad \|\mu^{(n)}\|^2_{l^2(SL_2(\mathbf{Z}_q))} = \frac{1}{|SL_2(\mathbf{Z}_q)|} \sum_j \lambda_j^n.$$

In general, this implication between spectral gap and rapid mixing (3.19) cannot be reversed; the problem is that λ_2 only directly influences one term in the summation on the right-hand side of (3.20), and so upper bounds on the left-hand side do not translate efficiently to upper bounds on λ_2. However, there is an algebraic miracle, which happens in the case of groups such as $SL_2(\mathbf{Z}_q)$, that allows one to reverse the implication.

---

[27]One can also see that the random walk on this Cayley graph takes a long time (about $O(N^2)$ steps) before it mixes to be close to the uniform distribution; with expander graphs on a set of N vertices, mixing instead occurs in time $O(\log N)$, thanks to the spectral gap.

Lemma 3.4.4 (Frobenius lemma). *Let q be prime. Then every non-trivial finite-dimensional unitary representation of $SL_2(\mathbf{Z}_q)$ has dimension at least $(q-1)/2$.*

Proof. Observe that $SL_2(\mathbf{Z}_q)$ can be generated by parabolic elements, so given a non-trivial representation $\rho : SL_2(\mathbf{Z}_q) \to U(V)$, there exists a parabolic element a whose representation $\rho(a)$ is non-trivial. By a change of basis we may take

$$a = \begin{pmatrix} 1 & 1 \\ 0 & 1 \end{pmatrix}.$$

On the one hand, we have $a^q = 1$ and hence $\rho(a)^q = 1$; thus all eigenvalues of $\rho(a)$ are q^{th} roots of unity. On the other hand, $\rho(a)$ is non-trivial, so at least one of the eigenvalues of $\rho(a)$ differs from 1. Thirdly, conjugating a by diagonal matrices in $SL_2(\mathbf{Z}_q)$, we see that a is conjugate to a^m whenever m is a quadratic residue mod q, and so the eigenvalues of $\rho(a)$ must be stable under the operation of taking m^{th} powers. On the other hand, there are $(q-1)/2$ quadratic residues. Putting all this together we see that $\rho(a)$ must take at least $(q-1)/2$ distinct eigenvalues, and the claim follows. $\square$

Remark 3.4.5. For our purposes, the exact value of $(q-1)/2$ is irrelevant; any multiplicity which grows like a power of q would suffice.

Applying this lemma to the eigenspace of λ_2, we obtain

Corollary 3.4.6. *The second eigenvalue λ_2 of the operation $f \mapsto f * \mu$ appears with multiplicity at least $(q-1)/2$.*

Combining this corollary with (3.20), one can now reverse the previous implication and obtain a spectral gap $\lambda_2 \leq 1-c$ as soon as one gets a mixing estimate (3.19) for some $n = O(\log q)$ and some sufficiently small ε.

The task is now to obtain the mixing estimate (3.19). The quantity $\|\mu^{(n)}\|_{l^2(SL_2(\mathbf{Z}_q))}$ starts at 1 when $n = 0$ and decreases with n. If we assume (as we may) that S generates a free group, then it is not hard to see that $\mu^{(n)}$ expands rapidly for $n \ll \log q$ (because all the words generated by S will be distinct until one encounters the "wrap-around" effect of taking residues modulo q). Using this one can get a preliminary mixing bound

$$\|\mu^{(n)}\|_{l^2(SL_2(\mathbf{Z}_q))} \leq q^{-\delta}$$

for some absolute constant $\delta > 0$ and some $n = O(\log q)$. Also, since S modulo q generates all of $SL_2(\mathbf{Z}_q)$, we know that the probability measure $\mu^{(n)}$ is not trapped inside any proper subgroup H of $SL_2(\mathbf{Z}_q)$; indeed, using the classification of subgroups of $SL_2(\mathbf{Z}_q)$ (or some general "escape from subvarieties" machinery of [**EsMoOh2005**]) one can show that

$$\mu^{(n)}(H) \leq q^{-\delta}$$

for any such subgroup, and some $n = O(\log q)$. The result now follows from iterating the following lemma, which is the heart of the argument:

Lemma 3.4.7 (l^2 flattening lemma). *Let ν be a symmetric probability measure on $SL_2(\mathbf{Z}_q)$ which is a little bit dispersed in the sense that*

$$\|\nu\|_{l^2(SL_2(\mathbf{Z}_q))} \le q^{-\delta}$$

for some $\delta > 0$, and is not concentrated in a subvariety in the sense that $\nu * \nu(H) \le q^{-\delta}$ for any proper subgroup H of $SL_2(\mathbf{Z}_q)$. Suppose also that ν is not entirely flat in the sense that

$$\|\nu\|_{l^2(SL_2(\mathbf{Z}_q))} \ge q^{-3/2+\delta}$$

(note that the minimal l^2 norm for a probability measure is comparable to $q^{-3/2}$, attained for the uniform distribution). Then $\nu * \nu$ is "flatter" than ν in the sense that

$$\|\nu * \nu\|_{l^2(SL_2(\mathbf{Z}_q))} \le q^{-\varepsilon}\|\nu\|_{l^2(SL_2(\mathbf{Z}_q))}$$

for some $\varepsilon > 0$ depending on δ.

In the special case when ν is the uniform distribution on some set A, the flattening lemma is very close to the following theorem of Helfgott [**He2008**]:

Theorem 3.4.8 (Product theorem). *Let q be a prime. Let A be a subset of $SL_2(\mathbf{Z}_q)$ which is not too large in the sense that $|A| \le q^{3-\delta}$ for some $\delta > 0$, and which is not contained in any proper subgroup H of $SL_2(\mathbf{Z}_q)$. Then $|A \cdot A \cdot A| \ge |A|^{1+\varepsilon}$ for some $\varepsilon > 0$ depending on δ.*

Indeed, by using some standard additive combinatorics, in particular (a non-commutative version of) the Balog-Szemerédi-Gowers lemma (which can be found for instance in [**TaVu2006**]), which connects "statistical" multiplication, such as that provided by convolution $\mu, \nu \mapsto \mu * \nu$, with "combinatorial" multiplication, coming from the product set operation $A, B \mapsto A \cdot B$, one can show that these two statements are in fact equivalent to each other.

The product theorem is a manifestation of certain "non-linear" or "non-commutative" behaviour in the group $SL_2(\mathbf{Z}_p)$; see Section 2.3 of *Structure and Randomness* for a bit more discussion on this. For now, let me just say that Helfgott's proof of this uses a variety of algebraic and combinatorial computations exploiting the special structure of $SL_2(\mathbf{Z}_p)$ (especially how commutativity or non-commutativity of various elements in this group interact with the trace of various combinations of these elements), as well as the following sum-product estimate [**BoKaTa2004**], [**BoKo2003**]:

Theorem 3.4.9 (Sum-product theorem). *Let q be prime. Let A be a subset of $\mathbf{Z}_q$ which is not too large in the sense that $|A| \le q^{1-\delta}$ for some $\delta > 0$. Then $|A + A| + |A \cdot A| \ge |A|^{1+\varepsilon}$ for some $\varepsilon > 0$ depending only on δ.*

There are now some quite elementary proofs of this theorem, but I will not discuss them here (see e.g. [**Ta2008c**] for further discussion). I should note, though, that the bulk of the Bourgain-Gamburd-Sarnak work is pre-occupied with establishing a suitable extension of this sum-product theorem to the case when q is not prime, in a manner which is uniform in the number of prime factors; this turns out to be a surprisingly difficult task.

Notes. This talk first appeared at `terrytao.wordpress.com/2008/11/20`, and was given as the final talk in my series of four Marker Lectures in Penn State University in November of 2008. Thanks to Luca Trevisan and Jonathan vos Post for corrections.

Bibliography

[Aa1997] J. Aaronson, *An introduction to infinite ergodic theory*, Mathematical Surveys and Monographs 50, American Mathematical Society, Providence, RI, 1997.

[AgKaSa2004] M. Agrawal, N. Kayal, N. Saxena, *PRIMES is in P*, Annals of Mathematics **160** (2004), no. 2, 781–793.

[AlSoVa2003] A. Alfonseca, F. Soria, A. Vargas, *A remark on maximal operators along directions in* $\mathbf{R}^2$, Math. Res. Lett. **10** (2003), no. 1, 41–49.

[Al1999] N. Alon, *Combinatorial Nullstellensatz*, Recent trends in combinatorics (Mátraháza, 1995), Combin. Probab. Comput. **8** (1999), no. 1-2, 7–29.

[AlGr1992] S. Altschuler, M. Grayson, *Shortening space curves and flow through singularities*, J. Differential Geom. **35** (1992), no. 2, 283–298.

[AmGe1973] W. O. Amrein, V. Georgescu, *On the characterization of bound states and scattering states in quantum mechanics*, Helv. Phys. Acta **46** (1973/74), 635–658.

[AuTa2008] T. Austin, T. Tao, *On the testability and repair of hereditary hypergraph properties*, preprint.

[AvGeTo2008] J. Avigad, P. Gerhardy, H. Towsner, *Local stability of ergodic averages*, preprint.

[BaHaPi2001] R. C. Baker, G. Harman, J. Pintz, *The difference between consecutive primes. II*, Proc. London Math. Soc. (3) **83** (2001), no. 3, 532–562.

[Ba2008] S. Ball, *On sets of points in a finite affine plane containing a line in every direction*, preprint.

[Ba1996] J. Barrionuevo, *A note on the Kakeya maximal operator*, Math. Res. Lett. **3** (1996), no. 1, 61–65.

[Ba2008] M. Bateman, *Kakeya sets and directional maximal operators in the plane*, preprint.

[BaKa2008] M. Bateman, N. Katz, *Kakeya sets in Cantor directions*, Math. Res. Lett. **15** (2008), no. 1, 73–81.

[BeFo1996] F. Beleznay, M. Foreman, *The complexity of the collection of measure-distal transformations*, Ergodic Theory Dynam. Systems **16** (1996), no. 5, 929–962.

[BeCaTa2006] J. Bennett, A. Carbery, T. Tao, *On the multilinear restriction and Kakeya conjectures*, Acta Math. **196** (2006), no. 2, 261–302.

[Be1987] V. Bergelson, *Weakly mixing PET*, Ergodic Theory Dynam. Systems **7** (1987), no. 3, 337–349.

[BeHoKr2005] V. Bergelson, B. Host, B. Kra, *Multiple recurrence and nilsequences*. With an appendix by Imre Ruzsa. Invent. Math. **160** (2005), no. 2, 261–303.

[BeHoMcCPa2000] V. Bergelson, B. Host, R. McCutcheon, F. Parreau, *Aspects of uniformity in recurrence*, Colloq. Math. **84/85** (2000), 549–576.

[BeLe1996] V. Bergelson, A. Leibman, *Polynomial extensions of van der Waerden's and Szemerédi's theorems*, J. Amer. Math. Soc. **9** (1996), no. 3, 725–753.

[BeLe1999] V. Bergelson, A. Leibman, *Set-polynomials and polynomial extension of the Hales-Jewett theorem*, Ann. of Math. **150** (1999), no. 1, 33–75.

[BeLe2007] V. Bergelson, A. Leibman, *Distribution of values of bounded generalized polynomials*, Acta Math. **198** (2007), no. 2, 155–230.

[BeLeLe2007] V. Bergelson, A. Leibman, E. Lesigne, *Intersective polynomials and polynomial Szemeredi theorem*, preprint. `arxiv:0710.4862`

[BeTaZi2009] V. Bergelson, T. Tao, T. Ziegler, *An inverse theorem for the uniformity seminorms associated with the action of F_p^∞*, preprint.

[Be1995] I. Berkes, *On the almost sure central limit theorem and domains of attraction*, Probability Theory and Related Fields **102** (1995), 1–17.

[Be1919] A. Besicovitch, *Sur deux questions d'intégrabilité des fonctions* J. Soc. Phys. Math. **2** (1919), 105–123.

[Be1928] A. Besicovitch, *On Kakeya's problem and a similar one*, Mathematische Zeitschrift **27** (1928), 312–320.

[BeBeBoMaPo2008] L. Bessiéres, G. Besson, M. Boileau, S. Maillot, J. Porti, *Weak collapsing and geometrisation of aspherical 3-manifolds*, preprint.

[BlHoMa2000] F. Blanchard, B. Host, A. Maass, *Topological complexity*, Ergodic Theory Dynam. Systems **20** (2000), no. 3, 641–662.

[Bl1993] A. Blass, *Ultrafilters: where topological dynamics = algebra = combinatorics*, Topology Proc. **18** (1993), 33–56.

[BlMa2008] A. Blokhuis and F. Mazzocca, *The finite field Kakeya problem*, Building Bridges Between Mathematics and Computer Science, Bolyai Society Mathematical Studies, Vol. 19, Martin Grötschel, Gyula O. H. Katona (eds.), Springer, 2008.

[Bo1987] E. Bombieri, *Le grand crible dans la théorie analytique des nombres*, 2nd ed., Astérisque 18, Paris 1987.

[BoDa1966] E. Bombieri, H. Davenport, *Small differences between prime numbers*, Proc. Roy. Soc. Ser. A **293** (1966), 1–18.

[Bo1991] J. Bourgain, *Besicovitch type maximal operators and applications to Fourier analysis*, Geom. Funct. Anal. **1** (1991), no. 2, 147–187.

[Bo1999] J. Bourgain, *On the dimension of Kakeya sets and related maximal inequalities*, Geom. Funct. Anal. **9** (1999), no. 2, 256–282.

[Bo2001] J. Bourgain, *Λ_p-sets in analysis: results, problems and related aspects*, Handbook of the geometry of Banach spaces, Vol. I, pp. 195–232, North-Holland, Amsterdam, 2001.

[Bo2005] J. Bourgain, *New encounters in combinatorial number theory: from the Kakeya problem to cryptography*, Perspectives in analysis, pp. 17–26, Math. Phys. Stud., 27, Springer, Berlin, 2005.

[BoGa2008] J. Bourgain, A. Gamburd, *Uniform expansion bounds for Cayley graphs of $SL_2(\mathbf{F}_p)$*, Ann. of Math. **167** (2008), no. 2, 625–642.

[BoGaSa2006] J. Bourgain, A. Gamburd, P. Sarnak, *Sieving and expanders*, C. R. Math. Acad. Sci. Paris **343** (2006), no. 3, 155–159.

[BoKaTa2004] J. Bourgain, N. Katz, T. Tao, *A sum-product estimate in finite fields, and applications*, Geom. Funct. Anal. **14** (2004), no. 1, 27–57.

[BoKo2003] J. Bourgain, S. Konyagin, *Estimates for the number of sums and products and for exponential sums over subgroups in fields of prime order*, C. R. Math. Acad. Sci. Paris **337** (2003), no. 2, 75–80.

[Br1993] J. W. Bruce, *A really trivial proof of the Lucas-Lehmer test*, Amer. Math. Monthly **100** (1993), no. 4, 370–371.

[Br1915] V. Brun, *Über das Goldbachsche Gesetz und die Anzahl der Primzahlpaare*, Archiv for Math. Og Naturvid. B34 (1915).

[Br2004] R. Bryant, *Gradient Kahler Ricci solitons*, preprint.

[BuZw] N. Burq, M. Zworski, *Control theory and high frequency eigenfunctions*, slides available at http://math.berkeley.edu/ zworski/bz1.pdf

[CaTo2008] E. Cabezas-Rivas, P. Topping, *The canonical shrinking soliton associated to a Ricci flow*, preprint.

[Ca1958] E. Calabi, *An extension of E. Hopf's maximum principle with an application to Riemannian geometry*, Duke Math. J. **25** (1958), 45–56.

[Ca1996] H.-D. Cao, *Existence of gradient Kähler-Ricci solitons*, Elliptic and parabolic methods in geometry (Minneapolis, MN, 1994), pp. 1–16, A K Peters, Wellesley, MA, 1996.

[CaZh2006] H.-D. Cao, X.-P. Zhu, *A complete proof of the Poincaré and geometrization conjectures—application of the Hamilton-Perelman theory of the Ricci flow*, Asian J. Math. **10** (2006), no. 2, pp. 165–492.

[CaFr1967] J. W. S. Cassels, A. Fröhlich (eds.), *Algebraic number theory*, Academic Press Inc. [Harcourt Brace Jovanovich Publishers], London, 1986, Reprint of the 1967 original.

[Ch1970] J. Cheeger, *Finiteness theorems for Riemannian manifolds*, Amer. J. Math. **92** (1970), 61–74.

[ChGr1971] J. Cheeger, D. Gromoll, *The splitting theorem for manifolds of nonnegative Ricci curvature*, J. Differential Geometry **6** (1971/72), 119–128.

[ChGr1972] J. Cheeger, D. Gromoll, *On the structure of complete manifolds of nonnegative curvature*, Ann. of Math. **96** (1972), 413–443.

[Ch1966] J. R. Chen, *On the representation of a large even integer as the sum of a prime and the product of at most two primes*, Kexue Tongbao **17** (1966), 385–386.

[ChLuTi2006] X. Chen, P. Lu, G. Tian, *A note on uniformization of Riemann surfaces by Ricci flow*, Proc. Amer. Math. Soc. **134** (2006), no. 11, 3391–3393.

[Ch1991] B. Chow, *The Ricci flow on the 2-sphere*, J. Differential Geom. **33** (1991), no. 2, 325–334.

[ChCh1995] B. Chow, S.-C. Chu, *A geometric interpretation of Hamilton's Harnack inequality for the Ricci flow*, Math. Res. Lett. **2** (1995), no. 6, 701–718.

[ChCh1996] B. Chow, S.-C. Chu, *A geometric approach to the linear trace Harnack inequality for the Ricci flow*, Math. Res. Lett. **3** (1996), no. 4, 549–568.

[CCGGIIKLLN2008] B. Chow, S.-C. Chu, D. Glickenstein, C. Guenther, J. Isenberg, T. Ivey, D. Knopf, P. Lu, F. Luo, L. Ni, *The Ricci flow: Techniques and applications. Part II. Analytic aspects.* Mathematical Surveys and Monographs, 144. American Mathematical Society, Providence, RI, 2008.

[ChKn2004] B. Chow, D. Knopf, *The Ricci flow: an introduction*, Mathematical Surveys and Monographs, 110. American Mathematical Society, Providence, RI, 2004.

[ChLuNi2006] B. Chow, P. Lu, L. Ni, *Hamilton's Ricci flow*, Graduate Studies in Mathematics, 77. American Mathematical Society, Providence, RI; Science Press, New York, 2006.

[Ch1984] M. Christ, *Estimates for the k-plane transform*, Indiana Univ. Math. J. **33** (1984), no. 6, 891–910.

[CoMi1997] T. Colding, W. Minicozzi, *Harmonic functions on manifolds*, Ann. of Math. **146** (1997), no. 3, 725–747.

[CoMi2005] T. Colding, W. Minicozzi, *Estimates for the extinction time for the Ricci flow on certain 3-manifolds and a question of Perelman*, J. Amer. Math. Soc. **18** (2005), no. 3, 561–569.

[CoMi2007] T. Colding, W. Minicozzi, *Width and finite extinction time of Ricci flow*, preprint.

[Co1977] A. Cordoba, *The Kakeya maximal function and the spherical summation multipliers*, Amer. J. Math. **99** (1977), no. 1, 1–22.

[CoFe1977] A. Cordoba, R. Fefferman, *On the equivalence between the boundedness of certain classes of maximal and multiplier operators in Fourier analysis*, Proc. Nat. Acad. Sci. U.S.A. **74** (1977), no. 2, 423–425.

[CoSC1993] T. Coulhon, L. Saloff-Coste, *Isopérimétrie pour les groupes et les variétés*, Rev. Mat. Iberoamericana **9** (1993), no. 2, 293–314.

[Cr1936] Harald Cramér, *On the order of magnitude of the difference between consecutive prime numbers*, Acta Arithmetica **2** (1936), 23–46.

[Cu1971] F. Cunningham Jr., *The Kakeya problem for simply connected and for star-shaped sets*, Amer. Math. Monthly **78** (1971), 114–129.

[Da1971] R. Davies, *Some remarks on the Kakeya problem*, Proc. Cambridge Philos. Soc. **69** (1971), 417–421.

[DaPuRo1961] M. Davis, H. Putnam, J. Robinson, *The decision problem for exponential Diophantine equations*, Ann. Math. **74** (1961), 425–436.

[DeT1983] D. DeTurck, *Deforming metrics in the direction of their Ricci tensors*, J. Differential Geom. **18** (1983), no. 1, 157–162.

[Dr1983] S. W. Drury, L^p *estimates for the X-ray transform*, Illinois J. Math. **27** (1983), no. 1, 125–129.

[Dv2008] Z. Dvir, *On the size of Kakeya sets in finite fields*, preprint.

[EcHu1991] K. Ecker, G. Huisken, *Interior estimates for hypersurfaces moving by mean curvature*, Invent. Math. **105** (1991), no. 3, 547–569.

[Ei2006] M. Einsiedler, *Ratner's theorem on* $SL(2, \mathbf{R})$*-invariant measures*, Jahresber. Deutsch. Math.-Verein. **108** (2006), no. 3, 143–164.

[EiMaVe2007] M. Einsiedler, G. Margulis, A. Venkatesh, *Effective equidistribution for closed orbits of semisimple groups on homogeneous spaces*, preprint.

[El1958] R. Ellis, *Distal transformation groups*, Pacific J. Math. **8** (1958), 401–405.

[En1978] V. Enss, *Asymptotic completeness for quantum mechanical potential scattering. I. Short range potentials*, Comm. Math. Phys. **61** (1978), no. 3, 285–291.

[Er1940] P. Erdős, *The difference of consecutive primes*, Duke Math. J. **6** (1940), 438–441.

[ErTu1936] P. Erdős, P. Turán, *On some sequences of integers*, J. London Math. Soc. **11** (1936), 261–264.

[EsMoOh2005] A. Eskin, S. Mozes, H. Oh, *On uniform exponential growth for linear groups*, Invent. Math. **160** (2005), no. 1, 1–30.

[Ev1998] L. C. Evans, *Partial differential equations*, Graduate Studies in Mathematics, 19. American Mathematical Society, Providence, RI, 1998.

[Fa2000] I. Farah, *Approximate homomorphisms. II. Group homomorphisms*, Combinatorica **20** (2000), no. 1, 47–60.

[FiMa2005] D. Fisher, G. Margulis, *Almost isometric actions, property (T), and local rigidity*, Invent. Math. **162** (2005), no. 1, 19–80.

[Fo1970] J. Folkman, *Graphs with monochromatic complete subgraphs in every edge coloring*, SIAM J. Appl. Math. **18** (1970), 115–124.

[Fr1973] G. Freiman, Foundations of a structural theory of set addition. Translated from the Russian. Translations of Mathematical Monographs, Vol 37. American Mathematical Society, Providence, RI, 1973.

[Fu1961] H. Furstenberg, *Strict ergodicity and transformation of the torus*, Amer. J. Math. **83** (1961), 573–601.

[Fu1963] H. Furstenberg, *The structure of distal flows*, Amer. J. Math. **85** (1963), 477–515.

[Fu1977] H. Furstenberg, *Ergodic behavior of diagonal measures and a theorem of Szemerédi on arithmetic progressions*, J. Analyse Math. **31** (1977), 204–256.

[Fu1981] H. Furstenberg, *Recurrence in ergodic theory and combinatorial number theory*, Princeton University Press, Princeton, NJ, 1981.

[FuKa1979] H. Furstenberg, Y. Katznelson, *An ergodic Szemerédi theorem for commuting transformations*, J. Analyse Math. **34** (1978), 275–291 (1979).

[FuKa1985] H. Furstenberg, Y. Katznelson, *An ergodic Szemerédi theorem for IP-systems and combinatorial theory*, J. Analyse Math. **45** (1985), 117–168.

[FuKa1991] H. Furstenberg, Y. Katznelson, A density version of the Hales-Jewett theorem, *J. d'Analyse Math.* **57** (1991), 64–119.

[FuWe1978] H. Furstenberg, B. Weiss, *Topological dynamics and combinatorial number theory*, J. Analyse Math. 34 (1978), 61–85 (1979).

[FuWe1996] H. Furstenberg, B. Weiss, *A mean ergodic theorem for* $\frac{1}{N}\sum_{n=1}^{N} f(T^n x)g(T^{n^2} x)$, Convergence in ergodic theory and probability (Columbus, OH, 1993), pp. 193–227, Ohio State Univ. Math. Res. Inst. Publ., 5, de Gruyter, Berlin, 1996.

[GaHa1986] M. Gage, R. Hamilton, *The heat equation shrinking convex plane curves*, J. Differential Geom. **23** (1986), no. 1, 69–96.

[Ga1976] P. X. Gallagher, *On the distribution of primes in short intervals*, Mathematika **23** (1976), no. 1, 4–9.

[Ga2002] A. Gamburd, *On the spectral gap for infinite index "congruence" subgroups of* $SL_2(\mathbf{Z})$, Israel J. Math. **127** (2002), 157–200.

[GeLe1993] P. Gérard, E. Leichtnam, *Ergodic properties of eigenfunctions for the Dirichlet problem*, Duke Math. J. **71** (1993), no. 2, 559–607.

[Ge2008] P. Gerhardy, *Proof mining in topological dynamics*, Notre Dame J. Formal Logic **49** (2008), 431–446.

[Gi1987] J.-Y. Girard, *Proof theory and logical complexity*, Vol. I, Bibliopolis, Naples, 1987.

[Gl2000] E. Glasner, *Structure theory as a tool in topological dynamics. Descriptive set theory and dynamical systems* (Marseille-Luminy, 1996), pp. 173–209, London Math. Soc. Lecture Note Ser., 277, Cambridge Univ. Press, Cambridge, 2000.

[Gl2003] E. Glasner, *Ergodic theory via joinings*, Mathematical Surveys and Monographs, 101. American Mathematical Society, Providence, RI, 2003.

[GoGrPiYi2006] D. A. Goldston, S.W. Graham, J. Pintz, C. Y. Yıldırım, *Small gaps between products of two primes*, preprint.

[GoMo1987] D. Goldston, H. Montgomery, *Pair correlation of zeros and primes in short intervals*, Analytic number theory and Diophantine problems (Stillwater, OK, 1984), 183–203, Progr. Math., 70, Birkhäuser Boston, Boston, MA, 1987.

[GoPoYi2005] D. Goldston, J. Pintz, C. Yıldırım, *The path to recent progress on small gaps between primes*, preprint.

[GoPoYi2005a] D. Goldston, J. Pintz, C. Yıldırım, *Primes in tuples I*, preprint.

[GoPoYi2007] D. Goldston, J. Pintz, C. Yıldırım, *Primes in tuples II*, preprint.

[Go1998] W. T. Gowers, *A new proof of Szemerédi's theorem for progressions of length four*, GAFA **8** (1998), no. 3, 529–551.

[Go2001] W. T. Gowers, *A new proof of Szemerédi's theorem*, Geom. Funct. Anal. **11** (2001), no. 3, 465–588.

[Go2008] T. Gowers, *Decompositions, approximate structure, transference, and the Hahn-Banach theorem*, preprint.

[Gr1995] A. Granville, *Harald Cramér and the distribution of prime numbers*, Harald Cramér Symposium (Stockholm, 1993). Scand. Actuar. J. **1995**, no. 1, 12–28.

[Gr2005] B. Green, *Finite field models in additive combinatorics*, Surveys in Combinatorics 2005, London Math. Soc. Lecture Notes 327, pp. 1–27.

[Gr2006] B. Green, *Three topics in additive prime number theory*, preprint.

[GrTa2006] B. Green, T. Tao, *Restriction theory of the Selberg sieve, with applications*, Journal de Théorie des Nombres de Bordeaux **18** (2006), 137–172.

[GrTa2007] B. Green, T. Tao, *The distribution of polynomials over finite fields, with applications to the Gowers norms*, preprint.

[GrTa2008] B. Green, T. Tao, *The primes contain arbitrarily long arithmetic progressions*, Annals Math. **167** (2008), 481–547.

[GrTa2009] B. Green, T. Tao, *Linear equations in primes*, to appear, Annals of Math.

[GrTa2009a] B. Green, T. Tao, *New bounds for Szemeredi's theorem, I: Progressions of length 4 in finite field geometries*, preprint.

[GrTa2009b] B. Green, T. Tao, *New bounds for Szemeredi's theorem, II: A new bound for $r_4(N)$*, preprint.

[GrTa2009c] B. Green, T. Tao, *The quantitative behaviour of polynomial orbits on nilmanifolds*, preprint.

[GrTa2009d] B. Green, T. Tao, *An inverse theorem for the Gowers $U^3(G)$ norm*, preprint.

[GrTa2009e] B. Green, T. Tao, *Quadratic uniformity of the Möbius function*, preprint.

[GrTa2009f] B. Green, T. Tao, *The Möbius function is asymptotically orthogonal to nilsequences*, preprint.

[Gr1961] L. W. Green, *Spectra of nilflows*, Bull. Amer. Math. Soc. **67** (1961), 414–415.

[GrLeRo1972] R. L. Graham, K. Leeb, B. L. Rothschild, *Ramsey's theorem for a class of categories*, Advances in Math. **8** (1972), 417–433.

[GrRoSp1980] R. Graham, B. Rothschild, J. H. Spencer, *Ramsey theory*, John Wiley and Sons, New York, 1980.

[GrSo2007] A. Granville, K. Soundararajan, *An uncertainty principle for arithmetic sequences*, Ann. of Math. (2) **165** (2007), no. 2, 593–635.

[Gr1981] M. Gromov, *Groups of polynomial growth and expanding maps*, Inst. Hautes Études Sci. Publ. Math. No. **53** (1981), 53–73.

[Gr2003] M. Gromov, *Isoperimetry of waists and concentration of maps*, Geom. Funct. Anal. **13** (2003), no. 1, 178–215.

[Gr1975] L. Gross, *Logarithmic Sobolev inequalities*, Amer. J. Math. **97** (1975), no. 4, 1061–1083.

[Gr2006] L. Gross, *Hypercontractivity, logarithmic Sobolev inequalities, and applications: a survey of surveys*, Diffusion, quantum theory, and radically elementary mathematics, pp. 45–73, Math. Notes, 47, Princeton Univ. Press, Princeton, NJ, 2006.

[GuLe1973] R. Gulliver, F. Lesley, *On boundary branch points of minimizing surfaces*, Arch. Rational Mech. Anal. **52** (1973), 20–25.

[Gu2008] L. Guth, *The endpoint case of the Bennett-Carbery-Tao multilinear Kakeya conjecture*, preprint.

[Gu1988] R. Guy, *The strong law of small numbers*, The American Mathematical Monthly **95** (1988), 697–712.

[HaJe1963] A. W. Hales, R. I. Jewett, *Regularity and positional games*, Trans. Amer. Math. Soc. **106** (1963), 222–229.

[Ha1982] R. Hamilton, *Three-manifolds with positive Ricci curvature*, J. Differential Geom. **17** (1982), no. 2, 255–306.

[Ha1986] R. Hamilton, *Four-manifolds with positive curvature operator*, J. Differential Geom. **24** (1986), no. 2, 153–179.

[Ha1988] R. Hamilton, *The Ricci flow on surfaces*, Mathematics and general relativity (Santa Cruz, CA, 1986), pp. 237–262, Contemp. Math., 71, Amer. Math. Soc., Providence, RI, 1988.

[Ha1993] R. Hamilton, *The Harnack estimate for the Ricci flow*, J. Differential Geom. **37** (1993), no. 1, 225–243.

[Ha1993b] R. Hamilton, *The formation of singularities in the Ricci flow*, Surveys in differential geometry, Vol. II (Cambridge, MA, 1993), pp. 7–136, Int. Press, Cambridge, MA, 1995.

[Ha1995] R. Hamilton, *A compactness property for solutions of the Ricci flow*, Amer. J. Math. **117** (1995), no. 3, 545–572.

[Ha1997] R. Hamilton, *Four-manifolds with positive isotropic curvature*, Comm. Anal. Geom. **5** (1997), no. 1, 1–92.

[Ha1999] R. Hamilton, *Non-singular solutions of the Ricci flow on three-manifolds*, Comm. Anal. Geom. **7** (1999), no. 4, 695–729.

[HaSi1985] R. Hardt, L. Simon, *Area minimizing hypersurfaces with isolated singularities*, J. Reine Angew. Math. **362** (1985), 102–129.

[HaLi1923] G. H., Hardy, J. E. Littlewood, *Some problems of 'Partitio Numerorum', III. On the expression of a number as a sum of primes.*, Acta Math. **44** (1923), 1–70.

[Ha1973] L. D. Harmon, *The recognition of faces*, Scientific American **229** (1973), 71–82.

[Ha2008] A. Hassell, *Ergodic billiards that are not quantum unique ergodic*, preprint. With an appendix by Andrew Hassell and Luc Hillairet.

[He2008] H. A. Helfgott, *Growth and generation in* $SL_2(\mathbb{Z}/p\mathbb{Z})$, Ann. of Math. **167** (2008), no. 2, 601–623.

[He1991] E. Heller, *Wavepacket dynamics and quantum chaology*, Chaos et physique quantique (Les Houches, 1989), pp. 547–664, North-Holland, Amsterdam, 1991.

[Hi1951] G. Higman, *A finitely generated infinite simple group*, J. London Math. Soc. **26** (1951), 61–64.

[Hi1969] S. Hildebrandt, *Boundary behavior of minimal surfaces*, Arch. Rational Mech. Anal. **35** (1969), 47–82.

[Hi1974] N. Hindman, *Finite sums from sequences within cells of a partition of N*, J. Combin. Thy. Ser. A **17** (1974), 1–11.

[Hi] J. Hirschfeld, *The nonstandard treatment of Hilbert's fifth problem*, Trans. Amer. Math. Soc. **321** (1990), no. 1, 379–400.

[HoKr2005] B. Host, B. Kra, *Nonconventional ergodic averages and nilmanifolds*, Ann. of Math. (2) **161** (2005), no. 1, 397–488.

[Il1997] K. Ilinski, *The physics of finance*, Proceedings of Budapest's conference on Econophysics (July 1997).

[Iv1993] T. Ivey, *Ricci solitons on compact three-manifolds*, Differential Geom. Appl. **3** (1993), no. 4, 301–307.

[Iw1978] H. Iwaniec, *Almost-primes represented by quadratic polynomials*, Invent. Math. **47** (1978), no. 2, 171–188.

[Jo1991] J. Jost, *Two-dimensional geometric variational problems*, Pure and Applied Mathematics (New York). A Wiley-Interscience Publication. John Wiley & Sons, Ltd., Chichester, 1991.

[Ka1999] N. H. Katz, *Remarks on maximal operators over arbitrary sets of directions*, Bull. London Math. Soc. **31** (1999), no. 6, 700–710.

[Ka2005] N. H. Katz, *On arithmetic combinatorics and finite groups*, Illinois J. Math. **49** (2005), no. 1, 33–43.

[KaLaTa2000] N. Katz, I. Łaba, T. Tao, *An improved bound on the Minkowski dimension of Besicovitch sets in R^3*, Ann. of Math. (2) **152** (2000), no. 2, 383–446.

[KaTa1999] N. Katz, T. Tao, *Bounds on arithmetic projections, and applications to the Kakeya conjecture*, Math. Res. Lett. **6** (1999), no. 5-6, 625–630.

[KaTa2002] N. Katz, T. Tao, *Recent progress on the Kakeya conjecture*, Proceedings of the 6th International Conference on Harmonic Analysis and Partial Differential Equations (El Escorial, 2000). Publ. Mat. 2002, Vol. Extra, pp. 161–179.

[KaTa200b] N. Katz, T. Tao, *New bounds for Kakeya problems*, Dedicated to the memory of Thomas H. Wolff, J. Anal. Math. **87** (2002), 231–263.

[Ka1981] J. Kazdan, *Another proof of Bianchi's identity in Riemannian geometry*, Proc. Amer. Math. Soc. **81** (1981), no. 2, 341–342.

[KaWa1974] J. Kazdan, F. W. Warner, *Curvature functions for compact 2-manifolds*, Ann. of Math. (2) **99** (1974), 14–47.

[Ke1995] M. Keane, *The essence of large numbers*, Algorithms, Fractals, and Dynamics (Okayama/Kyoto, 1992), pp. 125–129, Plenum, New York, 1995.

[Ke1999] U. Keich, *On L^p bounds for Kakeya maximal functions and the Minkowski dimension in R^2*, Bull. London Math. Soc. **31** (1999), no. 2, 213–221.

[KeRo1969] H. Keynes, J. Robertson, *Eigenvalue theorems in topological transformation groups*, Trans. Amer. Math. Soc. **139** (1969), 359–369.

[KiVi2008] R. Killip, M. Visan, *Nonlinear Schrodinger equations at critical regularity*, available at http://www.math.uchicago.edu/ mvisan/ClayLectureNotes.pdf

[KlRo2007] S. Klainerman, I. Rodnianski, *A Kirchoff-Sobolev parametrix for the wave equation and applications*, J. Hyperbolic Differ. Equ. **4** (2007), no. 3, 401–433.

[Kl2007] B. Kleiner, *A new proof of Gromov's theorem on groups of polynomial growth*, preprint.

[KlLo2006] B. Kleiner, J. Lott, *Notes on Perelman's papers*, preprint.

[Kn1929] H. Kneser, *Geschlossene Flächen in dreidimensionalen Mannigfaltigkeiten*, Jahresbericht der Deut. Math. Verein. **38** (1929), 248–260.

[KoSc1997] N. Korevaar, R. Schoen, *Global existence theorems for harmonic maps to non-locally compact spaces*, Comm. Anal. Geom. **5** (1997), no. 2, 333–387.

[Ko2006] D. Kotschick, *Monopole classes and Perelman's invariant of four-manifolds*, preprint.

[Kow2006] E. Kowalski, *Écarts entre nombres premiers succesifs (d'après Goldston, Pintz, Yıldırım)*, Séminaire Bourbaki, 58éme année, 2005–2006, no. 959.

[Kr2006] B. Kra, *From combinatorics to ergodic theory and back again*, International Congress of Mathematicians. Vol. III, pp. 57–76, Eur. Math. Soc., Zürich, 2006.

[La2008] I. Łaba, *From harmonic analysis to arithmetic combinatorics*, Bull. Amer. Math. Soc. (N.S.) **45** (2008), no. 1, 77–115.

[LaTa2001] I. Łaba, T. Tao, *An improved bound for the Minkowski dimension of Besicovitch sets in medium dimension*, Geom. Funct. Anal. **11** (2001), no. 4, 773–806.

[Le1998] A. Leibman, *Polynomial sequences in groups*, J. Algebra **201** (1998), no. 1, 189–206.

[Le2002] A. Leibman, *Polynomial mappings of groups*, Israel J. Math. **129** (2002), 29–60.

[Le2005] A. Leibman, *Pointwise convergence of ergodic averages for polynomial sequences of translations on a nilmanifold*, Ergodic Theory Dynam. Systems **25** (2005), no. 1, 201–213.

[Le2005b] A. Leibman, *Pointwise convergence of ergodic averages for polynomial actions of $\mathbf{Z}^d$ by translations on a nilmanifold*, Ergodic Theory Dynam. Systems **25** (2005), no. 1, 215–225.

[LiYa1986] P. Li, S.-T. Yau, *On the parabolic kernel of the Schrödinger operator*, Acta Math. **156** (1986), no. 3-4, 153–201.

[LiWa2002] M.-C. Liu, T. Wang, *On the Vinogradov bound in the three primes Goldbach conjecture*, Acta Arith. **105** (2002), no. 2, 133–175.

[Lo2008] J. Lott, *Optimal transport and Perelman's reduced volume*, preprint.

[LoMeSa2008] S. Lovett, R. Meshulam, A. Samorodnitsky, *Inverse conjecture for the Gowers norm is false*, STOC'08.

[Ma1985] H. Maier, *Primes in short intervals*, Michigan Math. J. **32** (1985), no. 2, 221–225.

[Ma1951] A. I. Mal'cev, *On a class of homogeneous spaces*. Izv. Akad. Nauk. SSSR. Ser. Mat. **13** (1949), 9–32. Translated in: Amer. Math. Soc. Transl. **1951**, no. 39, 33 pp.

[Ma1970] Y. Matiyasevich, *Enumerable sets are Diophantine*, Dokl. Akad. Nauk SSSR, **191** (1970), 279–282. English translation: Soviet Mathematics. Doklady **11** (1970), 354–358.

[Ma2003] J. Matousek, *Using the Borsuk-Ulam theorem*, Lectures on topological methods in combinatorics and geometry. Written in cooperation with Anders Björner and Günter M. Ziegler. Universitext. Springer-Verlag, Berlin, 2003.

[McM1978] D. McMahon, *Relativized weak disjointness and relatively invariant measures*, Trans. Amer. Math. Soc. **236** (1978), 225–237.

[MeYa1980] W. Meeks, S-T. Yau, *Topology of three-dimensional manifolds and the embedding problems in minimal surface theory*, Ann. of Math. (2) **112** (1980), no. 3, 441–484.

[Me2003] A. Melas, *The best constant for the centered Hardy-Littlewood maximal inequality*, Ann. of Math. (2) **157** (2003), no. 2, 647–688.

[Mi1962] J. Milnor, *A unique factorization theorem for 3-manifolds*, Amer. J. Math. **84** (1962), 1–7.

[Mi1968] J. Milnor, *A note on curvature and fundamental group*, J. Diff. Geom. **2** (1968), 1–7.

[Mi2003] J. Milnor, *Towards the Poincaré conjecture and the classification of 3-manifolds*, Notices Amer. Math. Soc. **50** (2003), no. 10, 1226–1233.

[Mi2006] J. Milnor, *The Poincaré conjecture. The millennium prize problems*, pp. 71–83, Clay Math. Inst., Cambridge, MA, 2006.

[MoTa2004] G. Mockenhaupt, T. Tao, *Restriction and Kakeya phenomena for finite fields*, Duke Math. J. **121** (2004), no. 1, 35–74.

[Mo1995] N. Mok, *Harmonic forms with values in locally constant Hilbert bundles*, Proceedings of the Conference in Honor of Jean-Pierre Kahane (Orsay, 1993), J. Fourier Anal. Appl. **1995**, Special Issue, 433–453.

[Mo1952] E. Moise, *Affine structures in 3-manifolds. V. The triangulation theorem and Hauptvermutung*, Ann. of Math. (2) **56** (1952), 96–114.

[MoZi1955] D. Montgomery, L. Zippin, *Topological transformation groups*. Reprint of the 1955 original. Robert E. Krieger Publishing Co., Huntington, NY, 1974.

[Mo2007] J. Morgan, *The Poincaré conjecture*, International Congress of Mathematicians. Vol. I, pp. 713–736, Eur. Math. Soc., Zürich, 2007.

[MoTi2007] J. Morgan, G. Tian, *Ricci flow and the Poincaré conjecture*, Clay Mathematics Monographs, 3. American Mathematical Society, Providence, RI; Clay Mathematics Institute, Cambridge, MA, 2007.

[MoTi2008] J. Morgan, G. Tian, *Completion of the proof of the geometrization conjecture*, preprint.

[Mo1948] C. Morrey, *The problem of Plateau on a Riemannian manifold*, Ann. of Math. (2) **49** (1948), 807–851.

[Mo1964] J. Moser, *A Harnack inequality for parabolic differential equations*, Comm. Pure Appl. Math. **17** (1964), 101–134

[Mu2006] R. Müller, *Differential Harnack inequalities and the Ricci flow*. EMS Series of Lectures in Mathematics. European Mathematical Society (EMS), Zürich, 2006.

[Mu1960] J. Munkres, *Obstructions to the smoothing of piecewise-differentiable homeomorphisms*, Ann. of Math. (2) **72** (1960) 521–554.

[Na2007] A. Naber, *Noncompact shrinking 4-solitons with nonnegative curvature*, preprint.

[NaStWa1978] A. Nagel, E. Stein, S. Wainger, *Differentiation in lacunary directions*, Proc. Nat. Acad. Sci. U.S.A. **75** (1978), no. 3, 1060–1062.

[Ne1954] B. H. Neumann, *An essay on free products of groups with amalgamations*, Philos. Trans. Roy. Soc. London. Ser. A. **246** (1954), 503–554.

[Ni2004] L. Ni, *The entropy formula for linear heat equation*, J. Geom. Anal. **14** (2004), no. 1, 87–100.

[NiWa2007] L. Ni, N. Wollach, *On a classification of the gradient shrinking solitons*, preprint.

[Pa1969] W. Parry, *Ergodic properties of affine transformations and flows on nilmanifolds*, Amer. J. Math. **91** (1969), 757–771.

[Pe1994] G. Perelman, *Proof of the soul conjecture of Cheeger and Gromoll*, J. Differential Geom. **40** (1994), 209–212.

[Pe2002] G. Perelman, *The entropy formula for the Ricci flow and its geometric applications*, preprint, math.DG/0211159.

[Pe2003] G. Perelman, *Ricci flow with surgery on three-manifolds*, preprint, math.DG/0303109.

[Pe2003b] G. Perelman, *Finite extinction time for the solutions to the Ricci flow on certain three-manifolds*, preprint, math.DG/0307245.

[Pe2006] P. Petersen, *Riemannian geometry*, 2nd ed., Graduate Texts in Mathematics, 171. Springer, New York, 2006.

[PeWy2007] P. Petersen, W. Wylie, *On the classification of gradient Ricci solitons*, preprint.

[Ra1937] R. A. Rankin, *The difference between consecutive primes*, J. Lond. Math. Soc. **13** (1938), 242–247.

[Ra1962] R. A. Rankin, *The difference between consecutive prime numbers. V*, Proc. Edinburgh Math. Soc. **13** (1962/1963), 331–332.

[ReTrTuVa2008] O. Reingold, L. Trevisan, M. Tulsiani, S. Vadhan, *New proofs of the Green-Tao-Ziegler dense model theorem: An exposition*, preprint.

[Ri1859] B. Riemann, *Über die Anzahl der Primzahlen unter einer gegebenen Grösse*, Monatsberichte der Berliner Akademie, 1859.

[Ro1953] K. F. Roth, *On certain sets of integers*, J. London Math. Soc. **28** (1953), 245–252.

[Ru1969] D. Ruelle, *A remark on bound states in potential-scattering theory*, Nuovo Cimento A **61** (1969), 655–662.

[Ru1994] I. Z. Ruzsa, *Generalized arithmetical progressions and sumsets*, Acta Math. Hungar. **65** (1994), no. 4, 379–388.

[SaUh1981] J. Sacks, K. Uhlenbeck, *The existence of minimal immersions of 2-spheres*, Ann. of Math. (2) **113** (1981), no. 1, 1–24.

[SaSu2008] S. Saraf, M. Sudan, *Improved lower bound on the size of Kakeya sets over finite fields*, preprint.

[Sa1998] Y. Saouter, *Checking the odd Goldbach conjecture up to 10^{20}*, Math. Comp. **67** (1998), no. 222, 863–866.

[SaXu1991] P. Sarnak, X. Xue, *Bounds for multiplicities of automorphic representations*, Duke Math. J. **64** (1991), no. 1, 207–227.

[Sc1998] W. Schlag, *A geometric inequality with applications to the Kakeya problem in three dimensions*, Geom. Funct. Anal. **8** (1998), no. 3, 606–625.

[Sc1962] I. Schoenberg, *On the Besicovitch-Perron solution of the Kakeya problem*, Studies in mathematical analysis and related topics, pp. 359–363, Stanford Univ. Press, Stanford, CA, 1962.

[Sc1916] I. Schur, *Über die Kongruenz $x^m + y^m = z^m$ (mod p)*, Jber. Deutsch. Math.-Verein. **25** (1916), 114–116.

[Sc1980] J. T. Schwartz, *Fast probabilistic algorithms for verification of polynomial identities*, J. ACM **27** (1980), 701–717.

[Se1965] A. Selberg, *On the estimation of Fourier coefficients of modular forms*, Proc. Sympos. Pure Math., Vol. VIII, pp. 1–15, Amer. Math. Soc., Providence, RI, 1965.

[Sh1996] N. Shah, *Invariant measures and orbit closures on homogeneous spaces for actions of subgroups generated by unipotent elements*, Lie groups and ergodic theory (Mumbai, 1996), pp. 229–271, Tata Inst. Fund. Res. Stud. Math., 14, Tata Inst. Fund. Res., Bombay, 1998.

[Sh1997] Y. Shalom, *Expanding graphs and invariant means*, Combinatorica **17** (1997), no. 4, 555–575.

[Sh1998] Y. Shalom, *The growth of linear groups*, J. Algebra **199** (1998), no. 1, 169–174.

[Sh1989] W.-X. Shi, *Deforming the metric on complete Riemannian manifolds*, J. Differential Geom. **30** (1989), no. 1, 223–301.

[Sm1961] S. Smale, *Generalized Poincaré's conjecture in dimensions greater than four*, Ann. of Math. (2) **74** (1961), 391–406.

[So2006] K. Soundararajan, *Small gaps between prime numbers: The work of Goldston-Pintz-Yıldırım*, preprint.

[So2007] K. Soundararajan, *The distribution of prime numbers*, Equidistribution in Number Theory, An Introduction, NATO Science Series II: Mathematics, Physics and Chemistry, Springer Netherlands, 2007.

[St1961] E. M. Stein, *On limits of sequences of operators*, Ann. of Math. (2) **74** (1961), 140–170.

[St1970] E. M. Stein, *Topics in harmonic analysis related to the Littlewood-Paley theory*. Annals of Mathematics Studies, No. 63, Princeton University Press, Princeton, NJ; University of Tokyo Press, Tokyo, 1970.

[StSt1983] E. M. Stein, J.-O. Strömberg, *Behavior of maximal functions in R^n for large n*, Ark. Mat. **21** (1983), no. 2, 259–269.

[St1969] S. A. Stepanov, *The number of points of a hyperelliptic curve over a finite prime field*, Izv. Akad. Nauk SSSR Ser. Mat. **33** (1969) 1171–1181.

[Sz1969] E. Szemerédi, *On sets of integers containing no four elements in arithmetic progression*, Acta Math. Acad. Sci. Hungar. **20** (1969), 89–104.

[Sz1975] E. Szemerédi, *On sets of integers containing no k elements in arithmetic progression*, Acta Arith. **27** (1975), 299–345.

[Ta] T. Tao, *An ergodic transference theorem*, unpublished. Available at http://www.math.ucla.edu/~tao/preprints/Expository/limiting.dvi.

[Ta2] T. Tao, *Perelman's proof of the Poincaré conjecture: A nonlinear PDE perspective*, unpublished. Available at http://arxiv.org/abs/math/0610903.

[Ta3] T. Tao, *The dyadic pigeonhole principle in harmonic analysis*, unpublished. Available at http://www.math.ucla.edu/~tao/preprints/Expository/pigeonhole.dvi.

[Ta2001] T. Tao, *From rotating needles to stability of waves: emerging connections between combinatorics, analysis, and PDE*, Notices Amer. Math. Soc. **48** (2001), no. 3, 294–303.

[Ta2005] T. Tao, *A new bound for finite field Besicovitch sets in four dimensions*, Pacific J. Math. **222** (2005), no. 2, 337–363.

[Ta2006] T. Tao, *A quantitative ergodic theory proof of Szemerédi's theorem*, Electron. J. Combin. **13** (2006), no. 99, 1–49.

[Ta2006b] T. Tao, *The Gaussian primes contain arbitrarily shaped constellations*, J. d'Analyse Math. **99** (2006), 109–176.

[Ta2007] T. Tao, *The ergodic and combinatorial approaches to Szemerédi's theorem*, Centre de Recerches Mathématiques, CRM Proceedings and Lecture Notes, Vol. 43, 2007, pp. 145–193.

[Ta2007b] T. Tao, *Structure and randomness in combinatorics*, Proceedings of the 48th annual symposium on Foundations of Computer Science (FOCS), 2007, pp. 3–18.

[Ta2007c] T. Tao, *A correspondence principle between (hyper)graph theory and probability theory, and the (hyper)graph removal lemma*, J. d'Analyse Math. **103** (2007), 1–45.

[Ta2008] T. Tao, *Norm convergence of multiple ergodic averages for commuting transformations*, Ergodic Theory and Dynamical Systems **28** (2008), 657–688.

[Ta2008b] T. Tao, *Global regularity of wave maps IV. Absence of stationary or self-similar solutions in the energy class*, preprint.

[Ta2008c] T. Tao, *Global regularity of wave maps V. Large data local wellposedness in the energy class*, preprint.

[Ta2008b] T. Tao, *Structure and randomness: Pages from year one of a mathematical blog*, American Mathematical Society, Providence, RI, 2008.

[Ta2008c] T. Tao, *The sum-product phenomenon in arbitrary rings*, preprint.

[TaVaVe1998] T. Tao, A. Vargas, L. Vega, *A bilinear approach to the restriction and Kakeya conjectures*, J. Amer. Math. Soc. **11** (1998), no. 4, 967–1000.

[TaVu2006] T. Tao, V. Vu, *Additive combinatorics*, Cambridge University Press, Cambridge, 2006.

[TaVu2008] T. Tao, V. Vu, *Random matrices: Universality of ESDs and the circular law*, preprint.

[TaZi2008] T. Tao, T. Ziegler, *The primes contain arbitrarily long polynomial progressions*, Acta Math. **201** (2008), 213–305.

[Ta] A. Tarski, *Une contribution a la théorie de la mesure*, Fund. Math. **15** (1930), 42–50.

[Ti1972] J. Tits, *Free subgroups in linear groups*, J. Algebra **20** (1972), 250–270.

[To1959] V. A. Toponogov, *Riemann spaces with curvature bounded below*, Uspehi Mat. Nauk **14** (1959), 87–130. (Russian)

[To1964] V. A. Toponogov, *The metric structure of Riemannian spaces of non-negative curvature containing straight lines*, Sibirsk. Mat. Ž. **5** (1964), 1358–1369. (Russian)

[Ul1929] S. Ulam, *Concerning functions of sets*, Fund. Math. **14** (1929), 231–233.

[vA1942] H. van Alphen, *Generalization of a theorem of Besicovitsch*, Mathematica, Zutphen. B. **10** (1942), 144–157.

[vdDrWi1984] L. van den Dries, A. J. Wilkie, *Gromov's theorem on groups of polynomial growth and elementary logic*, J. Algebra **89** (1984), no. 2, 349–374.

[vdW1927] B. L. van der Waerden, *Beweis einer Baudetschen Vermutung*, Nieuw. Arch. Wisk. **15** (1927), 212–216.

[vKa1933] E. R. van Kampen, *On the connection between the fundamental groups of some related spaces*, American Journal of Mathematics **55** (1933), 261–267.

[Va1959] P. Varnavides, *On certain sets of positive density*, J. London Math. Soc. **39** (1959), 358–360.

[Vi1937] I. M. Vinogradov, *The method of trigonometrical sums in the theory of numbers*, Tr. Steklov. Math. Inst. **10** (1937). (Russian)

[Wa2000] M. Walters, *Combinatorial proofs of the polynomial van der Waerden theorem and the polynomial Hales-Jewett theorem*, J. London Math. Soc. (2) **61** (2000), no. 1, 1–12.

[Wh1961] J. H. C. Whitehead, *Manifolds with transverse fields in Euclidean space*, Ann. of Math. (2) **73** (1961), 154–212.

[Wo1968] J. Wolf, *Growth of finitely generated solvable groups and curvature of Riemannian manifolds*, J. Diff. Geom. **2** (1968), 421–446.

[Wo1995] T. Wolff, *An improved bound for Kakeya type maximal functions*, Rev. Mat. Iberoamericana **11** (1995), no. 3, 651–674.

[Wo1999] T. Wolff, *Recent work connected with the Kakeya problem*, Prospects in mathematics (Princeton, NJ, 1996), pp. 129–162, Amer. Math. Soc., Providence, RI, 1999.

[Ye] R. Ye, *Lecture notes*, available at `http://www.math.ucsb.edu/ỹer/ricciflow.html`.

[Ye2008] R. Ye, *On the l-function and the reduced volume of Perelman. I*, Trans. Amer. Math. Soc. **360** (2008), no. 1, 507–531.

[Ye2007] S. Yekhanin, *Towards 3-query locally decodable codes of subexponential length*, Proceedings of the thirty-ninth annual ACM symposium on Theory of computing (STOC), 2007, pp. 266–274.

[Ze2004] S. Zelditch, *Note on quantum unique ergodicity*, Proc. Amer. Math. Soc. **132** (2004), 1869–1872.

[Zi2007] T. Ziegler, *Universal characteristic factors and Furstenberg averages*, J. Amer. Math. Soc. **20** (2007), no. 1, 53–97.

[Zh2007] Q. Zhang, *A uniform Sobolev inequality under Ricci flow*, preprint.

[Zh2008] Q. Zhang, *Heat kernel bounds, ancient κ solutions and the Poincaré conjecture*, preprint.

[Zi1976] R. Zimmer, *Extensions of ergodic group actions*, Illinois J. Math. **20** (1976), no. 3, 373–409.

Index